Power Systems

Electrical power has been the technological foundation of industrial societies for many years. Although the systems designed to provide and apply electrical energy have reached a high degree of maturity, unforeseen problems are constantly encountered, necessitating the design of more efficient and reliable systems based on novel technologies. The book series Power Systems is aimed at providing detailed, accurate and sound technical information about these new developments in electrical power engineering. It includes topics on power generation, storage and transmission as well as electrical machines. The monographs and advanced textbooks in this series address researchers, lecturers, industrial engineers and senior students in electrical engineering.

Power Systems is indexed in Scopus

Avinash Aithal · Andrew Wainwright · Steve Atkins

Distribution System Operation: Flexibility Services

Summary of Key Outputs From the Open Networks Programme

The UK's Journey of Establishing a Fully Integrated Local Flexibility Market

Avinash Aithal
Open Networks, Innovation and Electricity Systems
Energy Networks Association
London, UK

Andrew Wainwright
Scottish and Southern Electricity Networks
Reading, UK

Steve Atkins
Scottish and Southern Electricity Networks
London, UK

ISSN 1612-1287 ISSN 1860-4676 (electronic)
Power Systems
ISBN 978-3-031-92904-5 ISBN 978-3-031-92905-2 (eBook)
https://doi.org/10.1007/978-3-031-92905-2

© The Editor(s) (if applicable) and The Author(s) 2025. This book is an open access publication.

Open Access This book is licensed under the terms of the Creative Commons Attribution 4.0 International License (http://creativecommons.org/licenses/by/4.0/), which permits use, sharing, adaptation, distribution and reproduction in any medium or format, as long as you give appropriate credit to the original author(s) and the source, provide a link to the Creative Commons license and indicate if changes were made.
The images or other third party material in this book are included in the book's Creative Commons license, unless indicated otherwise in a credit line to the material. If material is not included in the book's Creative Commons license and your intended use is not permitted by statutory regulation or exceeds the permitted use, you will need to obtain permission directly from the copyright holder.
The use of general descriptive names, registered names, trademarks, service marks, etc. in this publication does not imply, even in the absence of a specific statement, that such names are exempt from the relevant protective laws and regulations and therefore free for general use.
The publisher, the authors and the editors are safe to assume that the advice and information in this book are believed to be true and accurate at the date of publication. Neither the publisher nor the authors or the editors give a warranty, expressed or implied, with respect to the material contained herein or for any errors or omissions that may have been made. The publisher remains neutral with regard to jurisdictional claims in published maps and institutional affiliations.

This Springer imprint is published by the registered company Springer Nature Switzerland AG
The registered company address is: Gewerbestrasse 11, 6330 Cham, Switzerland

If disposing of this product, please recycle the paper.

Foreword

ENA members came together with energy regulator Ofgem and the Department for Business, Energy, and Industrial Strategy (BEIS) in 2017 to consider how the UK's electricity distribution networks could harness the growing volumes of distributed generation and demand to develop markets for flexibility. At the time there was little understanding of terms such as 'flexibility' and 'distribution system operation' and early work involved defining these terms and achieving a common understanding across Great Britain.

Since then, distribution and transmission experts have come together under the banner of Open Networks. This programme, led by ENA, has sought to define the role of distribution system operation and develop new methodologies for planning and operations of distribution networks. Now, the UK has a world-leading flexibility market. We continue to deliver at pace and ensure we are truly 'Open Networks'.

It became clear that energy flexibility markets are a practical and low-cost way to improve energy resilience and reduce bills for householders as we transition to net zero. The entire energy sector has come together to unlock its full potential while ensuring transparency. Working with subject matter experts from across the industry, we aligned with government policies and created viable outcomes for customers.

Open Networks often challenged the status quo and united the entire value stream of flexibility, opening up dialogue with stakeholders to identify barriers to accessing the flexibility market. We focused on making it easier for flexibility service providers to participate in the flexibility market by harmonising user experience; from first qualifying to take part in the market through dispatching their services and settling payments, improving operational coordination to remove barriers to the successful delivery of services, improving transparency of system operator's decision making processes and the size and carbon impacts of our flexibility market, and improving operational coordination between networks and companies to remove barriers to the delivery of flexibility services.

Through a commitment to 'flexibility first', ENA members together with the industry have established a local flexibility market with record high volumes each year. In 2024, we recorded over 6 GW of tendered flexibility, with 80% of the contracted flexibility coming from non-fossil fuel-based energy sources and a total of 11.7 GWh dispatched over past year. This is equivalent to supplying electricity to approximately 32,500 UK households for an entire year, or powering 3900 electric vehicles for 10,000 miles. The transition to a smart and flexible networks is in full swing.

This book is a testament to the breadth and depth of the work undertaken by Open Networks since its inception. ENA are proud to have worked with industry colleagues to progress both flexibility markets and distribution system operation into business as usual within network companies.

This isn't the end of the journey for our smart, flexible energy future. Rather, this book represents the end of a chapter. Our work will lead onto further industry developments through the market facilitator, having set strong foundations for the markets of the future.

Lawrence Slade
Chief Executive
Energy Networks Association
London, UK

Preface

Energy Networks Association (ENA) is a not-for-profit industry body representing the companies which operate the energy networks in the UK and Ireland. ENA helps its members meet the challenge of delivering energy to communities across the UK and Ireland safely, sustainably, and reliably. ENA's mission is to facilitate collaboration across our sector, share best practices, promote and protect our industry's reputation, seek to influence those who have an interest in what our members do, and support its members achieve their objectives.

ENA Members

As the authoritative and credible voice of the networks, ENA leads a number of cross-industry initiatives on behalf of its members to address key challenges across the energy sector and to facilitate low carbon flexibility. The Open Networks programme was one of these strategic initiatives leading the transition to a smart and flexible energy system and getting networks ready for net zero.

As more homes and businesses adopt low carbon technologies (LCTs), flexibility markets enable consumption and generation patterns to be aligned with the availability of clean energy and network needs. Between 2017–2025, ENA's Open Networks has been dedicated to accelerating the transition to a smart and flexible grid, so that we can make the most of our electricity networks while bringing people and businesses along on the net zero journey. The Open Networks programme was instrumental in informing the transition to DSO and establishing the UK as the world leader in the development of local flexibility markets, which could save consumers £10 bn/year in energy costs by 2050 and reduce total costs of the net zero transition by up to £70 bn.

Driven by the government's Smart Systems and Flexibility Plan(s) and supported by the Office of Gas and Electricity Markets (Ofgem), we worked to make the energy flexibility market easier to take part in, more coordinated, and more transparent. Open Networks developed the UK's local flexibility market from the ground up, starting from developing a UK-wide business case and market principles to the formulation of procedures and technical specifications. The programme has successfully united the entire value chain of flexibility markets from across sectors, by bringing together over 500 delegates from energy network companies, the regulator, government, and the wider industry.

The book presents detailed learnings from the development of local flexibility markets, from inception to implementation starting from the programme's launch in 2017 till its closure in 2025. Power utilities worldwide that are looking to establish similar flexibility markets can benefit from our efforts, ensuring that the net zero transition is inclusive, transparent, and affordable for all.

London, UK	Avinash Aithal
Reading, UK	Andrew Wainwright
London, UK	Steve Atkins

Contributions and Acknowledgements ENA's Open Networks programme embodies the commitment of the UK's network companies to come together as an industry to progress key climate challenges and to ensure that the sum of our efforts is greater than individual parts. The authors would like to acknowledge the contribution of all the members that came together under the banner of ENA's Open Networks. The development of the local flexibility market or this book would not have been possible if not the invaluable contributions of all the contributors listed below.

Abisola Dapo-Akinpelu	Adam Bain	Adam Brown	Adam Sims
Adam Towl	Adrian Sellar	Ahmed Mohamed	Aileen McLeod
Alan Brown	Alan Kelly	Alan Minton	Alasdair Gaw
Alex de Santos Aranda	Alex Green	Alex Haffner	Alex Howison
Alex Munnery	Alex Travell	Alexandra Williams	Ali Harper
Ali Resali Ahadi	Ali Reza Ahmadi	Alison Scott	Amir Aziz
Amy Hamilton	Amy Weltevreden	Anda Baumerte	Andrea Ballanti
Andres Moreno	Andrew Cupples	Andrew Spencer	Andrew Urquhart
Andy Rice	Andy Smith	Aneesa Parkar	Angela Schorah
Anne-Claire Leydier	Annie Robbins	Anthony Walsh	Aoife Bradish
Apostolos Koutras	Autumn Davies	Bahij Youssef	Barry Hatton
Ben Godfrey	Ben Smith	Bernard O'Sullivan	Bethan Winter
Beverley Hudson	Bless Kuri	Brian Hoy	Brian Wann
Calum Jardine	Cameron Dobbie	Campbell Graham	Cara Blockley
Carl Hashim	Carol Choi	Catherine Winning	Charlie Edwards
Charlon Balrey	Charlotte Grant	Chris Allanson	Chris Gamble
Chris Thompson	Christopher Kungu	Christopher Surgeoner	Christos Kaloudas
Christos Takoudis	Colin Mathieson	Colin Nicholl	Colin Thomson
Colm Murphy	Connor Innes	Corinna Jones	Cormac Bradley
Craig Campbell	Craig Graham	Damien Kelly	Dan Clements
Dan Noon	Daniel Burke	Daniel Clelland	Daniel Mellis
Daniel Robinson	Danielle Stewart	Dario Minissale	Dave Tuffery
David Bowman	David Boyer	David Boyland	David Darley
David Hill	David McDonald	David Overman	David Spencer
David Tuffery	David Van Kesteren	David Willmot	Dawn Parlett
Dean Pearson	Deborah MacPherson	Deepak Lala	Deirdre Macduff
Derryl Miranda	Dimitris Konstantinidis	Diyar Kadar	Dovydas Dyson
Drew Sambridge	Duncan Hughes	Duncan Oliphant	Duncan Olive
Ed O'Carroll	Edward Gill	Eleanor Horn	Elin Williams
Elizabeth Porter	Emily Jones	Emily Leonard	Emily Smith

(continued)

(continued)

Emma Buckton	Emma Carr	Euan Norrington	Evangelos Karagiannis
Faith Natukunda	Farina Farrier	Fiona Fulton	Fiona Navesey
Fiona Odonnell	Franck Ngendakumana	Fraser MacIntryre	Gareth Hislop
Gareth Lyness	Garry MacDonald	Garry Turner	Gary Dolphin
Gary Huskinson	Gary O'Hare	Gary Stokes	Gavin Baillie
Gavin Stewart	Gemma Lampert	Gerry Boyd	Gibu Jacob
Gillian Williamson	Gordon Kelly	Graeme Keddie	Graham Campbell
Grant McBeath	Greg Dodd	Greg Farrell	Gregor Tait
Griffin John	Guy Shapland	Gwen MacIntyre	Hadi K.Khouzani
Hannah Cummins	Hannah Lewis	Hannah Wilson	Harriet Walsh
Hedd Roberts	Helen Boyle	Helen Hassan	Helen Jarva
Helen Priestley	Helen Sawdon	Henry Walker	Hope James
Howard W. Thomas	Hui Heng	Iain Dallas	Iain MacIntyre
Iain Miller	Iain Symon	Ian Bailie	Ian Coates
Ian Cooper	Ian Dunstan	Ian Pashley	Ian Povey
Ian Sandford	IasonIraklis Avramidis	Imran Mohammed	Irena Horobet
Jack Scoffham	Jake Pattison	James Abrahams	James Greenhalgh
James Hope	James Kelloway	James Watson	James Whiteford
James Yu	Janine Watson	Janire Casado	Jason Brogden
Jason Hicks	Jason Shreeve	Jennifer King	Jeeva Aithal
Jenny Rawlinson	Jeremy Meara	Jialiang Yi	Jim Cardwell
Jim McOmish	Jingchao Deng	Jo Dowling	Joana Dowling
Joanne Kerrigan	Jodie Garner-Jones	Joe Davey	Joe Horgan
Joe Nolan	Joel Martin	Joel McCreery	John Heywood
John Leighton	John Orr	John Spurgeon	John Stapleton
John West	Jon Berry	Jon Wisdom	Jonny Pollock
Joseph Kavanagh	Joseph Mitchell	Joseph Nolan	Josh Watson
Joshua Atkins	Julian Leslie	Julie Jackson	Julio Perez Olvera
Kaloudas Christos	Kanan Ganakesavan	Kara Burke	Karl Watson
Katharine Clench	Kavita Patel	Keith Evans	Keith Houston
Keith Owen	Kieran Davis	Kiran Jassal	Kyle Murchie
Laura Brown	Laura Dunn	Laura Henry	Laura McArthur
Le Fu	Leigh Lipton	Lisa O'Neil	Lisa O'Neill
Liz Sidebotham	Lois Clark	Lorna Millington	Lottie Wheatcroft
Luke Harker	Lynne McDonald	Maciej Fila	Magda Paluch
Mai Cao	Malcolm Arthur	Malcolm Barnacle	Malcolm Grisdale
Marc Vincent	Mark Beasley	Mark Goudie	Mark Herring

(continued)

Acknowledgements and Declarations

(continued)

Mark McCabe	Mark Nicholson	Mark Perry	Mary Black
Matt Bent	Matt Rivett	Matt Watson	Matt Webb
Matt White	Matthew Hamilton	Matthew Hindle	Matthew Paige-Stimson
Maurice Lynch	Max Peeters	Melanie Bryce	Michael Alexander
Michael Clarke	Michael Green	Michael Mclaughlin	Michael Oxenham
Michael Rieley	Michael Sowcroft	Michelle Cullis	Mike Gordon
Mike Harding	Mike Scowcroft	Minjiang Chen	Mohsen Beauville
Natasha Antill	Neal Wade	Neil Bennett	Neil Carter
Neil Sandison	Neil Stovold	Nia Lowe	Niall Murphy
Nicholas Harvey	Nicholas Sullivan	Nick Evans	Nick George
Nicola Bruce	Nicolas Manea	Nicole Butterworth	Nigel Turvey
Nisha Doshi	Odilia Bertetti	Oli Ricketts	Oli Spink
Orla Martin	Padraig Lyons	Paris Hadjiodysseos	Paul Fidler
Paul Fitton	Paul Fitzgerald	Paul Hartshorne	Paul Jewell
Paul McGimpsey	Paul Simpkin	Paul Sullivan	Peter Aston
Peter Coyne	Peter Gaskin	Peter Glover	Peter Kocen
Peter Turner	Phil Lawson	Philip Halsey	Philippa Williamson
Priya Mothilal Bhagavathy	Qi Tang	Rachel Donoghue	Ralph Eyre-Walker
Randolph Brazier	Rashmi Radhakrishnan	Rebecca Cailes	Rebecca Lees
Rebecca Threlfall	Reece Breen Begadon	Rhiannon Marsh	Rhys Penman
Rhys Williams	Richard Arnold	Richard Derby	Richard Jones
Richard Mandeya	Richard Smith	Richard Wilson	Richard Woodward
Rita Shaw	Rob Nickerson	Robert Brereton	Roddy Wilson
Roger Hey	Ross Anderson	Ross Easton	Ross Thompson
Roy Barnes	Ruby Smtih	Russell Bryans	Ryan Kavanagh
Ryan McAvoy	Ryan Place	Sam Do	Sam Turner
Sanjeev Loi	Sara Afifi	Sara Scarrott	Selma Awadallah
Shivananda Pukhrem	Sima Davarzani	Simanand Gandhi Jeyaraj	Simon Brooke
Simon Jesson	Simon Killoran-Codling	Simon Lewis	Sofia Cobo de Guzman
Sorcha Schnittger	Sotiria Kordi	Sotiris Georgiopoulos	Stathis Mokkas
Stefeni Cura	Steph Wootton	Stephen Dunlop	Steve Backhouse
Steve Cox	Steve Davenport	Steve Dugmore	Steve Field
Steve Halsey	Steve Miller	Steve Mockford	Steve Mould
Steve Shaw	Steven Gough	Steven Vallance	Stewart Reid
Stuart Easterbrook	Stuart Fowler	Susan Beveridge	Susanne Laing

(continued)

(continued)

Sven Hoffman	Tariq Hakeem	Tee Shengji	Thomas Koller
Thomas MacDowell	Tim Cox	Tim Manandhar	Tony Bright
Tony Hearne	Tracy Joyce	Trung Tran	Vibish Johnson
Vicci Page	Vicky Chiles	Victori Fleming-Williams	Victoria Gosling
Wayne Mullins	Wendy Mantle	Will Bowen	Will Monnaie
Will Seward	Yujia Du	Zahin Rahim	Zivanayi Musanhi

Competing Interests The authors have no competing interests to declare that are relevant to the content of this manuscript.

Contents

Part I Background and Context

1 Need for Distribution System Operation 3
   1.1 Introduction 3
      1.1.1 DSO Definition 3
   1.2 DNO to DSO Transition 4
   1.3 Suggested Further Reading 10
   References 11

2 Market Principles 13
   2.1 Flexibility First Commitment 13
   2.2 Market Principles 14
   2.3 Suggested Further Reading 15

Part II Local Flexibility Services

3 Flexibility Products 19
   3.1 Legacy DSO Flexibility Products 19
      3.1.1 Flexibility Service Product Parameters 19
   3.2 Realignment of Flexibility Products 21
      3.2.1 Investigation Approach 22
      3.2.2 Common Product Parameters Characterisation 25
   3.3 New DSO Flexibility Products 27
      3.3.1 New DSO Flexibility Product Definitions 29
   3.4 Implementation of New DSO Products 29
      3.4.1 Tracking Volumes of the DSO Products Procured 36
   3.5 Suggested Further Reading 37
   Reference 37

4 Forecasting and Development Plans (Pre-procurement) 39
   4.1 Distribution Future Energy Scenario (DFES) 39
      4.1.1 DFES and FES Alignment 42

		4.1.2	The Best View Scenario	45
	4.2	Network Development Plans		47
		4.2.1	NDP Form of Statement (FoS)	48
		4.2.2	Network Headroom Report	49
		4.2.3	Network Development Reporting	56
		4.2.4	NDP Methodology Reporting	57
	4.3	Selection and Evaluation of Options (Common Evaluation Methodology)		58
		4.3.1	Pre-tender Stage	62
		4.3.2	Tender Stage	63
		4.3.3	Post-tender Stage	64
	4.4	Common Evaluation Methodology (CEM) Tool		66
		4.4.1	Assessment Options	66
		4.4.2	Economic Assessment	69
	4.5	Illustrative Example		72
		4.5.1	Background and Network Need Description	72
		4.5.2	Application of the CEM Tool	74
	4.6	Suggested Further Reading		77
	References			77

Part III Deployment of Flexibility Services

5	**Procurement Process**			81
	5.1	Publication of Requirements		82
	5.2	Pre-qualification Process		82
		5.2.1	Review of Pre-qualification Process	83
		5.2.2	Aligned Pre-qualification Criteria	83
		5.2.3	Accepted Deviations to the Templates	99
	5.3	Tendering and Contracting		99
		5.3.1	Procurement Timelines	99
	5.4	Standard Agreement (Contract)		102
		5.4.1	Bilateral Versus Framework Approach	105
		5.4.2	Adoption of Service-Based Schedules	107
		5.4.3	Standard Agreement Version Tracker	109
		5.4.4	Future Considerations	112
	5.5	Suggested Further Reading		112
	References			113
6	**Dispatch of Flexibility Services**			115
	6.1	Phases of Dispatch		116
	6.2	Areas of Development		120
	6.3	Suggested Further Reading		120
	References			120

Contents

7 Settlement ... 123
- 7.1 Performance/Payment Calculations ... 124
  - 7.1.1 Availability Payments (for Turn-Up/Turn-Down Services) ... 125
  - 7.1.2 Utilisation Payments (for Turn-Up/Turn-Down Services) ... 127
  - 7.1.3 Utilisation Calculations (for Peak Reduction Services) ... 130
- 7.2 Metering Specifications ... 132
  - 7.2.1 Metering Granularity ... 132
  - 7.2.2 Metering Data Requested ... 132
  - 7.2.3 Metering Accuracy Standards ... 132
  - 7.2.4 Site Meter Location ... 134
- 7.3 Billing and Invoicing ... 134
  - 7.3.1 Process Steps ... 134
- 7.4 Baselining ... 137
  - 7.4.1 Baseline Methodology Groups ... 138
  - 7.4.2 Use of Baselining Methodology ... 142
- 7.5 Suggested Further Reading ... 145
- References ... 145

Part IV Interactions Across Markets

8 Interoperability of Flexibility Dispatch Systems ... 149
- 8.1 Application Program Interface (API) ... 149
  - 8.1.1 Stakeholder Views ... 150
  - 8.1.2 Review of Architectural Requirements ... 151
  - 8.1.3 Data Representation Requirements ... 160
  - 8.1.4 Security Requirements ... 162
  - 8.1.5 Testing Requirements ... 167
- 8.2 Delivering Interoperable Dispatch ... 168
  - 8.2.1 OpenADR 2.0 and 3.0 ... 169
  - 8.2.2 Other Standards and Approaches Considered ... 170
  - 8.2.3 Next Steps ... 176
- 8.3 Suggested Further Reading ... 179
- References ... 179

9 Flexible Connections and Flexibility Services ... 181
- 9.1 Principle of Flexible Connections ... 181
- 9.2 Flexible Connection (ANM) ... 182
  - 9.2.1 Types of Flexible Connections ... 183
  - 9.2.2 Applying for a Flexible Connection (ANM) ... 183
  - 9.2.3 Curtailment of Flexible Connections (Pre-SCR) ... 184
- 9.3 Flexible Connections (ANM) Versus Flexibility Services ... 188
- 9.4 Suggested Further Reading ... 189
- References ... 190

10	**Conflict Management (Primacy Rules)**		191
	10.1	Defining Primacy Rules	192
		10.1.1 High Level Primacy Rules	193
		10.1.2 Consolidated Use Cases	195
	10.2	Cost Benefit Analysis	200
		10.2.1 Cost Benefit Analysis Approach	200
		10.2.2 Cost Benefit Analysis Results	201
	10.3	Implementing Primacy Rules (Use Case)	203
		10.3.1 Transmission Constraint Management and DSO Flexibility Services	203
		10.3.2 Process and Data Flows for Primacy Rules Deployment	204
		10.3.3 Learnings and Considerations for Implementation of Primacy Rules	213
	10.4	Suggested Further Reading	217
11	**Stackability of Flexibility Services**		219
	11.1	Primacy Versus Stackability	219
	11.2	Revenue Stacking of Services	219
		11.2.1 Flexibility Service Provider Perspective	222
	11.3	Stacking Assessment for DSO Services	223
		11.3.1 Jumping of Services	224
		11.3.2 Splitting of Services	228
		11.3.3 Co-delivery of Service	229
		11.3.4 Other Key Findings for Service Stacking	230
	11.4	Individual DSO Service Assessments	232
		11.4.1 Peak Reduction	233
		11.4.2 Scheduled Utilisation	234
		11.4.3 Operational Utilisation	236
		11.4.4 Scheduled Availability + Operational Utilisation	239
		11.4.5 Variable Availability + Operational Utilisation	242
	11.5	Stackability Consideration for New Services (Stackable by Design)	245
	11.6	Baselining Considerations When Service Stacking	247
		11.6.1 Baselining Methodologies and Considerations for Service Stacking	247
	11.7	Suggested Further Reading	251
	Reference		252

Part V Additional Resources

12	**Carbon Impact Assessment of Flexibility Service**		255
	12.1	Comparison Between Carbon Reporting Methodologies	256
	12.2	Key Observations from Comparison	257
		12.2.1 Purpose of Each Methodology	257

		12.2.2 Calculation Approach	264
	12.3	Approach Adopted by DSOs for Carbon Reporting	266
		12.3.1 Carbon Impact Reporting Methodology	267
		12.3.2 Report Submission Format	271
	12.4	Suggested Further Reading	274
	References		276
13	**Industry Jargon Buster**		277
	13.1	Industry Codes and Sources of Detailed Definitions	277
	13.2	GB Network Voltage Levels	279
	13.3	Industry Terms with Common English Description	279
	References		320

Appendix ... 321

General Terms and Conditions ... 331

Flexibility Services Service Terms—Company Active Services ... 351

Annexes to Flexibility Services Service Terms—Company Active Services ... 361

Forms and Templates to Flexibility Services Service Terms—Company Active Services ... 363

About the Authors

Dr Avinash Aithal is Chartered Engineer and Fellow of the IET with a Masters in Electrical Energy Systems and a Ph.D. in Smart Grids, both from Cardiff University, UK. Avinash is Head of the Open Networks at Energy Networks Association (ENA), functioning as the programme director and technical principal for the programme. Avinash is also a member of UKRI's Science Engineering and Technology board, utilising his expertise to champion multidisciplinary and emerging scientific research, laying the foundations of a smart grid in Great Britain. Avinash has more than 15 years of experience in the energy sector, working with EA Technology, Alstom and GE, prior to his role at ENA.

Andrew Wainwright is Fellow of the IET with over 30 years in the electricity sector working across distribution and transmission entities. Andrew has been involved with Open Networks for eight years and a member of its Steering Group for around six years. Andrew currently leads SSEN's work on strategic development and is responsible for the publication of its Distribution Future Energy Scenarios (DFES) and Distribution Networks Options Assessment (DNOA). Previously Andrew has led the NESO's work into the Distribution System Operation (DSO) transition and been responsible for implementation of many Open Networks policies through the NESO's Regional Development Programmes.

Steve Atkins spent 17 years in developing sales and marketing strategies for the financial services industry before leaving to set up his own renewables installation business. He joined SSEN in 2014 to look after Major Commercial Connections in the south before joining the Future Networks team in 2017 to develop its DSO strategy. He was part of the ENA's Open Networks Steering Group and chaired the programme's workstream on DSO Transition. Following the implementation of SSEN's DSO Directorate, he led on local authority engagement as part of the Whole System team. He is currently part of SSE's Corporate Directorate where he leads on stakeholder relations for its SSEN Distribution business.

Abbreviations

ANM	Active Network Management
API	Application Programming Interface
BAFO	Best and Final Offer
BEIS	Department for Business, Energy, and Industrial Strategy
BSP	Bulk Supply Point
CBA	Cost Benefit Analysis
CEM	Common Evaluation Methodology
CEP	Clean Energy Package
CI	Customer Interruption
CIM	Common Information Model
CML	Customer Minutes Lost
CMZ	Constraint Management Zones
CPP	Consumer Protection Party
CUSC	Connection and Use of System Code
DCC	Data Communications Company
DCUSA	Distribution Connection and Use of System Agreement
DER	Distributed Energy Resources
DESNZ	Department for Energy Security and Net Zero
DFES	Distribution Future Energy Scenario
DFS	Demand Flexibility Service
DNO	Distribution Network Operator/Operation
DNOA	Distribution Network Options Assessment
DSO	Distribution System Operator/Operation
DUKES	Digest of UK Energy Statistics
EDL	Electronic Dispatch Logging
ENA	Energy Networks Association
ENWL	Electricity North West Limited
EU ETS	European Union Emissions Trading System
EV	Electric Vehicle
FES	Future Energy Scenario
FoS	Form of Statement

FPN	Final Physical Notifications
FSP	Flexibility Service Provider
GHG	Greenhouse Gas
GSP	Grid Supply Points
GWP	Global Warming Potential
ICO	Information Commissioner's Office
IDNO	Independent DNO
IPCC	Intergovernmental Panel on Climate Change
ITT	Invitation to Tender
LAEP	Local Area Energy Plans
LCT	Low Carbon Technology
LES	Local Energy Systems
LHEES	Local Heat and Energy Efficiency Strategies
LIFO	Last In First Out
LTDS	Long-Term Development Statements
MBMA	Meter Before—Meter After
NCSC	National Cyber Security Centre
NDP	Network Development Plans
NESO	National Energy System Operator
NGED	National Grid Electricity Distribution
NIR	National Inventory Report
NIS	Network and Information Systems
NOA	Network Options Assessment
NPg	Northern Powergrid
NPV	Net Present Value
OFGEM	Office of Gas and Electricity Markets
PKI	Public Key Infrastructure
PQQ	Pre-Qualification Questionnaire
RDP	Regional Development Programme
RDSP	Relevant Digital Service Provider
RIIO	Revenue = Incentives + Innovation + Outputs
SCR	Significant Code Review
SDA	Same-Day Adjustment
SDO	Standards Development Organisations
SECR	Streamlined Energy and Carbon Reporting
SGAM	Smart Grid Architecture Model
SGANM	Single Generation Active Network Management
SLC	Standard Licence Condition
SMP	Single Markets Platform
SO	System Operation/Operator
SPEN	SP Energy Networks
SSEN	Scottish and Southern Electricity Networks
STOR	Short Term Operating Reserve
TCM	Transmission Constraint Management
TIM	Totex Incentive Mechanism

TO	Transmission Owner
UKPN	UK Power Networks
UMEI	Universal Market Enabling Interface
UML	Universal Modelling Language
UNFCCC	United Nations Framework Convention on Climate Change
USEF	Universal Smart Energy Framework
UVDB	Utilities Vendor Database
WAAPI	Wider Access API
WPD	Western Power Distribution (now National Grid Electricity Distribution)

List of Figures

Fig. 1.1	Traditional versus smart flexible energy system	4
Fig. 1.2	Flexibility market arrangements (World B)—coordinated DSO–ESO procurement & dispatch	9
Fig. 1.3	Flexibility market arrangements (World E)—flexibility coordinator (market facilitator)	10
Fig. 3.1	Visualisation of flexibility product parameters	20
Fig. 3.2	Steps for the Flexibility Products alignment proposals	25
Fig. 3.3	Flexibility products in operation (with a rough old versus new product mapping)	35
Fig. 3.4	Flexibility procurement volumes ("open networks flexibility figures")	37
Fig. 4.1	High level process of capacity provision using flexibility services	40
Fig. 4.2	DFES as part of annual DSO planning processes	41
Fig. 4.3	Common DFES methodology framework	44
Fig. 4.4	Different network reports and their range of timeframes	46
Fig. 4.5	Different parts of the NDP FoS	48
Fig. 4.6	Network planning end-to-end process	58
Fig. 4.7	LTDS, DFES and NDP timeline	60
Fig. 4.8	High level process of capacity provision using Flexibility Services	62
Fig. 4.9	High level process for pre-tender stage	63
Fig. 4.10	High level process for tendering flexibility services	64
Fig. 4.11	High level process for ongoing contract monitoring and review	65
Fig. 5.1	Typical procurement milestones timelines (indicative)	103
Fig. 5.2	Timeline of evolution from bilateral to framework approach	106
Fig. 5.3	Schedule structure (in Ver 2.0 and above)	108
Fig. 5.4	Published versions of Standard Agreement for Flexibility Services	110

Fig. 7.1	Indicative visual representation impact on payment with changing delivery percentage (Turn-up/Turn-down)	131
Fig. 7.2	Processes steps in the settlement process across all DSOs	135
Fig. 7.3	Illustration of morning of adjustment for a weather-sensitive asset	141
Fig. 8.1	Simple dispatch architectural diagram	152
Fig. 8.2	Common dispatch management systems	153
Fig. 8.3	Region specific dispatch systems	155
Fig. 8.4	A Single operator's sharded dispatch system	156
Fig. 8.5	Provider-sharded architecture with multiple operators	157
Fig. 8.6	Market, dispatch and settlement functionary combined	159
Fig. 8.7	High level description of the scope of the planned work	178
Fig. 9.1	Approach to determine curtailment of flexible connections (ANM)	186
Fig. 9.2	Curtailment index	187
Fig. 10.1	Illustration of flexibility conflict use cases	197
Fig. 10.2	Process flows for Rule (1a): DSO priority-information shared ahead of time	207
Fig. 10.3	Process flows for Rule (1b): DSO priority—Closer to real time information sharing	208
Fig. 10.4	Process flows for Rule (2a): management of planned outages—additional coordination of planned outages	210
Fig. 10.5	Process flows for Rule (2b): management of planned outages—Closer to real time planned outage cancellation	211
Fig. 12.1	Calculation methodology—reporting boundary	275

List of Tables

Table 3.1	Historical nomenclature and high-level descriptors of the GB flexibility products	20
Table 3.2	Descriptions of DSO defined parameters for the legacy flexibility products	22
Table 3.3	Description of FSP defined parameters for the legacy flexibility products	23
Table 3.4	Legacy flexibility products parameters	24
Table 3.5	Common product parameters—structure, availability and utilisation	26
Table 3.6	List of GB's new flexibility products (for procurement beyond 2023)	28
Table 3.7	Peak Reduction product definitions	30
Table 3.8	Scheduled Utilisation product definitions	30
Table 3.9	Operational utilisation product definitions	31
Table 3.10	Scheduled Availability + Operational Utilisation product definitions	32
Table 3.11	Variable Availability + Operational Utilisation product definitions	33
Table 3.12	Early indications of DSOs' plans for procuring new flexibility products	36
Table 4.1	Distribution system operation roles and activities, prescribed by Ofgem [1]	40
Table 4.2	Network headroom report parameters	50
Table 4.3	Network headroom report headings	55
Table 4.4	Scope of the network development report	56
Table 4.5	Range of stakeholders expected to use NDPs	59
Table 4.6	Links to individual Network companies' DFES	61
Table 4.7	Links to individual Network companies' NDP	61
Table 4.8	Key areas of the CEM	67
Table 4.9	Standard inputs from the Ofgem CBA used for CEM	70
Table 4.10	Use cases for CEM methodology	73

Table 4.11	Data and information for the CEM Tool for the ceiling/guide price or price range	75
Table 4.12	Confirmed data and information for the CEM Tool evaluation	76
Table 5.1	Number of pre-qualification criteria questions per DSO (pre-alignment)	83
Table 5.2	Stakeholder Feedback on PQQ (pre-alignment)	84
Table 5.3	Standardised commercial PQQ template (31 questions)	85
Table 5.4	Standardised Technical PQQ Template (35 Questions)	91
Table 5.5	Noted deviations from the PQQ template	100
Table 5.6	Historic assessment weightage	101
Table 5.7	Scoring criteria and recommended weightage	101
Table 5.8	Pros and cons of procurement alignment between DSO and NESO	104
Table 7.1	Payment structures for new flexibility products	124
Table 7.2	Signage used in calculations for different asset types	125
Table 7.3	Parameters used in availability calculations	126
Table 7.4	Parameters used in utilisation calculations	128
Table 7.5	Aligned metering parameters	133
Table 7.6	Grace factors and performance multipliers	136
Table 7.7	Common DSO baseline methodology descriptions	143
Table 7.8	NESO baselining methodologies for balancing services	144
Table 9.1	Comparison of uses for typical Flexible Connections (ANM) and Flexibility Services	189
Table 10.1	Flexibility conflict use cases	199
Table 10.2	Likely data that will need to be exchanged during the Rules implementation	212
Table 11.1	Distinction between primacy and stackability	220
Table 11.2	Distinction between different categories of stackability	221
Table 11.3	Key for flexibility service stacking tables (applicable to splitting and jumping)	224
Table 11.4	Ability to jump different services	225
Table 11.5	Ability to split different services	226
Table 11.6	Service stacking summary for Peak Reduction	235
Table 11.7	Stacking summary for Scheduled Utilisation	237
Table 11.8	Stacking summary for Operational Utilisation	240
Table 11.9	Stacking summary for Scheduled Availability + Operational Utilisation	243
Table 11.10	Stacking summary for variable availability + operational utilisation	246
Table 12.1	NESO carbon intensity of GB electricity system	256
Table 12.2	EU ETS monitoring and reporting	257
Table 12.3	The GHG protocol	258
Table 12.4	The green book and supplementary guidance	259
Table 12.5	IPCC guidelines for national GHG inventory	260

Table 12.6	PAS 2080	261
Table 12.7	Pro low carbon	262
Table 12.8	Comparison of purpose of methodologies	263
Table 12.9	Conversion factor sources	269
Table 12.10	SLC 31E technology categories by the relevant technology categories	272
Table 12.11	Template for total carbon impact divided by the total energy delivered	273
Table 13.1	Wider Industry Code and Contractual Documents	278
Table 13.2	Main voltage levels used for GB distribution and transmission	280
Table 13.3	Industry terms with common English description	281

Part I
Background and Context

Chapter 1
Need for Distribution System Operation

1.1 Introduction

The energy industry is changing. In response to the need to decarbonise our energy usage, new decentralised forms of electricity generation have evolved, supported by smart grid technologies. These technologies are also changing the way we use electricity, leading both residential and business users to become increasingly active in their energy usage, as shown in Fig. 1.1. The three challenges of decarbonisation, decentralisation and digitisation have the potential to create whole system opportunities by transforming the way distribution networks behave and creating new flexibility market opportunities for potential service providers. These markets will enable Flexibility Services to compete alongside traditional investment options for all relevant network reinforcements or upgrades of significant value, and to make the most cost-effective investment decisions in the future. Through these services and the broader benefits of more active networks and customers we can deliver:

- Lower carbon energy at the lowest overall cost for Customers
- Opportunities for Customers to realise value from services and new technology
- More sustainable energy markets and networks.

1.1.1 DSO Definition

The transition from the traditional distribution networks operation to distribution system operation was imperative to build real momentum into the development work required to enable the UK's energy networks to deliver a more secure, affordable, low-carbon energy system.

A key aspect of this transition was considering how the industry needed to evolve to account for a need for increased active management of distribution networks. Distribution system operation was seen as a critical new role that will facilitate new

- Lower carbon energy at the lowest overall cost for Customers
- Opportunities for Customers to realise value from services and new technology
- More sustainable energy markets and networks.

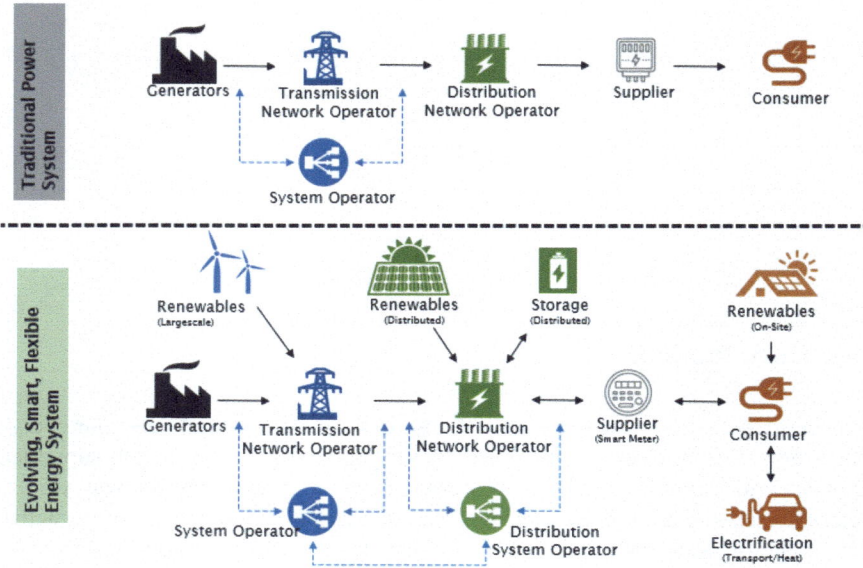

Fig. 1.1 Traditional versus smart flexible energy system

system operation roles for distribution networks including the needs of a (then future) neutral market facilitator. A key part of the transition was working with the industry to provide a definition of DSO, and its associated functions.

A DSO was therefore defined as

> "A Distribution System Operator (DSO) securely operates and develops an active distribution system comprising networks, demand, generation and other flexible Distributed Energy Resources (DER). As a neutral facilitator of an open and accessible market it will enable competitive access to markets and the optimal use of DER on distribution networks to deliver security, sustainability and affordability in the support of whole system optimisation. A DSO enables customers to be both producers and consumers; enabling customer access to networks and markets, customer choice and great customer service."

1.2 DNO to DSO Transition

In 2018, the assessment of "future worlds" was initiated by ENA's Open Networks programme through extensive consultation with the industry on a number of potential future market models. These models considered how future industry structures could best deliver flexibility markets providing services from DER for both national and

1.2 DNO to DSO Transition

regional (transmission and distribution) requirements. The "future worlds" represented the potential evolution of Industry structure to meet the challenges of decarbonisation, decentralisation and digitisation and to facilitate arrangements that work for future Customers, service providers and business models.

The list of key 'actors' that were represented in the consultation are listed below:

DNO/DSO: Distribution Network Operators (DNOs) are regulated entities that own and operate electricity distribution networks over a defined geographic area. Historically these networks have been passive in nature, but with increasing volumes of DER they are becoming increasingly active. This together with smart grid technologies is creating opportunities for DNOs to realise consumer value and develop SO functions and become a DSO.[1] The scale of functionality of this DSO entity would vary depending on the Future World that is ultimately implemented. (Independent DNOs (IDNOs) are covered as a separate actor.).

ESO: The Electricity System Operator (ESO) is the actor responsible for the design and operation of the transmission network in Great Britain This role is now fulfilled by the newly established National Energy System Operator (NESO), in Great Britain. This neutral party already fully performs the role of a SO at a transmission level. Historically this has required engagement with transmission connected parties and a few large DERs. However, as more active parties connect to distribution networks the NESO has established relationships with more DER and also with third parties such as commercial Aggregators.

Independent DNO: An Independent DNO (IDNO) is a network company licensed by the energy Regulator to own, develop, operate and maintain (including fault repair service) local electricity distribution networks. IDNO networks are directly connected to the DNO networks, or indirectly via another IDNO. IDNOs are regulated in the same way as DNOs, however, the IDNO licence does not have all the conditions of a full DNO licence.

Transmission Owner: A Transmission Owner (TO) is responsible for investing, building and maintaining their electricity transmission network. The TO provides network customers with a safe, secure and reliable network ensuring they receive high quality network services at value for money.

Flexibility Coordinator (Market Facilitator): One of the future scenarios identified required the creation of a new role of Flexibility Coordinator that did not exist at the time (its functions being picked up by the DSO or NESO in other Future Worlds). A Flexibility Coordinator is the responsible party for the management of the central hub(s) on which SOs request Flexibility Services and service providers offer their products. In this World, the Flexibility Coordinator would be responsible for the

---

[1] The organisational split of DNO and DSO functions vary across the network companies in GB, with some companies building capabilities within the existing business while others have opted for a formal business separation. For convenience, the regulated entity that own and operate electricity distribution networks together with system operation function is collectively designated as DSO throughout this book.

management of potential service conflicts. Whilst this actor could be a national monopoly party, there could also be several regional Flexibility Coordinators.

Gas: Gas represents an energy system from which useful gas energy resources can be extracted or recovered, either directly or by means of a conversion or transformation process (e.g. conversion of natural gas and derivatives into chemical energy). Gas can be stored either for use in a different location or a different time.

Government: In the context of future worlds, this actor represents both local, devolved and UK national Government bodies. This includes local authorities and local enterprise partnerships (LEP) which are formed by a variety of stakeholders such as employers, landlords, policy-makers, energy consumers and energy generators. They promote the social, economic and environmental well-being of their community. They participate in the implementation of national energy policy that delivers secure, clean and affordable energy supplies through the application of measures that reduce energy use, promote the extensive use of renewable sources and tackle fuel poverty. This includes the Department for Energy Security and Net Zero (DESNZ), who is responsible for the design and implementation of national energy policy.

Customer (Active and Passive Customers): Traditionally, customers have been passive entities taking power from networks for use in their homes and businesses. Through smart meters, microgeneration and storage, these parties will have the choice whether to become active in their use of electricity or remain passive. The Open Networks has defined both of these customer types.

Distributed Energy Resource: Distributed Energy Resources (DERs) are smaller scale power generation technologies (typically in the range of up to 10 MW and including electric energy storage facilities) and larger end use electricity consumers (e.g. industrial and commercial) with the ability to flex their demand (i.e. demand-side response) that are directly connected to the electricity distribution network. (System service providers and active participants including distributed generation, flexible demand providers, and distributed connected energy sources).

Aggregator: An aggregator is a company who acts as an intermediary between active parties such as DER and active Customers who can offer Flexibility Services, and SOs who wish to obtain such services for efficient management of networks. The aggregator groups distributed service providers into a single entity for the purpose of Flexibility Service provision. This can be through national aggregation or on a more regional basis.

Consumer Protection Party: A Consumer Protection Party (CPP) is a party responsible for acting as the voice of the consumer, for example Citizens Advice. Such a party may represent consumers in general or may focus on the needs and concerns of a defined subset, for example large energy users.

Data Communications Company: A Data Communications Company (DCC) is a party responsible for establishing and managing the data and communications network that connects smart meters to the business systems of energy Suppliers, network operators and other authorised service users of the network. The DCC is a monopoly company regulated by the GB energy regulator.

Heat: Heat represents an energy system from which useful heat energy resources can be extracted or recovered either directly or by means of a conversion or transformation process (e.g. conversion of heat exchanging fluids into thermal energy). Heat can be stored either for use in a different location or a different time.

Local Energy System: Local Energy Systems (LES) utilise peer-to-peer trading/ local energy market to the benefit of their participants (e.g. communities, companies, individuals). LES participants provide each other with energy and trade out the aggregate 'balance' in the wholesale electricity market. LES can provide Flexibility Services to Electricity System Operators (e.g. NESO, DSO) for electricity system balancing and transmission and distribution network constraint management. LES can include DER and active customers. LES can incorporate innovative energy distribution, management and metering, novel business models as well as clean transport systems. LES are emerging actors in the current world and includes community energy projects.

Regulator: The energy Regulator is responsible for regulating the electricity industry (Ofgem for Great Britain and Utility Regulator in for Northern Ireland). The energy Regulator carries out functions to protect the interests of current and future consumers of electricity supplied by authorised suppliers, wherever appropriate, by promoting effective competition between persons engaged in, or in commercial activities connected with, the generation, transmission, distribution or supply of electricity. The electricity Regulator works closely with industry in carrying out its functions such as licensing electricity suppliers, generators, transmission and distribution, setting the levels of return which the monopoly networks companies can make and deciding on changes to market rules.

Settlement Agent: A settlement agent is responsible for managing the settlement of payments to and from Flexibility Service providers. The settlement agent collects, validates, processes and aggregates metered data from service providers (generation and demand based services); sets up and maintains the systems that collect, securely store, and securely transmit the data necessary for settlement process; manages the settlement of payments by Flexibility Service providers; calculates payments and charges; and invoices and collects payments due.

Supplier: A supplier is a company that buys electricity in the wholesale market or directly from generators and sells it on to end use electricity consumers. The supplier sets the tariffs consumers pay for the electricity they use. Suppliers work in a competitive market where Customers can choose any Supplier to provide their electricity. Suppliers can also be active in flexibility markets; providing services through their Customers.

Supply Chain: The Supply Chain is responsible for the design, manufacture and supply of equipment and devices to the electricity industry.

Transmission Connected Demand: Transmission Connected Demand is large scale sources of demand (e.g. such as steelworks, refineries, or other major industrial demand) directly connected to the transmission network that supports the NESO balancing supply and demand, and helps manage transmission network constraints. Transmission Connected Demand can act as a source of flexibility to the NESO by reducing demand to make additional volumes of electricity available or by increasing demand to reduce imbalanced volumes of electricity.

Transmission Connected Generator: Transmission Connected Generators are large scale electricity generators (e.g. nuclear, gas powered, coal fired power stations, etc.) directly connected to the transmission network that support the NESO in balancing supply and demand and managing transmission network constraints. These generators act as a source of flexibility to the NESO by making additional volumes of electricity available or by reducing the volumes of electricity being generated.

As we move towards our smart, decentralised systems of the future, many of these actors continue to see significant change and evolution of their roles. For some, these changes would be largely consistent across the "Future Worlds", for others, there is considerable difference.

Using the information collated from the industry engagement, Five models of more rounded Future Worlds that consider not just the impact of flexibility markets but also the other requirements of a neutral market facilitator, whether a DNO acting as a DSO or some other party.

1. World A: DSO Coordinates—a World where the DSO takes a central role for all distribution connected parties acting as the neutral market facilitator for all DER and provides services on a locational basis to the ESO
2. **World B: Coordinated DSO-ESO Procurement and Dispatch—a World where DSO and ESO work together to efficiently manage networks through coordinated procurement and dispatch of flexibility resource (Fig. 1.2)**
3. World C: Price-Driven Flexibility—a World where changes developed through Ofgem's reform of electricity network access and forward-looking charges have improved access arrangements and forward-looking signals for Customers. This World has been built with flexibility arrangements as described in World B, but it is recognised that charging and access developments could be similarly progressed in other Worlds
4. World D: ESO Coordinate(s)—a World where the NESO takes a central role in the procurement and dispatch of Flexibility Services as the neutral market facilitator for DER with DSO's informing the NESO of their requirements
5. **World E: Flexibility Coordinator(s)—a World where a national third-party acts as the neutral market facilitator for DER providing efficient services to the DSO and/or ESO as required (Fig. 1.3).**

Note:
Since 2018, the development of the DSO and ESO market has predominantly followed World B: In this World, flexibility resources can provide services to multiple

1.2 DNO to DSO Transition

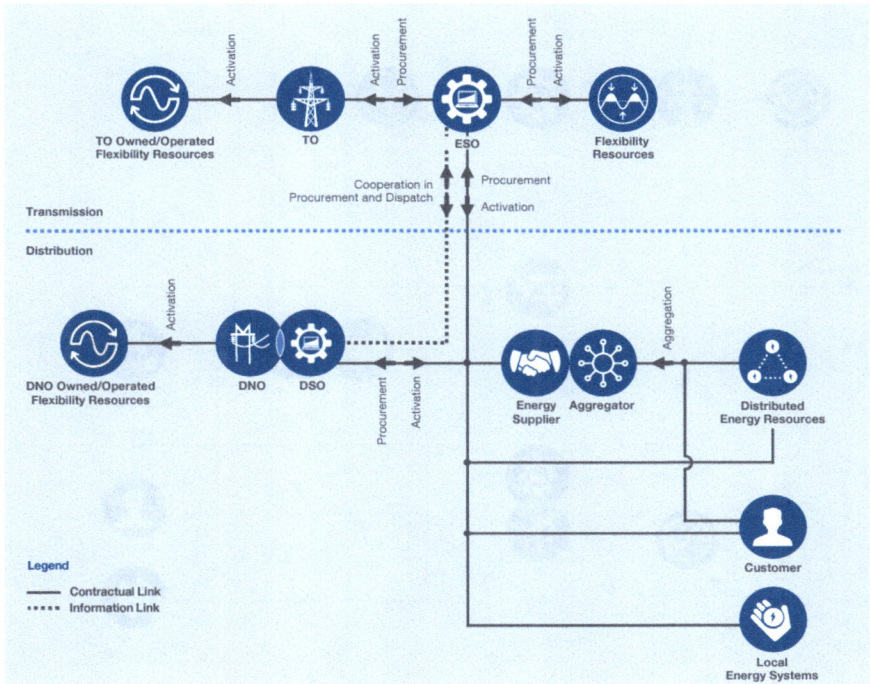

Fig. 1.2 Flexibility market arrangements (World B)—coordinated DSO–ESO procurement & dispatch

System Operators and are able to stack revenues from these differing System Operators. It was recognised that, on occasion, the needs of different System Operators will conflict, and it would be the joint responsibility of these System Operators to coordinate service procurement and dispatch activities. Figure 1.2 illustrates the flexibility market arrangements for this World.

In 2024 the GB regulator, Ofgem, has formalised the establishment of an independent market facilitator, with Elexon taking on the role. Whilst there are some key parallels to the World E, at the time of publication, details of the structure and market arrangements of the new market facilitator are still being developed [1, 2].

Details of the all the Future Worlds, impact assessment, cost benefit analysis and the detailed Smart Grid Architecture Model (SGAM) [3] can be accessed via the ENA Open Networks documents repository, links for which are included in Suggested Further Readings section.

These Future Worlds were developed through discussion with stakeholders at a series of workshops. There were a wide range of options that exist within these Future Worlds. Whilst some assumptions were made about each of these Worlds, it was noted at the time that further detailed policy development work would be required to develop the Worlds.

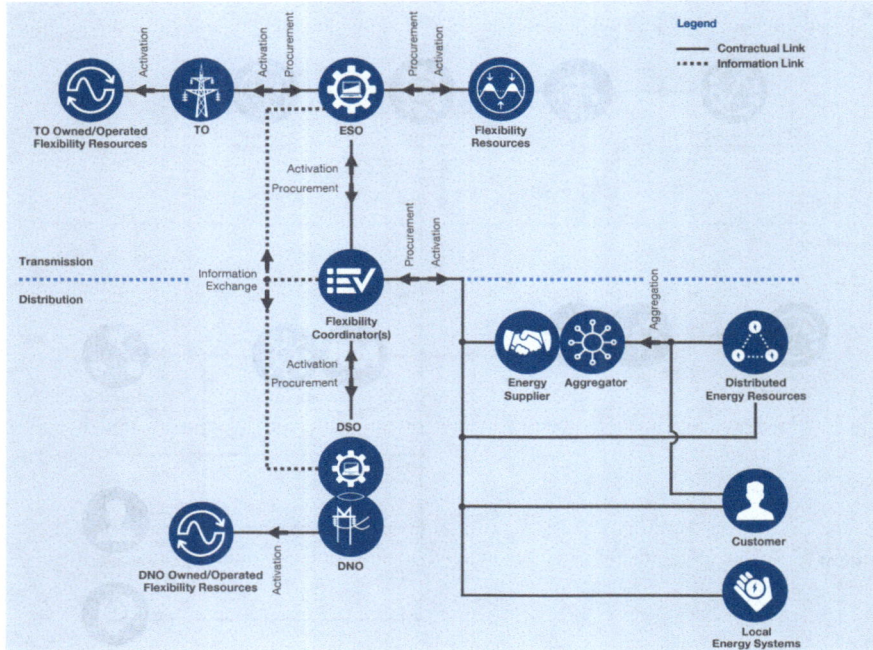

Fig. 1.3 Flexibility market arrangements (World-E)—flexibility coordinator (market facilitator)

It is noteworthy that there are some common principles that apply across all the Worlds. This includes the principle of neutral market facilitation. It also includes the need for network owners and operators to work together to ensure safe, secure and efficient design and operation of systems, including optimisation of existing network assets and management of network congestion and capacity.

1.3 Suggested Further Reading

Listed below are the links to Open Networks publications relevant to this chapter in reverse chronological order. Please note, some of the documents/spreadsheets may now by outdated or superseded as the topics have evolved.

1. DSO Transition Workstream Closeout Paper (2022).
2. DSO transition Conflicts of Interest and Unintended Consequences Register 2022 Final Update (2022).
3. Conflicts of Interest and Unintended Consequences webinar slides (2021).
4. DSO Implementation Plan Report (2021).
5. DSO Implementation Plan Appendices—detailing 8 DSO functions (2021).
6. Impact Assessment on Future DSO Worlds (Benefits) (2019).

7. DSO Transition—Integrated World Costs (2019).
8. DSO Transition Future Worlds Impact Assessment on (Costs) (2019).
9. Consultation DSO Transition Future Worlds Impact Assessment (2019).
10. DSO Transition Future World Impact Assessment Consultation Responses (2019).
11. DSO Transition Future Worlds Impact Assessment Report (2019).
12. DSO Transition—Future World Impact Assessment Report (independent assessment) (2019).
13. DSO Transition Future World Impact Assessment—Consultation Detailed Analysis (2019).
14. DSO Transition Future World Impact Assessment Consultation—ON Response (2019).
15. DSO Future World results (2019).
16. DSO Final Future World results—Sensitivity Publication (2019).
17. DSO Transition Pathway—Least Regrets Analysis 2019 PID Input (2018).
18. DSO Future Worlds—Least Regrets Tables Publication (2018).
19. Modelling DSO transition using the Smart Grid Architecture Model Publication (2018).
20. DSO Transition Pathway—Least Regrets Analysis PID Input (2018).

References

1. Ofgem (2024) Decision: market facilitator delivery body. [Online]. Available: https://www.ofgem.gov.uk/decision/decision-market-facilitator-delivery-body. Accessed 2024
2. Ofgem, "Market facilitator policy framework consultation," December 2024. [Online]. Available: https://www.ofgem.gov.uk/consultation/market-facilitator-policy-framework-consultation. Accessed 2024
3. IEC System Committee Smart Energy—SyC Smart Energy (2012) IEC SRD 63200–SGAM basics. [Online]. Available: https://syc-se.iec.ch/deliveries/sgam-basics/. Accessed 2021

Open Access This chapter is licensed under the terms of the Creative Commons Attribution 4.0 International License (http://creativecommons.org/licenses/by/4.0/), which permits use, sharing, adaptation, distribution and reproduction in any medium or format, as long as you give appropriate credit to the original author(s) and the source, provide a link to the Creative Commons license and indicate if changes were made.

The images or other third party material in this chapter are included in the chapter's Creative Commons license, unless indicated otherwise in a credit line to the material. If material is not included in the chapter's Creative Commons license and your intended use is not permitted by statutory regulation or exceeds the permitted use, you will need to obtain permission directly from the copyright holder.

Chapter 2
Market Principles

2.1 Flexibility First Commitment

A transition to smarter, more flexible energy system would see electricity move in multiple directions This would be underpinned by the principle of flexibility, giving people more control and choice over how Britain's homes, businesses and communities use their electricity and providing new competitive opportunities for them to participate in the energy market. The electricity networks were and continue to remain central to this transformation by integrating and enabling greater amounts of low-carbon and smart technology to help deliver new Flexibility Services markets.

In December 2018, Britain's local electricity network operators launched ENA's Flexibility first commitment.

Through this, all GB DNOs committed to openly test the market to compare relevant grid reinforcement and market flexibility solutions for all new projects of significant value.

As part of the flexibility commitment, DNOs were committed to:

- Opening up requirements for building significant new electricity network infrastructure to include smart Flexibility Services markets as part of day-to-day operations. This covered all new relevant projects of significant value, where local electricity operators face congestion in grid infrastructure that results from increased electricity demand and/or distributed energy projects being connected to the grid.
- Openly test the market to compare relevant grid reinforcement and market flexibility solutions for all new projects of any significant value.
- Working with Ofgem and other stakeholders to develop the forthcoming RIIO-2 price control framework to ensure that the financial incentives that network companies receive were fully aligned with the greater use of Flexibility Services and did not favour the building of new infrastructure where these services are more efficient.

Since that time, Open Networks has made significant strides forward as there has been a major uplift in tenders for Flexibility Services.

The remaining chapters in this book present the individual building blocks in establishing, implementing and growing the local flexibility markets, through the collation of learnings from ENA's Open Networks programme.

2.2 Market Principles

Building on the Flexibility First commitment, the DNOs, the Transmission Owners (TOs) and the national system operator committed to six key principles as Flexibility Services were being rolled out more widely and transitioning to a more efficient, smarter and low-carbon energy system fit for Britain's future. These principles apply to how DSOs will procure and use Flexibility Services to maximise benefits to households, businesses and communities acting as the foundation for enabling, supporting and growing flexibility markets across the country. This commitment ensured the transition delivers flexibility through open, transparent access to markets while maintaining secure and safe electricity supplies and delivering best value for customers, benefiting all customers; flexibility markets not only help reduce the cost of traditional network reinforcement and support decarbonisation progress towards "net zero", but they also create new revenue streams for all customers by enabling them to sell their services from flexible technologies into these new markets.

Extensive and inclusive stakeholder engagement was undertaken, sharing our flexibility developments and listening to wide reaching feedback at every step to arrive at these six market principles. The principles adopted at an early stage of Flexibility Services markets development would be fundamental in defining how the public, businesses and electricity networks interact into the future. The detail underpinning these six principles were since developed into consistent, tangible processes, procedures and agreed working methodologies by all participating electricity networks through the Open Networks programme and they will be important as new roles and responsibilities emerge including the Distribution System Operation and the Market Facilitator.

The six market principles are as follows:

Champion a level playing field: Market neutrality is a fundamental principle of operating Britain's energy network infrastructure. Flexibility Services shall be procured in a way that creates a level playing field for all energy technologies and services. All GB's electricity network companies (i.e. ENA electricity members) committed to facilitating and providing convergence and standardisation for customers in order to support this.

Ensure visibility and accessibility: Opportunities for Flexibility Services shall be highlighted where and when they exist, to play a role in ensuring a secure, consistent energy supply via electricity networks. Barriers shall be removed, enabling all customers to access multiple markets to provide services, for example where they can earn revenue from both the national market and local Flexibility Services markets. This will be undertaken consistently and easily and include sharing data with Flexibility Service providers to develop transparent markets.

Conduct procurement in an open and transparent manner: Remain committed to being open and transparent when deciding how and why services have been procured from different solutions in order to meet network needs, such as Flexibility Services from the market, smart grid solutions and traditional network reinforcement. Common methodologies shall be defined for all network operators to follow and be transparent about the criteria used in decision-making. The guiding principle underpinning all decisions is that the solution chosen must be most cost-effective for consumers, while meeting the needs of all customers, the system and the networks.

Provide clarity on the dispatch of services: Following transparency in the procurement process, a fair and clear approach shall be taken to the dispatch of Flexibility Services to meet electricity system or network needs by setting out the terms and methodology adopted. This includes any decision-making criteria underpinning the dispatch of services.

Provide regular, consistent and transparent reporting: Having committed to be transparent in processes and methods. The DSO shall also provide regular, consistent and transparent monitoring and reporting to provide confidence to the public and ensure all parties learn from what flexibility is used; including why and how this contributes to running energy networks in a smarter, more efficient way. All decisions and reasoning, such as traditional reinforcement compared to Flexibility Services options and cost–benefit analysis, will be clear and readily available. Network companies are committed to sharing these best practices across the wider industry.

Work together towards whole energy system outcomes: All electricity networks shall continue to work closely to facilitate coordinated and efficient arrangements which benefit households and businesses, including activities relating to the decarbonisation of heat and transport. This work shall be expanded to the wider energy industry, including the gas, heat, transport and waste sectors, to ensure that changes deliver the best outcomes for everyone on a whole energy system basis. (This applies to all six of the principles outlined above.)

2.3 Suggested Further Reading

Listed below are the links to Open Networks publications relevant to this chapter in reverse chronological order. Please note, some of the documents may now by outdated or superseded as the topics have evolved.

1. Flexibility Roadmap Implementation of local flexibility markets (2020).
2. Flexibility Market Principles (2019).
3. ENA Flexibility First Commitment confirmation (2018).

Open Access This chapter is licensed under the terms of the Creative Commons Attribution 4.0 International License (http://creativecommons.org/licenses/by/4.0/), which permits use, sharing, adaptation, distribution and reproduction in any medium or format, as long as you give appropriate credit to the original author(s) and the source, provide a link to the Creative Commons license and indicate if changes were made.

The images or other third party material in this chapter are included in the chapter's Creative Commons license, unless indicated otherwise in a credit line to the material. If material is not included in the chapter's Creative Commons license and your intended use is not permitted by statutory regulation or exceeds the permitted use, you will need to obtain permission directly from the copyright holder.

Part II
Local Flexibility Services

Chapter 3
Flexibility Products

When considering how best to solve network congestion using Flexibility Services, differing Flexibility products may be more suitable depending on the specific network scenario. These 'products' are traded as contracted services, generating monetary value for flexibility providers. These Flexibility products can primarily be differentiated by their intended utilisation condition and notice period.

3.1 Legacy DSO Flexibility Products

In 2018, four products were released initially as the common Distribution Flexibility Market Products, as shown in Table 3.1. These were established to meet the anticipated needs of the networks and the capabilities of Flexibility Service providers (FSPs) to provide these services, with the ambition that these products would capture the majority of use cases of distribution network flexibility needs that could be met by Flexibility Services.

3.1.1 Flexibility Service Product Parameters

Flexibility Products can be understood in the context of their technical and commercial parameters. This section of the chapter will introduce these parameters. A pictorial visualisation of Flexibility Product parameters is shown in Fig. 3.1.

1. **Aggregated:** A Flexibility Service made up from more than one DER or DER group.
2. **Available:** Means that the Flexibility Services from the Flexibility Provider have been made available to the DSO and could requested following a Utilisation Instructions.

Table 3.1 Historical nomenclature and high-level descriptors of the GB flexibility products

Product	Description
Sustain	The DSO procures, ahead of time, a pre-agreed change in input or output over a defined time period to prevent a network going beyond its firm capacity
Secure	The DSO procures, ahead of time, the ability to access a pre-agreed change in Service Provider input or output based on network conditions close to real-time
Dynamic	The DSO procures, ahead of time, the ability of a Service Provider to deliver an agreed change in output following a network abnormality
Restore	Following a loss of supply, the DSO instructs a provider to either remain off supply, or to reconnect with lower demand, or to reconnect and supply generation to support increased and faster load restoration under depleted network conditions

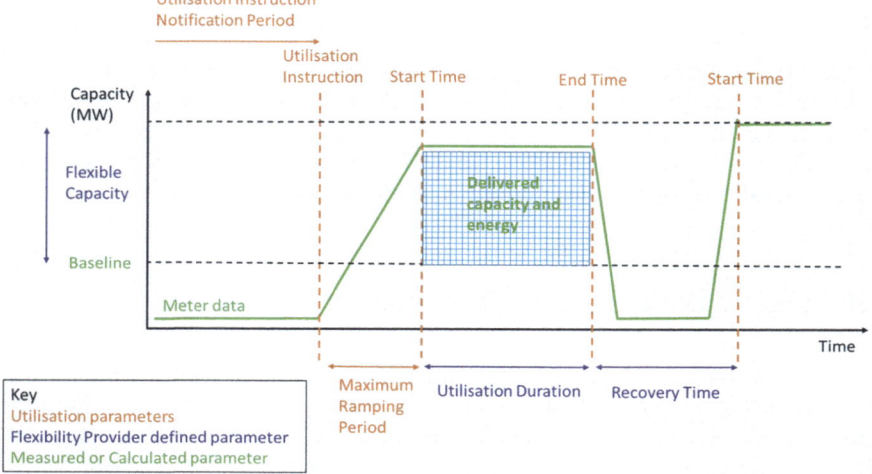

Fig. 3.1 Visualisation of flexibility product parameters

3. **Baseline Capacity:** The capacity against which the Flexible Capacity is measure.
4. **Contractible/Contracted:** Agreed/agreeable between the DSO and Flexibility Provider.
5. **Non-aggregated:** A Flexibility Service made of exactly one DER or DER group.
6. **Utilisation Instruction:** Means an instruction by the DSO to the Flexibility Provider to deliver Flexibility Service.
7. **Constraint management zone:** Means the area in which the Flexibility Services are to be delivered or the feeding area of the DERs being managed.

Flexibility Products are further defined through their commercial parameters (i.e. how they are remunerated). Flexibility Providers are paid a combination of:

8. **Availability Fee**: remuneration to Flexibility Providers to be in a state of readiness within Service Windows to deliver Flexibility Services following a Utilisation Instruction from DSO, paid for each MW of Flexibility made available per hour.
9. **Utilisation Fee**: remuneration to Flexibility Providers for each MWh of Energy Delivered (Flexibility) following a Utilisation Instruction, from DSO, capped at the level of delivery instructed through the Utilisation Instruction.

Availability Fees are payable for each MW of Flexibility made available per hour during the pre-agreed Service Window paid in £/MW/h. Utilisation Fees are payable for each MWh of Energy Delivered during a Utilisation event and are paid in £/MWh.

Note Availability Fees are for Capacity (MW) while Utilisation Fees are for Energy Delivered (MWh).

Description of DSO Defined Parameters for the Legacy Flexibility Products

The Flexibility products are characterised by parameters that provide technical characteristics to and for the Market Provider to describe their Flexibility Service need to the market. The list of parameters that are defines by the DSO and the FSPs are detailed in Tables 3.2 and 3.3 respectively.

Description of FSP Defined Parameters for the legacy Flexibility Products

Building on a gap analysis of individual DSO systems and capabilities at 2018, the range of various parameters were noted, as shown in Table 3.4 and remained in use till the realignment efforts in 2023/24 (discussed in following sections).

3.2 Realignment of Flexibility Products

As the GB distributed Flexibility Services market has matured, the use cases for the standardised distribution Flexibility Market Products have expanded. The four defined Products launched in 2018, have been used to facilitate a range of new and interesting markets, increasing market fluidity extensively. However, due to local, technical reasons and/or market reasons (such as Flexibility Service Providers (FSP) asset capabilities) as well as the high-level parameters used to originally define the products, some localised deviation in the utilisation approach were taken by the DSOs to aid liquidity. Responses to Open Networks stakeholder consultation 2022 suggested that FSPs would prefer more clarification and alignment of the different approaches.

Therefore, the products were then reassessed to investigate these differences and bring forward proposals to align DSO Flexibility Product definitions to enable and ensure that at least 80% of tendered flexibility should be tendered through common products, from the summer of 2024.

Table 3.2 Descriptions of DSO defined parameters for the legacy flexibility products

S. No.	Flexibility product parameter	Description
1	Minimum non-aggregated flexible capacity (MW)	The minimum non-aggregated flexible capacity a flexibility provider may make Available to the DSO
2	Minimum aggregated flexible capacity (MW)	The minimum Aggregated Flexible Capacity a Flexibility Provider may make available to the DSO
3	Minimum utilisation duration capability (Mins)	The minimum amount of time a Flexibility Provider must be able continually hold their Contacted Flexible Capacity, in minutes
4	Minimum utilisation (Mins)	The minimum amount of time a DSO will require the provision of a Flexibility Service from a Flexibility Provider, following a Utilisation Instruction
5	Maximum ramping period (Mins)	The maximum allowed time, once a Utilisation Instruction has been issued or becomes active, for a Flexibility Provider to reach their Contracted Flexible Capacity
6	Availability agreement period (Months/weeks)	The time period before a Flexibility Service is required by a DSO, in which the DSO and Flexibility Provider may agree the Flexibility Provider's Availability Window
7	Utilisation instruction notification period (Months/weeks)	The time period before a flexibility service is required by a DSO, in which a DSO may issue a utilisation instruction to a flexibility provider for the provision of a Flexibility Service
8	Utilisation instruction method (n/a)	The method by which an instruction is issued by the DSO to a flexibility provider for the delivery of a flexibility service
9	Metering requirements (Mins)	The time interval for the measurement of electrical energy by the Flexibility Provider, in minutes

3.2.1 Investigation Approach

A comprehensive review was carried out with a detailed discussion and subsequent agreement as to what was within and out with the scope of alignment. It progressed further to examine:

- The details of the Flexibility Products that had been procured by their organisations since their original release
- An internal review of the detail as to how each company was utilising the product to achieve its technical network requirement
- Areas of alignment
- Specification of Product Parameters

3.2 Realignment of Flexibility Products

Table 3.3 Description of FSP defined parameters for the legacy flexibility products

S. No.	Flexibility product parameter	Description
1	Flexible capacity (MW)	Is the power capacity, in megawatts (MW), that a Flexibility Provider can manage in order to provide a Flexibility Service. This is usually agreed at the contract stage
2	Maximum daily number of utilisation instructions (Nos)	The maximum number of times per day a DSO may issues a Utilisation Instruction to a Flexibility Provider
3	Recovery Time (Mins)	The minimum amount of time required between the end of a Flexibility Service delivery and the commencement of the next Flexibility Service
4	Minimum utilisation per utilisation instruction (Mins)	The minimum amount of time a Flexibility Provider can provide a Flexibility Service to a DSO for a single Utilisation Instruction
5	Maximum utilisation per utilisation instruction (Mins)	The maximum amount of time a Flexibility Provider can provide a Flexibility Service to a DSO for a single Utilisation Instruction
6	Maximum daily utilisation (Mins)	The maximum amount of time a Flexibility Provider shall provide Flexibility Services to a DSO in a single day
7	Maximum weekly utilisation (Mins)	The maximum amount of time a Flexibility Provider shall provide Flexibility Services to a DSO in a single week
8	Availability window (Mins)	A time period that Flexibility Provider makes a Flexibility Service Available to DSO

- Confirmation on the elements that cannot be aligned due to limitations relating to Flexibility Zones or Distributed Energy Resource (DER) parameters (Fig. 3.2).

Stakeholder engagement highlighted three main strands of alignment area for the Flexibility Products, they are.

- Payment Structure – how the service provision shall be recompensed
- Availability – the agreed product parameters that are available for use for the service provision
- Utilisation – the agreed product parameters that will be used to deliver the service provision

It was agreed that for each of these elements the area of alignment would be characterised under (what became known as) the Common Product Parameters (CPP). These are detailed later within this chapter.

There were a range of variables and parameters which are critical to the procurement of Flexibility Services that are not specific to how the Flexibility Product is defined but are crucial to how its need is assessed, how it is procured from the market and how it is measured, dispatched, and settled. Some of these are noted below.

Table 3.4 Legacy flexibility products parameters

	Legacy DSO Flexibility products				
Service Parameter	Sustain	Secure (Scheduled)	Secure (Dispatched)	Dynamic	Restore
When required?	Scheduled forecasted overload	Pre fault/peak shaving		Network abnormality/ Planned outage	Network abnormality
Risk to network	Low	Medium		High	High
Utilisation certainty	High	High		Low	Low
Frequency of use[a]	High	Medium		Low	Low
Minimum flexible capacity	0–50 kW				
Minimum utilisation duration capability	30 min				
Minimum utilisation	15–30 min				
Maximum ramping period	n/a	n/a	< 15 min	< 15 min	< 15 min
Availability agreement period		Contract stage	Week ahead	Contract stage (if applicable)	Contract stage (if applicable)
Utilisation Instruction notification period	Scheduled in advance[b]	Contract stage	Real time	Real time	Real time

[a] Frequency is location specific, defined at the point of procurement
[b] Utilisation requirements may differ to schedule and be instructed in real time

Alignment of some of these are discussed in the following chapters. However, a few of these are network situation/process/infrastructure specific:

- Network needs assessment (Network situation/process specific—Chap. 4)
- Price paid / Valuing Flexibility Service (Network situation/process specific—Chap. 4)
- Pre-Contract Processes (Chap. 5)
- Market Platform (Network process specific)
- Dispatch process and technology (API/Platform etc.) (Chap. 8)
- Settlement Process (Chap. 7)
- Baselining Methodology (Chap. 7).

3.2 Realignment of Flexibility Products

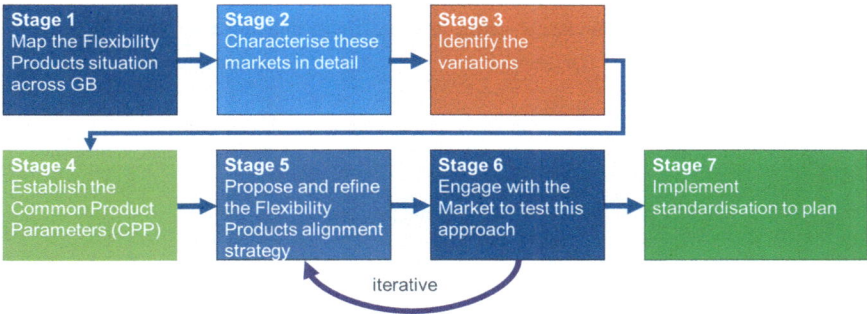

Fig. 3.2 Steps for the Flexibility Products alignment proposals

3.2.2 Common Product Parameters Characterisation

For each strand of the product definitions to be aligned i.e. Structure, Availability and Utilisation, a list of common parameters that described the products was established. These parameters provide economic and technical characteristics for the Market Provider to describe their Flexibility Service need to the market. These were then mapped together to provide the characterisation to facilitate the alignment of the Flexibility Products.

The alignment proposal was to retain the product structures as they were originally designed in 2018 but to be very explicit on the design up front by providing a small finite list of characteristics and options know as Common Product Parameters (CPP) to provide clarity for the Flexibility Service procurement from the outset.

Example

Common Product Parameter Name: Availability Payment Unit
Aligned Definition: The Units used for Availability Payments
Aligned Options available: £/MW/h,
£/MWh,
£/h

To capture the maximum range of technologies that can provide Flexibility Services while providing clarity of elements, such as speed of response; the new product definitions allowed a sub parameter option and have only one aspect that differs between them. As these share all other aspects of CPP, these were not classed as separate products. These aligned sub parameter options are presented in Table 3.5.

Table 3.5 Common product parameters—structure, availability and utilisation

Purpose	Parameter Name	Description	Options
Structure	Payment Structure	How the service is structured (i.e. what is the DSO asking of the FSP)	Utilisation Only Availability and Utilisation
	When Prices are set [procurement timescales]	Time before use that prices are determined	At Trade Operational
Availability	Availability Request Mechanism	How availability is requested from providers	Request initiated by DSO Request Initiated by FSP
	Availability Acceptance timing	When availability is accepted by the DSOs	At trade After trade
	Availability Refinement timing	Can the DSO refine the availability required? If so, when. This determines what availability payments are due	Yes No
	Availability Changes Allowed?	Can FSPs change their availability declaration post acceptance?	Yes No
	Minimum Aggregate Unit Size	The minimum volume requirement for provision of availability	e.g. 50 kW, 1 MW, N/A
	Partial Availability Acceptance Possible?	Whether the DSO can accepted a portion of the offered volume	Yes No
	Time Variable Availability Volumes Allowed	Can the FSP provide different volumes for availability for the different periods within the availability window?	Yes No
	Availability Payment Unit	The Units used for Availability Payments	£/MW/h £/MWh £/h
	Availability Period	The unit of time considered for Availability Instructions	Electricity Forward Agreement (EFA) blocks Settlement Periods Minutes
Utilisation	Utilisation Payment Unit	The Units used for Utilisation Payments	£/MWh £/MW/season
	Utilisation Period	The unit of time considered for Utilisation Instructions	Electricity Forward Agreement (EFA) blocks Settlement Periods Minutes

(continued)

Table 3.5 (continued)

Purpose	Parameter Name	Description	Options
	Delivery Expectation	How the FSP is expected to respond to a utilisation instruction	Continuous (a sustained delivery over the entire utilisation window) Peak Delivery (targeting the maximum response that can be delivered within the window) Maximum generation cap
	Maximum Response Time	Time from Utilisation Instruction to full output	At trade 15 min 2 min
	Payments during response time?	Are FSPs paid during the response time	Yes No
	Minimum Utilisation Time	The minimum time a unit can be instructed for	60 min (Hour) 30 min (Half Hour)
	Minimum Utilisation Volume	The minimum volume that can be instructed	Measured in kW
	Utilisation Instruction Timings	Timeframes in which utilisation Instructions are communicated	At trade Operational Real Time
	Partial Utilisation Instruction possible	Whether the DSO can instruct a portion of the available volume	Yes No
	Time Variable Utilisation Volumes Allowed	Can the DSO vary the utilisation instruction for the different periods within the availability window?	Yes No

3.3 New DSO Flexibility Products

A comprehensive stakeholder engagement was undertaken through consultation with FSPs, the regulator and wider industry stakeholders to understand the challenges for service providers with regards Flexibility Products descriptions. With reference to these challenges a wide-ranging review was carried out on historical utilisation of the products in the Flexibility Services market. This uncovered the specific market components that were driving differences in how the products were being utilised by the DSOs to facilitate delivery of the Flexibility Services.

A comprehensive standardisation exercise was undertaken to develop proposals for aligned definitions of key parameters with the aim of eliminating the differences. A consensus was achieved across all GB DSOs to launch five new/rebranded Flexibility Products for utilising DSO Flexibility Services beyond 2023, as listed in Table 3.6. These new products superseded the previous 2018 released standard products. Rigorous engagement with the Market Facilitation Platform providers was also undertaken to establish routes to integrate the proposals into the market procurement, dispatch and settlement systems.

To aid further understanding of the purpose of the products, some use case examples are presented, to show how the products can be used to provide Flexibility Services. Please note, this is not a definitive list of all use cases.

Peak Reduction

This product seeks a reduction in peak power utilised over time. This response can manage peaks in demand and could be provided by long-term energy efficiency activities. This product could be used where energy efficiency measures are planned that would reduce a sites overall electricity consumption across the year but specifically during high peak periods.

Scheduled Utilisation

In this product, the time that flexibility is delivered has been pre-agreed in advance with the provider. This product will primarily benefit FSPs that cannot respond in real-time or near to real-time. This service can be used by the Network Companies to manage seasonal peak demands and defer network reinforcement, for example.

Operational Utilisation

This product allows for the use case where the amount of flexibility delivered is agreed nearer to real time. This can be utilised to facilitate a change in demand profile from FSPs based on network conditions close to real-time. The assets will be dispatched for the required level of service that is required based upon actual network measurement data thus managing the cost.

A DSO may utilise this product in order to restore network supplies following an unplanned outage/fault where the regulatory funding does not allow for availability payments e.g. customer interruptions (CI).

Table 3.6 List of GB's new flexibility products (for procurement beyond 2023)

Product name	Payment structure
Peak reduction	Utilisation payment only
Scheduled utilisation	Utilisation payment only
Operational utilisation	Utilisation payment only
Operational Utilisation + Scheduled Availability	Availability and utilisation payment
Operational Utilisation + Variable Availability	Availability and utilisation payment

Operational Utilisation + Scheduled Availability

This product procures, ahead of time, the ability of an FSP to deliver an agreed change following a network abnormality. The availability will be defined at the point of procurement and cannot be modified once the contract has been agreed. The assets will be dispatched for the required level of service that is required based upon actual network measurement data, meaning that the DSO/NESO is only paying utilisation payments based upon the actual needs of the network.

An example use case for this product is when a DSO is planning for sufficiency of flexible services contracts based upon short-medium range forecasting of network constraints.

Operational Utilisation + Variable Availability

This product allows for DSOs and the NESO to procure a level of contracted capacity, but then refine the requirements in terms of availability closer to the event. The assets will be dispatched for the required level of service that is required based upon actual network measurement data, meaning that the DSO/NESO is only paying utilisation payments based upon the actual needs of the network.

An example use case for this product is when a DSO is planning for sufficiency of flexible services contracts based upon long range forecasting of network constraints.

3.3.1 New DSO Flexibility Product Definitions

Peak Reduction (Table 3.7)

Scheduled Utilisation (Table 3.8)

Operational Utilisation (Table 3.9)

Scheduled Availability + Operational Utilisation (Table 3.10)

Variable Availability + Operational Utilisation (Table 3.11).

3.4 Implementation of New DSO Products

At the time of publication, all GB DSOs had moved to the new aligned Flexibility Products. All DSOs started some procurement of the new products in parallel to existing products in 2023 and have since migrated to procuring only the standardised Flexibility Products for at least 80% of tendered flexibility (allowing 20% for innovation and trials of new products) by the end of 2024. Plans are underway to migrate existing contracts of legacy products. Figure 3.3 shows a technical summary of the new DSO products' implementation along with a rough mapping of the legacy products.

Table 3.7 Peak Reduction product definitions

Purpose	Common product parameter	Peak reduction
Structure	Payment structure	Utilisation Only
	When prices are set (procurement timescales)	At trade
Availability	Availability Request Mechanism	N/A
Utilisation	Utilisation Payment Unit	£/MWh
	Utilisation Period	Settlement Periods
	Delivery Expectation	Peak Delivery
	Maximum Response Time	N/A
	Payments during response time?	No
	Minimum Utilisation Time	30 min
	Minimum Utilisation Volume	End state: N/A Interim: differs per DSO
	Utilisation instruction timings	At trade
	Partial utilisation instruction possible	End State: Yes Interim: differs per DSO
	Time Variable Utilisation Volumes Allowed	End State: Yes Interim: differs per DSO

Table 3.8 Scheduled Utilisation product definitions

Purpose	Common product parameter	Scheduled utilisation	
Structure	Payment structure	Utilisation only	Utilisation only
	When prices are set (procurement timescales)	At trade	At trade
Availability	N/A	–	–
Utilisation	Utilisation payment unit	£/MWh	£/MWh
	Utilisation period	Settlement periods	Blocks
	Delivery expectation	Continuous	Continuous
	Maximum response time	N/A	N/A
	Payments during response time?	No	No
	Minimum utilisation time	30 min	30 min
	Minimum utilisation volume	End state: N/A Interim: differs per DSO	End state: N/A Interim: differs per DSO
	Utilisation instruction timings	At trade	At trade
	Partial utilisation instruction possible	End state: Yes Interim: differs per DSO	End state: Yes Interim: differs per DSO
	Time variable utilisation volumes allowed	End state: Yes Interim: differs per DSO	End state: Yes Interim: differs per DSO

3.4 Implementation of New DSO Products

Table 3.9 Operational utilisation product definitions

Purpose	Common product parameter	Operational utilisation		
Structure	Payment structure	Utilisation only	Utilisation only	Utilisation only
	When prices are set (procurement timescales)	At trade	At trade	At trade
Availability	N/A	–	–	–
Utilisation	Utilisation payment unit	£/MWh	£/MWh	£/MWh
	Utilisation period	Minutes	Minutes	Minutes
	Delivery expectation	Continuous	Continuous	Continuous
	Maximum response time	2 min	15 min	N/A
	Payments during response time?	No	No	No
	Minimum utilisation time	30 min	30 min	30 min
	Minimum utilisation volume	End state: N/A Interim: differs per DSO	End state: N/A Interim: differs per DSO	End state: N/A Interim: differs per DSO
	Utilisation instruction timings	**Real time**	**Real time**	**Operational—week ahead**
	Partial utilisation instruction possible	End state: Yes Interim: differs per DSO	End state: Yes Interim: differs per DSO	End state: Yes Interim: differs per DSO
	Time variable utilisation volumes allowed	End state: Yes Interim: differs per DSO	End state: Yes Interim: differs per DSO	End state: Yes Interim: differs per DSO

Each DSO provide more detailed implementation details as part of their Market Notification processes. Details of the products that will be procured by each network company will be published via the invitation to tender (ITT) process including the sub-parameter detail. Early indications of the 2025 plans (collated in 2024) for use of the aligned Flexibility Products are listed within Table 3.12. "Y" denotes instances where a DSO has either definitely confirmed their intention to procure or when there hasn't been a confirmation from the DSOs for not procuring a specific variant.

It is important to note that the use of the products will be driven by the network need, evolution of the flexibility market and the market platforms as well as technological development of new solutions deployed by FSPs and/or DSOs.

Following a review of the implementation plans, DSOs noted a few common product parameters will not be aligned from the very start. This is due to some differences in platform tools and dispatch systems utilised. Whilst these are critical

Table 3.10 Scheduled Availability + Operational Utilisation product definitions

Purpose	Parameter Name	Scheduled Availability + Operational Utilisation	
Structure	Payment Structure	Availability and Utilisation	Availability and Utilisation
	When prices are set (procurement timescales)	At trade	At trade
Availability	Availability Request Mechanism	Request initiated by DSO	Request initiated by DSO
	Availability Acceptance timing	At trade	At trade
	Availability Refinement timing	Not allowed	Not allowed
	Availability Changes Allowed?	No	No
	Minimum Aggregate Unit Size	End state: N/A Interim: differs per DSO	End state: N/A Interim: differs per DSO
	Partial Availability Acceptance Possible?	End State: Yes Interim: differs per DSO	End State: Yes Interim: differs per DSO
	Time Variable Availability Volumes Allowed	End State: Yes Interim: differs per DSO	End State: Yes Interim: differs per DSO
	Availability Payment Unit	£/MW/H	£/MW/H
	Availability Period	Settlement Periods	Settlement Periods
Utilisation	Utilisation Payment Unit	£/MWh	£/MWh
	Utilisation Period	Minutes	Minutes
	Delivery Expectation	Continuous	Continuous
	Maximum Response Time	**2 min**	N/A
	Payments during response time?	No	No
	Minimum Utilisation Time	30 min	30 min
	Minimum Utilisation Volume	End state: N/A Interim: differs per DSO	End state: N/A Interim: differs per DSO
	Utilisation instruction timings	**Real time**	**Operational—day ahead**
	Partial Utilisation Instruction possible	End State: Yes Interim: differs per DSO	End State: Yes Interim: differs per DSO
	Time Variable Utilisation Volumes Allowed	End State: Yes Interim: differs per DSO	End State: Yes Interim: differs per DSO

3.4 Implementation of New DSO Products

Table 3.11 Variable Availability + Operational Utilisation product definitions

Purpose	Parameter name	Variable Availability + Operational Utilisation			
Structure	Payment structure	Availability and utilisation	Availability and utilisation	Availability and utilisation	Availability and utilisation
	When prices are set (procurement timescales)	At trade	At trade	At trade	At trade
Availability	Availability request mechanism	Request initiated by DSO	Request initiated by DSO	Request initiated by DSO	Request initiated by DSO
	Availability acceptance timing	At trade	At trade	At trade	At trade
	Availability refinement timing	Week ahead	Week ahead	Week ahead	Month ahead
	Availability changes allowed?	No	No	No	No
	Minimum aggregate unit size	End state: N/A Interim: differs per DSO	End state: N/A Interim: differs per DSO	End state: N/A Interim: differs per DSO	End state: N/A Interim: differs per DSO
	Partial availability acceptance possible?	End state: Yes Interim: differs per DSO	End state: Yes Interim: differs per DSO	End state: Yes Interim: differs per DSO	End state: Yes Interim: differs per DSO
	Time variable availability volumes allowed	End state: Yes Interim: differs per DSO	End state: Yes Interim: differs per DSO	End state: Yes Interim: differs per DSO	End state: Yes Interim: differs per DSO
	Availability payment unit	£/MW/H	£/MW/H	£/MW/H	£/MW/H
	Availability period	Settlement periods	Settlement periods	Settlement periods	Settlement periods
Utilisation	Utilisation payment unit	£/MWh	£/MWh	£/MWh	£/MWh
	Utilisation period	Minutes	Minutes	Minutes	Minutes
	Delivery expectation	Continuous	Continuous	Continuous	Continuous

(continued)

Table 3.11 (continued)

Purpose	Parameter name	Variable Availability + Operational Utilisation			
Structure	Payment structure	Availability and utilisation	Availability and utilisation	Availability and utilisation	Availability and utilisation
	When prices are set (procurement timescales)	At trade	At trade	At trade	At trade
	Maximum response time	2 min	15 min	N/A	N/A
	Payments during response time?	No	No	No	No
	Minimum utilisation time	30 min	30 min	30 min	30 min
	Minimum utilisation volume	End state: N/A Interim: differs per DSO	End state: N/A Interim: differs per DSO	End state: N/A Interim: differs per DSO	End state: N/A Interim: differs per DSO
	Utilisation instruction timings	Real time	Real time	Operational—day ahead	Operational—week ahead
	Partial utilisation instruction possible	End state: Yes Interim: differs per DSO	End state: Yes Interim: differs per DSO	End state: Yes Interim: differs per DSO	End state: Yes Interim: differs per DSO
	Time variable utilisation volumes allowed	End state: Yes Interim: differs per DSO	End state: Yes Interim: differs per DSO	End state: Yes Interim: differs per DSO	End state: Yes Interim: differs per DSO

to the procurement of Flexibility Services, they are not specific to how the Flexibility Product is defined, but are specific to either the asset type providing the Flexibility Service or to the particular flexibility zone where the Service was needed. It was agreed that these would not be part of the alignment process.

These parameters are:

- Minimum Aggregate Unit Size
- Partial Availability Acceptance Possible
- Time Variable Availability Volumes Allowed

3.4 Implementation of New DSO Products

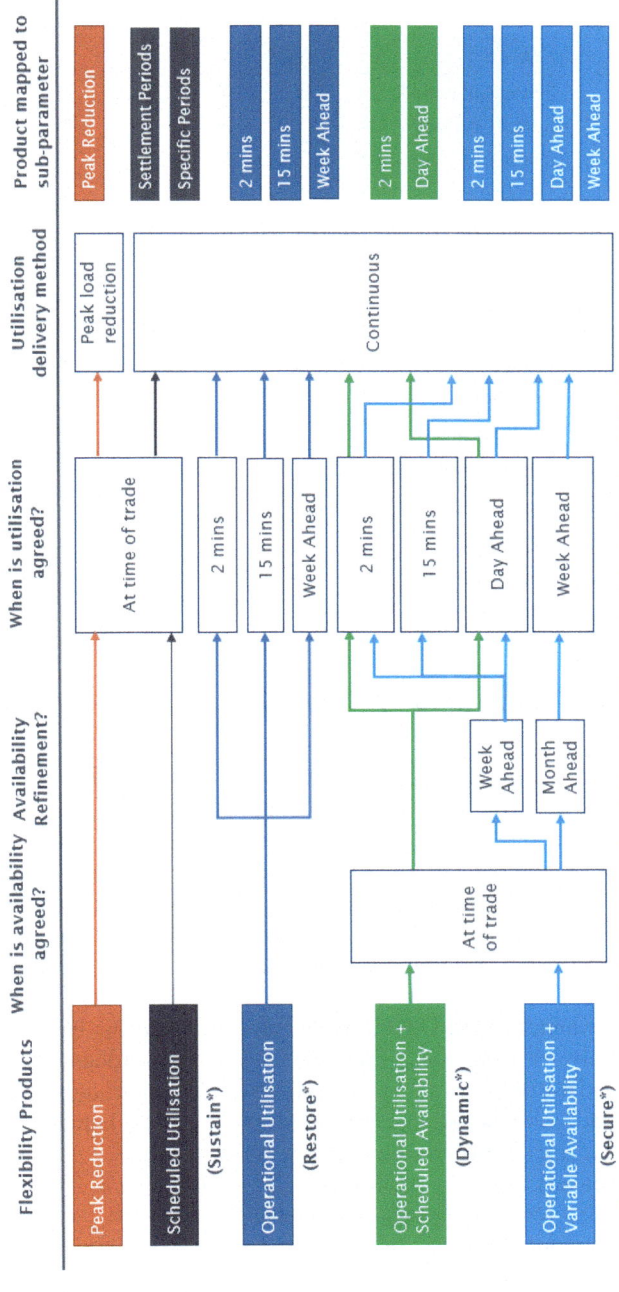

Fig. 3.3 Flexibility products in operation (with a rough old versus new product mapping)

Table 3.12 Early indications of DSOs' plans for procuring new flexibility products

Product name	Sub-parameter	DSO					
		NPg	SSEN	SPEN	NGED	UKPN	ENW
Peak reduction	N/A	Y	Y	Y	Y	Y	Y
Scheduled utilisation	Settlement periods	Y	Y	Y	Y	Y	Y
	Specific periods	Y	Y	N	Y	N	N
Operational utilisation	2 min	Y	Y	Y	Y	N	Y
	15 min	Y	Y	Y	Y	Y	Y
	Week ahead	Y	Y	N	N	N	N
Operational Utilisation + Scheduled Availability	2 min	Y	Y	Y	Y	N	Y
	Day ahead	Y	Y	Y	Y	Y	Y
Operational Utilisation + Variable Availability	2 min	Y	N	N	N	N	Y
	15 min	Y	N	Y	N	N	Y
	Day ahead	Y	Y	N	N	N	Y
	Week ahead	Y	Y	N	N	N	N

* Y includes "definitive and probable procurement plans" and N indicates "no plans to procure"

- Minimum Utilisation Volume
- Partial Utilisation Instruction Possible
- Time Variable Utilisation Volumes Allowed.

3.4.1 Tracking Volumes of the DSO Products Procured

Since 2018, Open Networks has tracked the growth of the local flexibility by collating the total tendered flexibility (the total volume of all Flexibility Products the networks are looking to procure) and the total contracted flexibility (the total volume of flexibility that was available at the right time and location to alleviate network constraints). These are tracked every twelve months. Figure 3.4 shows a summary of the market growth over the past seven years. Please refer to the 'Open Networks Flexibility figures' links in the Sect. 3.5 for more details (including per Product, per DSO, per year breakdown.)

Since 2022, the volumes of the DSO products procured can also be tracked and reviewed via the Network Companies Electricity Distribution Licence Condition 31E: Procurement and use of Distribution Flexibility Services reporting requirements [1]. This license condition mandates each DSO to produce, validate and submit a collated report of the Flexibility Products procured over a year, to the regulator (these reports are typically available on individual DSO websites). In 2022 reporting of the Open Networks Flexibility figures was adapted to align with the DSOs flexibility statement submission, moving from calendar year to regulatory year, and also including technology break up.

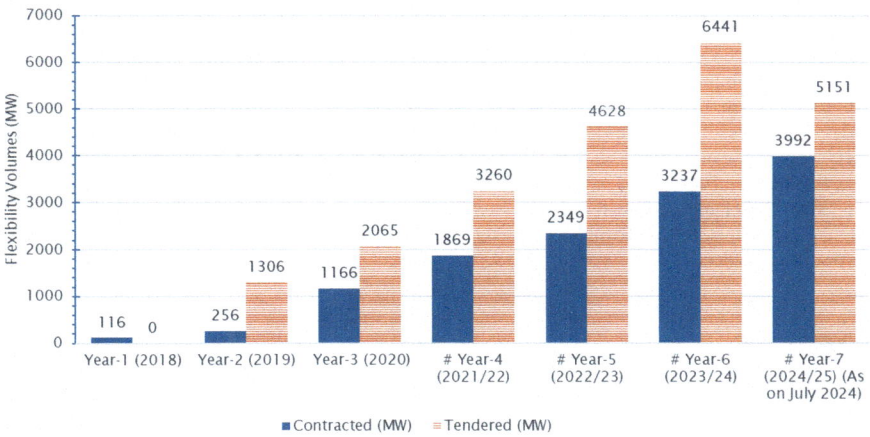

Fig. 3.4 Flexibility procurement volumes ("open networks flexibility figures")

3.5 Suggested Further Reading

Listed below are the links to Open Networks publications relevant to this chapter in reverse chronological order. Please note, some of the documents may now by outdated or superseded as the topics have evolved.

1. Flexibility Products Review and Alignment Report (2024).
2. Flexibility Products: Active Power Products Review (2022).
3. Flexibility Products: Active Power Service Definition and Implementation Plan Publication (2020).
4. Flexibility Figures 2024/25 Publication.
5. Flexibility Figures 2023/2024 Publication.
6. Flexibility Figures 2022/23 Publication.

Reference

1. Electricity Distribution Licence Condition 31E: Procurement and use of Distribution Flexibility Services reporting requirements. Ofgem (2022)

Open Access This chapter is licensed under the terms of the Creative Commons Attribution 4.0 International License (http://creativecommons.org/licenses/by/4.0/), which permits use, sharing, adaptation, distribution and reproduction in any medium or format, as long as you give appropriate credit to the original author(s) and the source, provide a link to the Creative Commons license and indicate if changes were made.

The images or other third party material in this chapter are included in the chapter's Creative Commons license, unless indicated otherwise in a credit line to the material. If material is not included in the chapter's Creative Commons license and your intended use is not permitted by statutory regulation or exceeds the permitted use, you will need to obtain permission directly from the copyright holder.

Chapter 4
Forecasting and Development Plans (Pre-procurement)

In Ofgem's RIIO-ED2 Business Plan Guidance [1] statement, published in September 2021, the three roles and five activities of distribution system operation, were defined as shown in Table 4.1.

Figure 4.1 illustrates the high-level process of network development involving Flexibility Services, where the three stages of pre-tender, tender and post-tender are shown within DSO Roles 1, 2 and 3 and where the Common Evaluation Methodology (CEM) Tool is used within these three stages. In addition, the areas that that have been defined by some DSOs as being within their network options assessment are shown.

In Pre-tender stage the key activities are forecasting and network needs identification. All DSOs generate and publish demand and generation forecasts in their Distribution Future Energy/Electricity Scenario (DFES) documents; and from May 2022 DSOs have developed and published their Network Development Plans (NDP) that detail their network's future needs for the next 10 years.

4.1 Distribution Future Energy Scenario (DFES)

Energy scenarios are produced as part of the annual long-term forecasting activities undertaken by all GB DSOs and the GB Electricity System Operator (NESO).

The Future Energy Scenarios (FES) is an annual process undertaken by NESO. It provides a set of scenario projections for Great Britain and focuses on the whole energy system, through the lens of how the energy system can be decarbonised. FES utilises information, insight and data from all sectors of the energy industry and is used as a fundamental part of the annual transmission network planning, security of supply and national system operability analysis. The FES is also regularly used beyond the regulated businesses, for example in consultancy, academia, public sector and investment planning.

Table 4.1 Distribution system operation roles and activities, prescribed by Ofgem [1]

Role	Activity
Role 1: Planning and network development	1.1 Plan efficiently in the context of uncertainty, taking in account of whole system outcomes, and promote planning data availability
Role 2: Network operation	2.1 Promote operational data visibility and data availability
	2.2 Facilitate efficient dispatch of distribution flexibility services
Role 3: Market development	3.1 Provide accurate, user friendly and comprehensive market information
	3.2 Embed simple, fair, and transparent rules for procuring distribution flexibility services

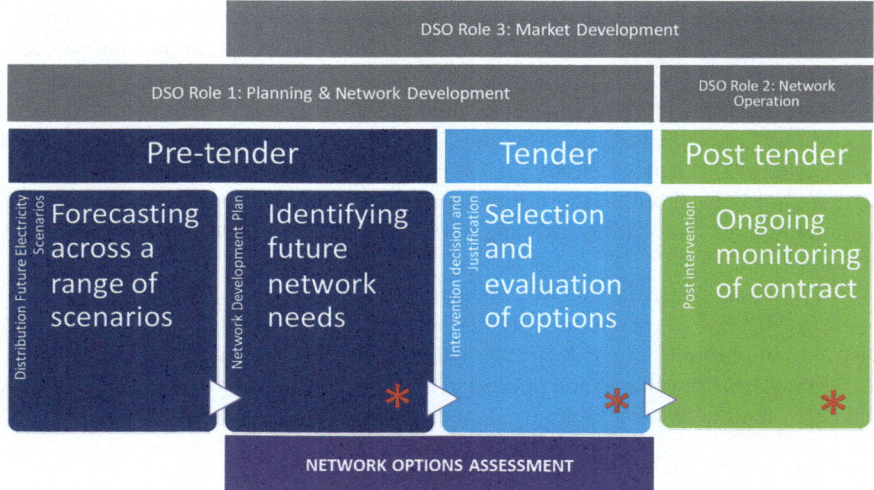

✱ Stages where the CEM Tool is utilised

Fig. 4.1 High level process of capacity provision using flexibility services

The DFES is an annual forecasting activity undertaken by DSOs across Great Britain which builds on the NESO's FES to understand projections at a distribution level. They provide granular scenario projections that incorporate regional factors and can be used at a local level for strategic planning of electricity distribution systems and networks. These projections are informed by local stakeholder engagement to understand the needs, plans and delivery progress of local authorities and other regional stakeholders, e.g., County Councils, Combined Authorities. The DFES provides an evidence base for DSOs to develop the business case necessary to support future investment, including regulated business plans.

4.1 Distribution Future Energy Scenario (DFES)

The transition of DNO planning processes to DSO functions requires DNOs to make well informed, optimal and transparent decisions to justify load related investment that includes among other interventions the Flexibility Services and conventional reinforcement. The DFES are a key component of the DSO planning process. As part of the annual DSO planning cycle, the long-term forecasts of electricity demand, distributed generation and storage from the DFES are used to inform:

- The Long-Term Development Statements (LTDS); and,
- From May 2022 the Network Development Plans (NDP) of DSOs.

The LTDS is a licence condition for all DSOs and presents the future distribution network requirements for the next five years. The NDP is a licence condition for all DSOs to provide stakeholders with transparency on network headroom. NDP is part of the European clean energy package [2], which was adopted in 2019 to help decarbonise EU's energy system in line with the European Green Deal objectives. The NDPs will detail future distribution network requirements for one to ten years beyond publication.

Alignment between DFES and FES and further DFES standardisation across all DSOs can better support whole system planning. It can also facilitate information and data exchanges between DSOs and the NESO to improve the energy scenarios. Alignment and standardisation also help stakeholders better understand and use the scenarios through improved consistency and transparency, see Fig. 4.2.

A key component of the DFES process is stakeholder engagement that includes local authorities (LAs), customers, energy communities and investors. Stakeholders provide valuable inputs to the DFES including both data directly provided by them, but also how implications of network development affect their decisions and connection behaviour. The learnings of this direct interaction with stakeholders put DSOs in a unique position to produce well informed and bottom up forecasts. The DFES should be also seen as an input to stakeholders' plans including Local Area Energy Plans (LAEPs), Climate Action Plan (CAPs) and Local Heat and Energy Efficiency Strategies (LHEES), LA & business decarbonisation plans and any type of planned

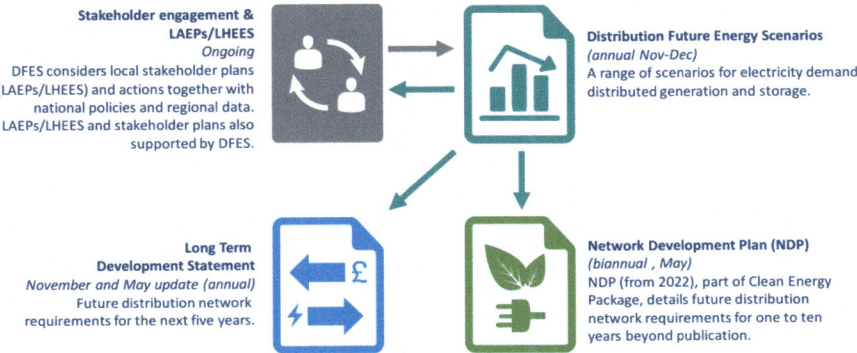

Fig. 4.2 DFES as part of annual DSO planning processes

developments. Granular data and insights on the local drivers for electricity demand and generation help stakeholders make well informed decisions. Focusing on LAEPs/CAPs/LHEES, the DFES can help LAs and other local stakeholders understand how future trends around the electrification of transport, heating, renewable generation and efficiency measures drive planned network developments, capacity provision and Flexibility Service opportunities presented in LTDS, NDP and Flexibility Service tenders. At the same time this bidirectional interaction between LAEP/CAPs/LHEES and DFES means that any developments that are part of LAEPs/CAPs/LHEES need to be shared with DSO forecasting and planning teams to inform DFES, which in turn inform network planning and load related investment.

4.1.1 DFES and FES Alignment

Both DFES and FES use a common scenario framework and definition of technologies, to allow for information exchange and comparison of datasets for all network and system operators. However, as described in the previous section, the DFES and FES serve two different primary purposes, i.e. DFES supports distribution planning and FES supports transmission planning & national system operability. This difference in the scenarios use in practice means that not all datasets between DFES and FES can be directly used or compared as the DFES and FES outputs need to be tailored and fit for purpose for the planning requirements at distribution and transmission level. Whilst a common scenario framework is used, regional variations in projections mean that the summation of DFES forecast ranges may not have identical alignment to the GB FES forecast range. A typical example to understand the alignment challenges is around the electricity peak demand forecasts from DFES and FES.

Both DSO and NESO planning provide the available network capacity to supply demand at distribution and transmission level, respectively. The distribution network planning standards require the use of true (or underlying or gross) demand, which is effectively the maximum demand on the network in case that local generators go off. Any adjustments for generation are made by planning standards (e.g., ER P2/8 [3]), but importantly this approach recognises the real risk in providing sufficient network capacity in local networks in case that one or a small number of local generators disconnect/go off and the network capacity needs to supply higher levels of demand. For transmission network planning the corresponding risk of disconnection of a larger number of small local generators is in practice negligible and what is required for cost efficient planning is to identify as peak demand the maximum measured demand that is suppressed by many small distributed generators. Beyond peak demand there are more examples of limitations for alignment that are due to the different uses of the DFES and FES outputs that aim to properly optimise planning in terms of reducing both costs and risks in DSO and NESO planning. For example, more diverse EV charging profiles should be used at transmission and sub-transmission voltage

4.1 Distribution Future Energy Scenario (DFES)

levels compared to lower voltage distribution levels to properly reflect the increase in demand diversification at higher voltage levels.

In 2020, the DSOs and NESO established an alignment process between DFES and FES, through the Open Networks Programme. The NESO and DSOs adopted the "Initial Alignment & Feedback Model" approach to increase the level of standardisation and improve information exchange. Based on this model the DFES across all DSOs and the NESO FES:

- Use a common scenario framework;
- Use the same high-level assumptions per scenario; and,
- Established the data exchange process of the whole system FES building blocks.

As part of the "Initial Alignment and Feedback Loop Model", all DSOs have agreed to share their whole system FES building blocks. These are effectively forecasting components for each transmission-distribution interface. The building blocks are reviewed on an annual basis. The 2021 building blocks can be grouped at high level as follows:

- Generation building blocks: these include installed capacities in MW for different distributed generation technologies including solar and wind generation, combined heat power plants, biomass and biofuel units, gas and diesel fuelled turbines, hydro generation and others.
- Demand building blocks: these include volumes of domestic and non-domestic customers, floor space projections for non-domestic customers and energy (electricity consumption) forecasts for baseline demand and different low carbon technologies such as electric vehicles (EVs) and heat pumps
- Low carbon technology (LCT) building blocks: these include volumes of LCTs including EVs and heat pumps, as well as volumes of EV chargers.
- Storage and flexibility building blocks: these include data around installed battery storage projections and data on domestic and non-domestic customer availability to provide Flexibility Services.

The common DFES methodology framework

All DSOs follow the same methodology framework to produce their DFES, as illustrated in Fig. 4.3. The agreed framework is a 6-step process where step 1 requires the definition of scenarios, which should include:

- The four common scenarios between DFES and FES (Steady Progression, System Transformation, Consumer Transformation and Leading the Way); and,
- A fifth "Best View" scenario that captures the highest certainty trends in the next 1 to 10 years horizon.

DSOs could have additional scenarios in their DFES. Even though this could increase stakeholder complexity, the use of additional scenarios could be particularly useful in cases that the other five scenarios cannot capture short and long-term uncertainties, e.g. due to accelerated decarbonisation.

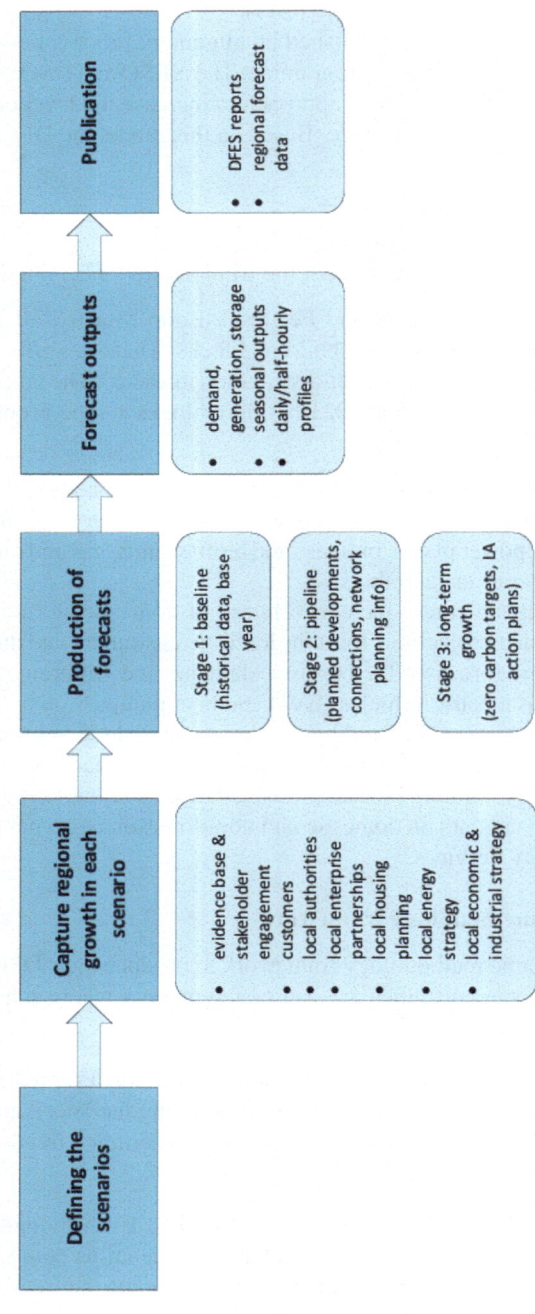

Fig. 4.3 Common DFES methodology framework

4.1 Distribution Future Energy Scenario (DFES)

The second step of the common methodology framework considers the DFES focus on capturing regional growth in the scenarios. Stakeholder engagement is critical to achieve this. This engagement includes a wide range of stakeholders from LAs and local enterprise partnerships to businesses, investors, local communities and other local bodies.

Next step is the production of the forecasts, which consists of three stages:

1. Stage 1—baseline: use of local measurements from substations and local generators to identify historical demand and generation, data processing to identify generation and other technology installations (e.g., heat pumps, battery storage)
2. Stage 2—pipeline: modelling of planned developments (incl. connections offers and acceptances) and other DSO planning information.
3. Stage 3—long term trends: consideration of LA action plans and Net Zero carbon targets to define long term demand and distributed generation growth

The forecasting outputs in the next step cover for all DSOs the necessary load related data to inform network planning and load related investment. This includes for all DSOs electricity demand, distributed generation and battery storage. Focusing on demand, the forecasting outputs are not purely about annual peak demand, but also for seasonal and daily variations. The final step of the common methodology framework is the publication of the DFES report and associated data. All DSOs present their scenarios, assumptions and messages based on the forecasting results in their DFES reports. All DSOs also publish more detailed and granular data that is accessible by stakeholders to help inform their decarbonisation and other plans.

The common DFES methodology framework improves consistency and standardisation, but it also provides sufficient flexibility to DSOs to optimise the use of DFES in network planning through:

- The facilitation of a competitive environment where learnings and best practices from developed tools and research findings can be shared between different DSOs. This competitive environment promotes transparency and sharing more data that improves stakeholder utility.
- Allowing individual DSOs capture particular characteristics in parts of their region, e.g. using additional scenarios to capture accelerated decarbonisation trends whilst allowing stakeholders understand how these compare with the common DFES-FES scenarios.
- Disincentivising static forecasting methodologies where importantly as more LCTs (e.g., EVs and heat pumps) novel modelling approaches would be required to capture customer behaviour.

4.1.2 The Best View Scenario

Further to the four common scenarios, DSOs have agreed (in 2021) to include a fifth Best View scenario in their DFES. The Best View scenario can be used together with the other DFES scenarios to produce the NDP and inform other network planning and reporting processes, e.g. LTDS. The main difference between the Best View scenario and the other DFES scenarios is that it will focus on high certainty in a 1 to 10 years horizon.

As a single scenario that reflects highest certainty, the Best View scenario can:

- Provide clarity and remove the complexity of multiple scenarios. This will help stakeholders understand local demand and generation trends over the short-term.
- Provide the highest certainty basis for assessing network impact and the need for interventions
- Support a "best view"/optimal Network Development Plan together with:

  other network factors including asset health
  all other DFES scenarios that have equal/lower certainty than "Best View" and can provide more insights on the uncertainty range in the > 10 years horizon.

As a guiding principles, the Best View scenario should:

- Be well understood through a transparent development methodology;
- Not allow a broad interpretation, but instead be well defined through an associated methodology that justifies it as the highest certainty scenario among all other DFES; and,
- Be consistent with wider scenario methodologies.

The Best View scenario development methodology incorporates the above mentioned guiding principles and builds on the current DFES methodology; Improvements are encouraged through decentralised thinking and competition across DSOs in developing and sharing learnings and methodologies. It builds on Ofgem's RIIO-ED2 guidance and Load Related Expenditure (LRE) framework and considers opportunities to explain regional sensitivities and justify the "best view" forecast of the future.

The Best View scenario is defined as the highest certainty scenario across all other DFES scenarios, focusing in specific on certainties that can be justified in a 1–10 years horizon acknowledging that longer term forecasts can be more uncertain. To produce the Best View scenario, each building block needs to be checked against three categories as shown in Fig. 4.4 to justify that the developed scenario reflects the highest certainty for the region.

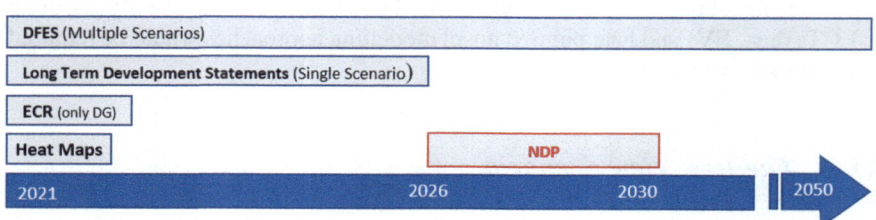

Fig. 4.4 Different network reports and their range of timeframes

1. Best View uses justification criteria for alignment with existing/announced policies
   - "best view" models national policies, which are currently to meet 2050 net zero
   - local policies can be used only if high certainty can be justified that they will prevail over national policies, e.g., to accelerate decarbonisation
2. Best View uses justification criteria for stakeholder engagement inputs.
   - "best view" considers connections pipeline and supporting evidence from local customer engagement or historical performance to model planned developments
   - regional strategic developments modelled in "best view" only if specific justification criteria can be presented, such as LA or UK government backing and secure funding
3. Best View uses justification criteria for regional and local characteristic inputs
   - local trends in "best view" are justified as the high certainty assumptions based on how regional and local characteristics can influence local customer/stakeholder actions differently from a national average (e.g., DFES trends affected by access to gas grid, planning permission requirements, socioeconomic conditions, housing stock and other local factors).

More information around the three categories of the justification criteria and information around how each DSO will produce a Best View scenario can be found in the Suggested Further Reading, Sect. 4.6.

4.2 Network Development Plans

Network planning and development is essential for ensuring the electricity network infrastructure supports customers' needs and is especially important due to the expected challenges as usage of electricity increases due to decarbonisation. Electrification of the heat and transport sectors in the form of Heat Pumps (HP) and Electric Vehicles (EV), connection of distribution energy resources, and incorporation of Flexibility Services are some of the expected challenges. DSOs publish network planning and development documents including Long Term Development Statements (LTDS) and Distribution Future Energy Scenarios (DFES). The aim of LTDS is to provide information on DSOs' existing network and availability of capacity in the short-term, while DFES provide long-term (until 2050) forecasts of future energy pathways in order to capture the envelope of uncertainties. This leaves a gap in information on DSOs' network development in the medium-term. To address this gap, Ofgem has introduced a new standard licence condition (SLC25B) [4] requiring DSOs to produce Network Development Plans (NDP) for a five-to-ten-year window.

The NDP should include the DSOs' plans for reinforcements, the use of Flexibility Services, and network capacity reporting. The new Licence condition implements elements of the Electricity Directive (EU) 2019/944 which is part of the Clean Energy for all Europeans Package [2, 5].

Article 32 of the Clean Energy Package mandates that distribution network operators publish a Network Development Plan (NDP) every two years to provide stakeholders with transparency on network constraints and needs for flexibility. The NDP is to present the 'best view' of planned asset based and flexible network developments over the five to ten-year period.

DSOs already report on distribution network capacity to empower customers and stakeholders, some of whom need information on immediate connection applications, some who may be looking for opportunities to locate developments to provide network services and others looking at long-term plans for our region. Different network reports, as shown in Fig. 4.4, have different purposes and therefore have a range of timeframes and content. The NDP aims to provide value to customers by aligning against existing and evolving reports. It is critical that each form of capacity report has a distinct scope so that stakeholders know where to find information, but the reports must also complement each other with consistency across each DSO's suite of documents.

The understanding of how these reports work together was used to develop a "form of statement (FoS)" for the NDP through the Open Networks Programme. The FoS serves as a consistent construct for all the DSOs to prepare and submit the NDPs.

4.2.1 NDP Form of Statement (FoS)

NDP FoS has been broken into three standalone reports as shown in Fig. 4.5, namely:

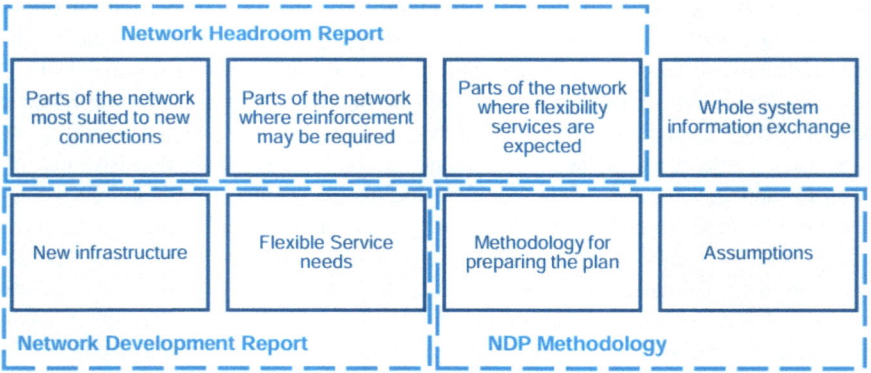

Fig. 4.5 Different parts of the NDP FoS

1. Network Development Report,
2. Network Headroom Report, and
3. NDP Methodology.

4.2.2 Network Headroom Report

The main objective of the Network Headroom Report element of the NDP is to indicate where it is anticipated that there will be network capacity to accommodate future connections, where further capacity may be necessary and where Flexibility Services may be required.

A standardised Network Capacity Report was developed by Open Networks in 2020 prior to awareness of the NDP requirements. However, the NDP encompasses reporting on network capacity and therefore previous learning and outputs were incorporated into the NDP FoS. All DSOs committed to the discretionary publication of a Network Capacity Report in 2021 with the objective of gathering stakeholder feedback to help refine the NDP FoS and on the understanding that it would be consumed within the NDP going forward. This part of the NDP will provide significant value through the additional information to major stakeholders and provide forward visibility of network opportunities. The scope and format of the Network Headroom Report part of the NDP is given in Table 4.2.

Date Range

The Network Headroom Report covers the date range to 2050, or aligning with the final year of the DFES forecast if that differs. This date range goes beyond the 5–10 years of the mandated NDP date range as justified on the following basis:

- Consideration to 2050 matches the DFES date range and so can reflect the uncertainty on long-term network impacts.
- Customers' adoption of low carbon technologies and transition to active networks into the long term means that there will be value in reporting network capacity beyond the current position and the short-term future covered by most of the existing network capacity reports.
- Proven value is presently obtained from network capacity indications for the current year in heat maps and Embedded Capacity Registers, and LTDS and calls for Flexibility Services covering the short-term future. Other stakeholders, including Local Authorities and large developers, are interested in network capacity beyond ten years in the future.

Note:

1. Restricting the capacity reporting to 10 years was also considered inappropriate because developers of potential flexible services require visibility beyond this especially because their development and construction phases can take a number of years. Also, adopting 10 years would not go much further than the five years currently covered by the LTDS.

Table 4.2 Network headroom report parameters

Scope of network headroom report	Deliverable
Date range	Every year to be covered individually between 1 and 10 years
	After the 10th year, this requirement moves to every five years up to 2050 or aligning with the final year of the DFES forecast;
Scenarios	DFES scenarios, plus a 'best view' scenario;
Network capacities and assessment methodology	Demand and generation headroom (available capacity) in MW per reported year per scenario
	Headroom calculations are considerate of financially approved network developments in delivery or planned for delivery, including asset-based enhancements and the use of flexibility services. This may include updates in network developments in the timeframe 0–5 years which were not included in the latest LTDS (November). If included, this must be stated in the accompanying notes and updated in the next LTDS (end May)
	Headroom calculations are considerate of thermal loading and fault level constraints as a minimum
Coverage	Capacity information to be provided for substations where the greatest voltage is greater than 20 kV, normally BSP and primary substations down to and including the primary secondary voltage, typically HV (20 kV, 11 kV or 6.6 kV)
Format and publication	The format of the network headroom report part of the NDP is tabular in nature, presented in Microsoft Excel or similar spreadsheet format. Interactivity can be added to the workbook to improve visualisation of the data
	Guidance shall be included to explain the scope of the data workbook, define each data element and give user instructions
	A contents and version control page is included to ensure that users are able to easily access data, accurately reference the report and view approvals. It also states the dates and versions of critical data sources including the LTDS and DFES
	Licensees shall endeavour to refresh the network headroom report with the latest Licensee's data annually, including the years in between publishing the whole NDP (which shall be published by 1st May every two years)
Information sources	Parameters for the existing network underlying the headroom calculations shall be based on the latest LTDS and incorporate a view of financially approved and planned interventions

(continued)

4.2 Network Development Plans   51

Table 4.2 (continued)

Scope of network headroom report	Deliverable
	Existing and future network demand and generation shall be based on the licensee's latest LTDS and DFES forecasts for demand and generation at the substation
	It is expected that the flexibility services incorporated in the NHR shall be in accordance with DSO flexibility procurement statements and reports or if not included in those reports, they must be stated in the accompanying notes. publication of flexibility procurement statements and reports is a new Standard Licence Condition 31E, and reporting detail is yet to be finalised, but will likely include the location and magnitude of contracted and prospective flexibility services

2. The Network Headroom Report is not intended to provide a customer about to submit a connection application with a latest indication of currently available capacity headroom to connect or to reflect the current status of connection offers, but to provide visibility of the distribution network's medium-term and longer-term capacity needs. It is expected that other more up to date sources of information on current network loading, capacity and headroom, such as heat maps, will be signposted for users seeking information on immediate connections. This is because the forecasted levels of demand and generation already reflect prospective new connections and so the given headroom values may already factor in the capacity needed for the customer's application. For example, from around 2023–2035, the further changes in the DFES generation scenarios predominantly reflect forecast connection projects which are not yet accepted for connection.

Reporting Granularity

The Network Headroom Report will be for every year for the first ten years, and every five years beyond that to the end of the date range. Five-year reports shall "snap" to the years ending in five or zero for simplicity and reflect the indicative nature of the report. The proposed years to be reported was justified on the following basis:

- Reporting every year for the first ten and every five years thereafter matches the needs of different stakeholders.
- Reporting network capacity for every year for only the first five years matching the current LTDS provision was discounted as it doesn't match with the NDP date range.
- Reporting every year for more than ten years was discounted because analysis workload is greater and not justified by increased benefits.
- Ten years matches the period that DSOs undertake detailed analysis to refine investment plans and roughly aligns with design and build timescales. By factoring

in the connection pipeline, forecasts have an element of assurance up to approximately ten years because this encompasses the typical build period of known projects in the pipeline and Embedded
- Capacity Register (formerly the System Wide Resource Register) reports on quoted and accepted connection offers.
- Flexibility stakeholders are interested in detailed short-term requirements, so they target developments in the right places.
- Timescale aligns with the DFES and reporting for each year matches the LTDS and other network capacity reports.
- Reporting for every five years after the first ten years matches the longer period and illustrates greater tolerance and divergence in DFES forecasts reflecting uncertainty in network needs.
- Reporting every five years to the end of the date range provides flexibility providers with indications of the longevity of network needs.
- The simple approach is efficient when considering multiple scenarios.
- Snapping to the years ending in five and zero shows the uncertainty in the results rather than appearing precise.
- The reason for discounting reporting every year to 2050 was the larger volume of data leading to an unmanageable report. Also, future uncertainty means that there is no benefit from reporting for every year up to 2050.

Forecast scenarios

The Network Headroom Report is based on the DFES scenarios, plus a 'best view' scenario. Inclusion of multiple scenarios goes beyond the requirements of the licence requirement and is justified on the basis of:

- Reporting data for multiple DFES scenarios is beneficial because they are used in DSOs planning of network developments included in the NDP. The DFES scenarios plus a best view scenario is appropriate because it matches the view from Ofgem that the NDP should be the DFES with purpose.
- Principles for distribution network forecasts requires the use of the most recent version of the LTDS when developing the NDP and potentially a single scenario.
- Although a single 'best view' scenario may avoid confusion in some stakeholder communities by simplifying the presentation to the highest certainty forecast, it shall be made clear that the 'best view' network development plan is also informed by other network factors including asset health and all other DFES scenarios that provide insights into the range of future uncertainties to avoid foreclosing development options.
- Forecasts with less certainty can help DSOs understand network requirements under more extreme conditions and so prepare the network for the next development stage.

4.2 Network Development Plans

Network coverage

The Network Headroom Report includes as a minimum the following components of distribution networks:

- Substations where the greatest voltage is more than 20 kV, normally:

    Bulk Supply Points, (BSPs) (typically 132/33 kV or 132/66 kV), and
    Primary substations (typically 33/11 kV or 33/6.6 kV).

- Exclude information on individual customers to comply with Licence condition 25B.6.

(In Scotland, 132/33 kV substations are known as Grid Supply Points (GSPs) rather than BSPs, due to the lower transmission/distribution boundary and would therefore be excluded from the network capacity reporting part of the NDP.) Where distribution networks are run interconnected with other DSOs, assumptions on how the interconnected networks have been modelled is included in the accompanying guidance document.

Network headroom reporting focuses on BSPs and primary substations for the following reasons:

- Aligns with the Licence condition clause 25B.3 requirement to cover the "11 kV network and above".
- Data is most readily available for BSPs and primary substations although it is recognised that some DSOs may report on capacity of lower voltage networks if they already have data or if it becomes available in the future.
- Range of network components matches published DFES forecasts.
- Detailed reporting of LV network capacity was discounted because there is currently insufficient accurate visibility of LV distribution/secondary substation loading. DFES publications don't yet present LV distribution/secondary substation forecasts and it was judged that reporting down to this level could lead to an unmanageable report.

It was decided that DSOs should not report on GSP capacities in the NDP as true understanding of the capacity of the transmission-distribution interface requires assessments involving the transmission network owner and National Energy Strategic Operator. Close collaboration on how the power system is operated, running arrangements and short-term asset ratings are required to study the distribution to transmission boundary, especially when considering combinations of arranged outages/faults for contingency analysis. Further consideration of how capacity could/should be reported across the transmission and distribution interface is required and needs input from the NESO, TOs and DSOs.

Assessment Parameters

The Network Headroom Report reflects thermal loading, fault level and voltage constraints if practical. Justification of this approach includes:

- The Network Headroom Report is based on thermal and fault level parameters to adequately reflect significant constraints on demand and generation headroom respectively.
- Consideration of fault level was included because it is a major constraint on generation connections. However, it is recognised that fault level indications must be accompanied with a clear description of the assumptions adopted in the assessment.
- Consideration of voltage rise, and drop was not mandated because of the strong dependency on where connections occur in the network and because voltage issues can be managed for example by restricting a generator's power factor. It is preferable that short-term analysis is based on studies which assess voltage where possible. Where detailed analysis of short-term conditions has shown voltage issues, then the Network Headroom Report may indicate capacity based on this constraint. Such reporting will highlight the issue to advertise, in advance of a formal flexible services tender, the potential for flexible solutions to alleviate voltage constraints.

Assessment Methodology

The Network Headroom Report is based on detailed (network modelling) analysis for the short term where practical, and simple tabular comparisons for the longer-term to 2050.

Capacity on distribution networks not only depends on local conditions, such as the rating of local assets, their configuration and local Flexibility Services, but also the capacity of interconnected networks. For example, the capacity of a primary substation to accommodate additional demand connections may depend upon the available capacity of the upstream circuits and BSP which supplies power to the primary substation. Another example is where capacity at a primary substation to accept additional generation depends on a restriction due to the fault level rating of equipment downstream of the primary substation.

The methodology for reporting network headroom in the Network Headroom Report reflects constraints across the wider network where possible. This is likely to be achieved through power system studies, but they may not be practicable in all cases, for example when future forecast loading is such that power system analysis of the present network is numerically impossible. Methodologies applied in the derivation of the NDP are explained to ensure good understanding of the sensitivities of reported network capacity headroom or deficit. It is important that stakeholders understand when the reported network capacity could be further limited by constraints on interconnected networks.

Network capacity assessments consider appropriate interactions between forecast generation and demand capacities. For example, demand assessments of primary demand capacity may consider demand to be offset by export from forecast additional small LV embedded generation but may not consider export from forecast HV generation to account for this being realised in a single unit which could be out of service.

4.2 Network Development Plans

Similarly, generation capacity assessments may consider the corresponding forecast demand at the time of peak generation export, to reflect the counterbalancing effect of future EVs, heat pumps, new domestic/I&C connections.

Network capacity headroom reporting within the NDP is based on detailed network analysis for the short-term because this greater level of detail is essential within planning to justify network investments and is justified as simpler tabular comparisons of loading versus firm capacity was deemed to be acceptable for the long term to 2050 as detailed analysis is not warranted due to the uncertainty. Also, the conditions in some scenarios may be so extreme that power flow analysis may not converge. Application of a tabular approach is not preferred for the first ten years as firm capacity can be an oversimplification not fully representative of complex networks.

Format and publication

The Network Headroom Report is presented in an Excel workbook format hosted on DSO.

Webpages as this approach is accessible to many users. A tabular format facilitates presentation of headroom of multiple indications of how much additional demand and generation can be accommodated on the network. Table headings shall be as shown in Table 4.3, with the headroom values being given for each scenario and year covered by the report.

The relevance of the reported parameters is described along with the underlying methodologies and assumptions in the NDP Methodology document. Contents include;

- Reference to the DFES document giving full descriptions of the background to the scenarios underlying the forecasts considered in the network capacity evaluations,
- Explanation of how headroom values have been calculated,
- Description of the studies employed to determine headroom values,
- Details of the network limitations considered in the evaluation, and
- Wha network interventions are included and when.

All nomenclature used in the Network Headroom Reporting is consistent with that used in the LTDS data tables. The network headroom values within the Network Headroom Report are evaluated based on data taken from the latest LTDS and DFES.

Table 4.3 Network headroom report headings

Substation Name	Voltage kV	BSP Group	GSP	Substation location	Demand Headroom MW	Generation headroom MW

4.2.3 Network Development Reporting

Alongside network capacity headroom reporting, network development reporting within the Network Development Report element of the NDP serves to provide the reader with valuable additional information on key projects set for delivery in terms of new infrastructure to be installed and upcoming flexible services to be employed. The information is provided with the objective of providing users with foresight whether network plans may impact on theirs and signpost requirements for Flexibility Services so users can target developments.

The scope of the Network Development Report, summarised in Table 4.4, is to provide a list of high-level plans for network interventions and flexible service requirements, specifically:

- For the years 1–10;
- Location of the intervention, covering whole network down to primary substation HV bars;
- Justification for the need for network developments, including the nature of any constraints and the created benefits;
- Development requirements for Flexibility Services and new infrastructure;
- Where a part of an interconnected network is expected to be constrained this may be highlighted as requiring further study to evaluate whole systems approaches, such as a Regional Development Plan;
- Where it resides on the delivery lifecycle (signposting, approved plan with secured financing, in delivery, planned for delivery etc.)

Reporting of network developments is complementary to existing reporting of planned network interventions in LTDS and Embedded Capacity Registers (ECRs) by focussing on anticipatory network interventions in addition to planned interventions likely to be delivered in the period five to ten years in the future.

The LTDS includes a Network Development sections providing the following details on network development proposals within five years of publication and for which finance has been secured:

Table 4.4 Scope of the network development report

Flexibility services	New infrastructure
• Magnitude • Year of intervention, likely duration i.e. number of years in the future • Location of the requirement • Nature of requirement/flexibility product	• Timing and high-level scope of intervention; construction duration (start & finish) • Details of connectivity • Asset quantities approx. circuit lengths, number of transformers etc. • Equipment ratings • Approximate locations, where appropriate

4.2 Network Development Plans

- Work to be carried out
- Expected timescale
- Impact on the distribution network.

The Network Development Report element of the NDP works well with this in that the focus is later, but it is likely that the NDP will reflect planned schemes in that they will be factored into the capacity underlying the assessments reported in the Network Headroom Report. The Network Development Report's focus on development plans for higher voltage networks and the biannual refresh is aligned with the years typically taken to design and construct 33 and 132 kV network reinforcements.

The format of the Network Development Report is as follows:

- Include an introduction to the purpose of the NDP in accordance with Licence condition 25B;
- Include accessible high-level descriptions of plans for network interventions and flexible service requirements;
- Have clear association of where the named schemes and services reside in terms of geography and network connectivity;
- Group development proposals by grid supply point;
- always accessible to many parties (likely to be in a pdf document format, possibly interactive);
- Use nomenclature consistent with the LTDS data tables and schematics. Alternatives would be reviewed with the ambition of future use of CIM;

The Network Development Report is prepared every other year.

4.2.4 NDP Methodology Reporting

The NDP must provide transparency in how it provided its outcomes. Each DSO must:

- Produce a methodology document to cover the end-to-end process
- Provide sufficient detail to allow stakeholders understand sensitivities and extrapolate NDP results
- Ensure format consists of readily accessible data in a manner coordinated with other network operators; Standalone document which is set up to not require significant updates each year.

The objective of the NDP is to be an integral part of DSO network planning and development, rather than simply being numbers produced for publication. Therefore, it is expected that the methodology used to prepare the data underlying the NDP shall be business as usual. Network assessment and planning practices shall be explained

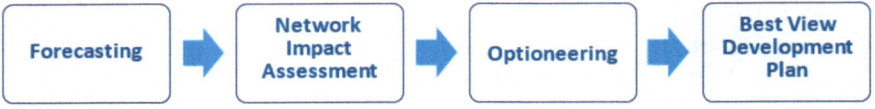

Fig. 4.6 Network planning end-to-end process

in sufficient detail to assist users understand developing plans by undertaking their own evaluations of the detailed information within the NDP Methodology.

The scope of the NDP Methodology to accompany the NDP includes:

- Description of the end to end process shown in Fig. 4.6.
- Assumptions, for example those on the export from existing and accepted generation connections;
- References to published data and network parameters; DFES/"best view" forecast methodologies, Network analysis and assessment methodologies;
- Standard network design and operation of all voltage levels and the nature of alternative network interventions including typical equipment ratings.

Stakeholder use of NDP

A detailed stakeholder engagement was undertaken to understand how to maximise the impact of the NDP. Table 4.5 collates how a range of stakeholders are expected to utilise the NDP and how this impacts the FoS (Fig. 4.5).

Timeline and Links (Tables 4.6 and 4.7).

4.3 Selection and Evaluation of Options (Common Evaluation Methodology)

All the GB DSO have committed to 'market testing' potential flexibility solutions as an alternative means of releasing capacity compared to traditional asset reinforcement. Initially each DSO developed their own methodology for decision making, and there had been a lack of standardisation of approach. The development of a common evaluation methodology was therefore developed, with an intention to provide transparency on how decisions are made to choose the most suitable solution to meet network needs between traditional network asset solutions (reinforcement) and procuring Flexibility Services from generators, storage operators or demand side response. This also addressed a key action outlined in the Ofgem and Department for Business, Energy & Industrial Strategy (BEIS) Open Letter[1] to the ENA in 2019.

In response, Open Networks committed to developing a common evaluation methodology (CEM) for network investment decisions, to be used by all DSOs from April 2021 for the remainder of RIIO ED1 and beyond. It was agreed that this work

---

[1] BEIS and Ofgem's Open Letter to Open Networks 2019.

4.3 Selection and Evaluation of Options (Common Evaluation Methodology)

Table 4.5 Range of stakeholders expected to use NDPs

Stakeholder	Utility (expected use and value)	What does this mean to the FoS?
Developers Property/Building, Generation, Industrial customers, Generation customers	• Understand future network constraints and plans for new capacity to signpost when and where connections would be most suitable (noting that indication of currently available capacity would not be by NDP but by other indications such as DSO heat maps)	Developers need detailed information in accessible format Likely not to be models, but some may be able to use them Location of assets to be included Flexible connection curtailment rates to be included Dispatch rates to be included for flex services
Local Authorities (LA)/ Government organisations	• Understand whether the electricity supply network will provide opportunities or would be a barrier to their advancement of their initiatives, especially decarbonisation or economic stimulus of typically larger areas and requiring greater capacities compared to individual customers	Accessible format They are interested in smaller capacities at lower voltages, EV connections, hubs etc. For local authority public sector software limitations (make it compatible with older versions of Excel) With relation to new housing developments—headroom on EHV networks as affected by larger connections Breakdown of where electrical infrastructure is within a LA
Interconnected electrical network operator e.g. IDNOs, TO and NESO	Understand future needs for considering alternative or whole system solutions Understand how constraints are to be managed to understand impacts on their network to adjust their assessments accordingly Identify synergies in their development for efficient delivery (can work be scheduled at the same time or to avoid conflicts in scheduling and associated risk to security of supply	Accurate evaluation of the capacity on any interconnected network is only possible with greater transparency of capacity on all networks This means that transfer of models would be convenient
Other network operators Transport Gas network Water network	Whole energy wider system analysis Policy optimisation Study the impact of electrification policy Co-ordinate or avoid overlap of works	Same format from all DSOs will facilitate whole GB analysis Accessible

(continued)

Table 4.5 (continued)

Stakeholder	Utility (expected use and value)	What does this mean to the FoS?
Flexibility service providers	Understanding network needs for existing/future service providers More notice of requirements and therefore time to prepare to participate Understanding the location, year of need, longevity and extent of the requirement	Developers need detailed information in accessible format Likely not to be models NCR would be considered for strategic and longer-term decisions and the Flex data details delivered via the Expressions of Interest. The flexibility opportunities would be sign posted in the NCR/NDP e.g. website
Community Energy	Understanding the location, year of need, longevity and extent of the requirement	Accessible format
Universities	Provide resources and understanding for further analysis	All assumptions and approaches to be well explained to support detailed evaluation and extrapolation of analysis and proposals

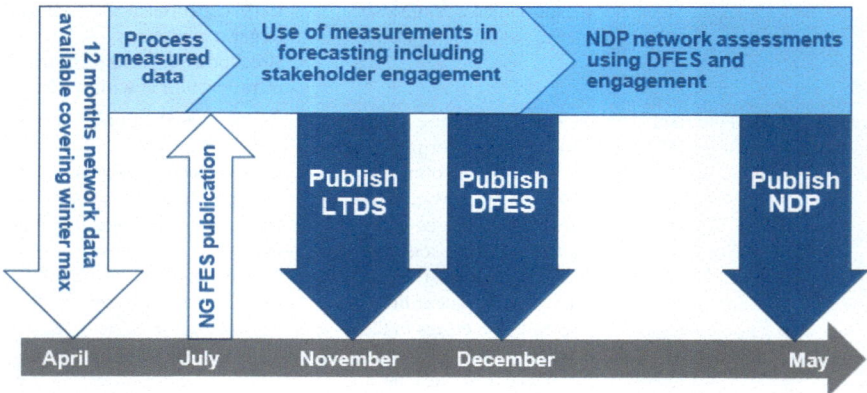

Fig. 4.7 LTDS, DFES and NDP timeline

would be progressed within the Open Networks project. The CEM would be used to decide which intervention to procure to mitigate a reinforcement need, whether that be a Flexibility Service, an asset reinforcement or an alternative innovative solution.

4.3 Selection and Evaluation of Options (Common Evaluation Methodology)

Table 4.6 Links to individual Network companies' DFES

Distribution Network	Link
Electricity North West	ENWL—Distribution Future Electricity Scenarios (DFES)
National Grid Electricity Distribution	NGED—Distribution Future Electricity Scenarios (DFES)
Northern Powergrid	NPg—Network Data
SP Energy Networks	SPEN—Distribution Future Electricity Scenarios (DFES)
Scottish and Southern Energy Networks	SSEN—Forecasting future needs of the network
UK Power Networks	UKPN—Distribution Future Electricity Scenarios (DFES)

Table 4.7 Links to individual Network companies' NDP

Distribution network	Link
Electricity North West	ENWL—Network development Plan
National Grid Electricity Distribution	NGED—Network Development Plan
Northern Powergrid	NPg—Network Data
SP Energy Networks	SPEN—Network development Plan
Scottish and Southern Energy Networks	SSEN—Strategic network planning process
UK Power Networks	UKPN—Long Term Development Statement (LTDS) & Network Development Plan (NDP)

The objective of the CEM was to develop a standard approach for the DSOs and create greater transparency. In turn, this should provide greater visibility and confidence amongst flexibility providers and help stimulate volumes and competition in the market, ultimately reducing costs for network customers.

Since the launch of the first version CEM in 2020, structured engagement was undertaken with users of the tool and third parties, and noted a need to enhance the model in two ways:

1. Develop the treatment and articulation of option value in the tool in order to ensure that the value of flexibility could be fully recognised, particularly under conditions of load growth uncertainty.
2. Expand on the calculations of carbon savings in the tool, making the inputs and calculations more explicit and standardised.

These enhancements were implemented during a second phase of work that concluded in 2021. The outcome of that work was the publication of the second version of the CEM tool.

A third version of the CEM tool has been released in 2024, introducing a 'Simple Ceiling Price' calculation, reflecting the latest Ofgem's Cost benefit Analysis (CBA) tool, and making other minor optimisations to the structure and calculations within

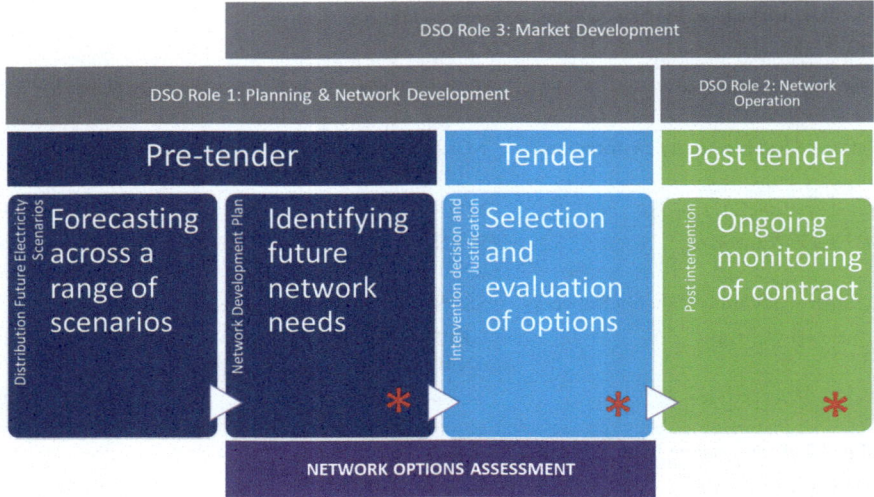

Fig. 4.8 High level process of capacity provision using Flexibility Services

the tool using key learnings from the networks utilisation of the tool to date. The following sections contain a description of the framework and key areas that make up the CEM.

Figure 4.8 illustrates the high-level process of network development using Flexibility Services, where the three stages of pre-tender, tender and post-tender are shown, highlighting where the CEM Tool is used within them. In addition, the areas that have been defined by DSOs as being within their network options assessment are shown.

The following section illustrates the generic high level process network development using Flexibility Services. Individual deployment of stages and associated steps may differ.

4.3.1 Pre-tender Stage

In this stage the key activities are forecasting and network needs identification. All DSOs generate and publish demand and generation forecasts in their Distribution Future Energy/Electricity Scenario (DFES) documents; and from May 2022 DSOs have developed and published their Network Development Plans (NDP) that detail their network's future needs for the next 10 years. Section 4.2.4 contains the links and details about the DFES and NDP to the webpages for each of these documents published by the GB DSOs.

4.3 Selection and Evaluation of Options (Common Evaluation Methodology)   63

Fig. 4.9 High level process for pre-tender stage

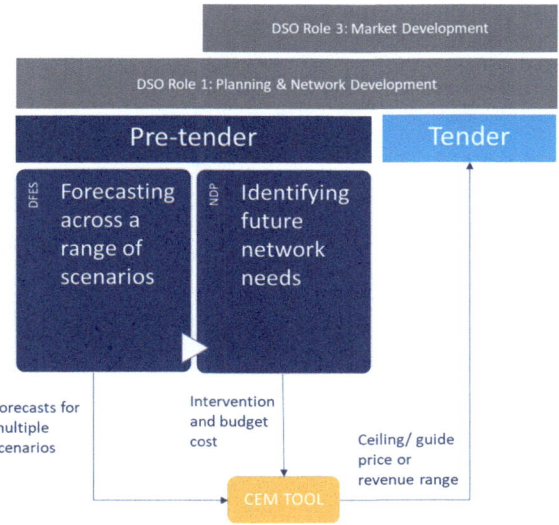

Figure 4.9 shows the inputs to the CEM Tool are (1) a range of scenarios (i.e. forecasts of future demand) for the network under consideration and (2) the budget costs and estimated timings of potential network intervention options being proposed to mitigate the network need. At this pre-tender stage the key output is the ceiling/guide price or a revenue range that is used in the tender process. This information is shared to illustrate the opportunities available to potential responders for participation; this approach has been adopted by DSOs following feedback from stakeholders in previous Open Networks' consultations. The aim of providing this information is to encourage early participation by showing the value of involvement thereby developing the flexibility market so that competition in flexibility provision is driven by price.

As shown the use of the CEM Tool in this pre-tender stage is to create the ceiling or guide price or revenue range for publication in the tender. This output from the CEM Tool is for the pre-tender stage, and is available on the Ceiling Price tab of the current CEM Tool.

4.3.2 Tender Stage

In this stage the key activity is the evaluation of the solution options and the selection of the optimal solution, using the CEM Tool.

Like the pre-tender stage, Fig. 4.10 shows the inputs to the CEM Tool are (1) a range of scenarios (i.e. forecasts of future demand) for the network under consideration and (2) the estimated costs and timings of potential network intervention options being proposed to mitigate the network need; with one of the options being

Fig. 4.10 High level process for tendering flexibility services

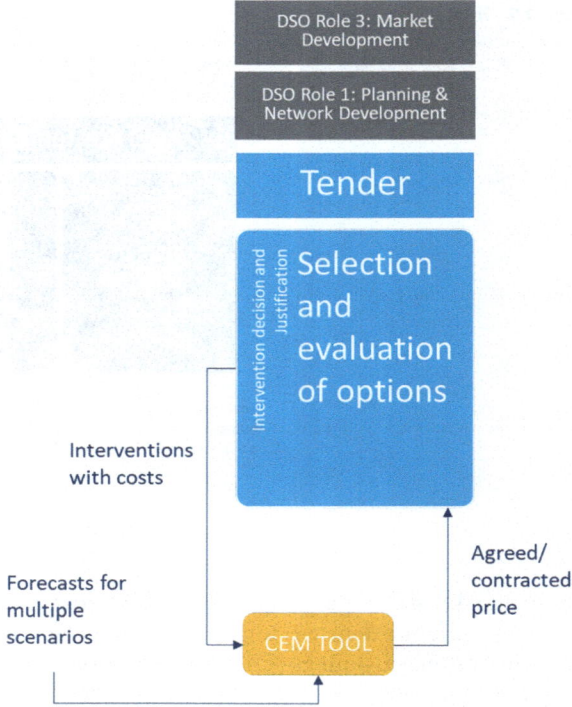

the costs of provision of the Flexibility Services from the tendering process. In this tender stage the key output is the selection of the solution to be employed to mitigate the network need. The DSOs will use a combination of the information from the 'Benefits by strategy' and 'Insights and Reporting' tabs, and where appropriate information from the 'Option value' tab within the CEM Tool.

Where Flexibility Services are designated as the optimal solution the DSO will finalise the contracting process with the Flexibility Services provider(s) and schedule for the services. In some cases, a Best And Final Offer (BAFO) stage may be included to ensure the lowest cost to consumers.

4.3.3 Post-tender Stage

Where Flexibility Services are defined as the optimal solution to mitigate the need it is important that any variances to the provision of Flexibility Services is monitored to ensure that the decision to utilise Flexibility Services remains value for money for customers compared with other network solutions. In this stage the key activities are the administrative monitoring of the service provider (including any post tender delivery milestones), the commercial monitoring of the services provided),

4.3 Selection and Evaluation of Options (Common Evaluation Methodology)

and the planning for future service provision. The administrative monitoring involves checking the financial status of the service provider on a regular basis, whilst the commercial monitoring is tracking the costs of provision of Flexibility Services, including the delivery performance of the flexibility provider, and the planning for future service provision is reviewing on an ongoing basis the expected service requirement and sharing this in advance with the Flexibility Service provider. These ongoing activities ensure that the viability and future use of Flexibility Services remains the optimal approach.

Figure 4.11 shows post-tender monitoring and review of the Flexibility Services contract and requirements. In this stage the DSO will monitor the utilisation of the agreed Flexibility Services against the scheduled current and future services regularly checking that the cost of service provision is within the bounds of the efficient economic decision initially made; the review cycle would be dependent upon the agreed contract length and the extent of the ongoing use of the Flexibility Services. For example, triggers may be put in place on the actual utilisation of the Flexibility Services compared against the scheduled utilisation.

This ongoing monitoring of the contract, the service provider and the utilisation of the Flexibility Services ensures the Flexibility Services solution remains value for money for customers compared with other network solutions.

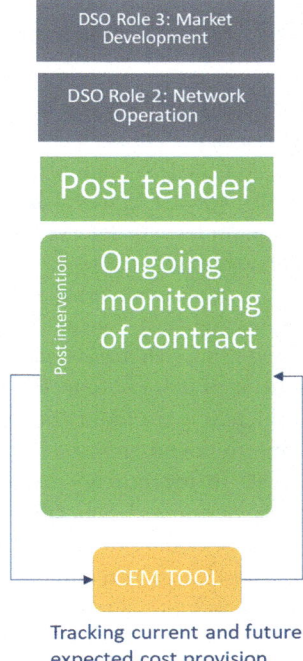

Fig. 4.11 High level process for ongoing contract monitoring and review

Distribution Network options assessment

The Distribution Network Options Assessment (DNOA) is a publication, which outlines reasons behind investment decisions made by a DSO in order to deal with constraints that arise across their license areas. DSOs are now utilising DNOA as the overall process of developing and evaluating a range of intervention/investment options to mitigate a network need. The DNOA suite of documents is published annually and provides transparency to the industry on the decisions DSOs are taking to meet the future capacity needs.

4.4 Common Evaluation Methodology (CEM) Tool

The CEM tool is built to enable DSOs to make investment decisions when comparing Flexibility Products to traditional network interventions. This section describes how the methodology and tool can be used to evaluate the Flexibility Products (discussed in Sect. 3.3), as well as options for alleviating export constraints where curtailment of renewables is occurring. Key areas withing the excel-based tool is shown in Table 4.8.

4.4.1 Assessment Options

Consistent with the Ofgem CBA, DSOs must clearly identify the range of solution options that were considered to meet the specific network need. For each investment decision, the DSO should clearly explain in supporting commentary boxes and worksheets in the CEM tool, what assumptions have been used and which regulations the minimum level of intervention relates to, as well as any calculations that have been done external to the tool.

A section is included in the CEM tool for DSOs to identify and clearly list the options they have considered for each investment decision. This list of options should include those that have been considered and rejected before full costing, and the shortlist of those options that have been considered and costed, with a clear rationale for including/excluding them, which is to be summarised (i.e. a few lines or bullets) in the comment box.

One of the primary use cases for this tool is to evaluate investment in Flexibility Services. When utilising the methodology for flexibility, the model aligns with the standard definitions for Flexibility Products as defined by Open Networks (In Sect. 3.3). The methodology assumes that the Flexibility Products are compared to the baseline scenario of network investment.

The model is built as a cost and benefit comparison tool for all DSOs to utilise when making network investment decisions on an asset by asset level basis. Given that some network interventions will meet more than one network need, there may

4.4 Common Evaluation Methodology (CEM) Tool

Table 4.8 Key areas of the CEM

Key area	Description
Options the model is set to consider	
Outlines the purpose of the methodology and the key use cases for DSOs to put the methodology and tool to use	
Defining the service requirement	
Load growth scenarios	As DSOs are assessing their network needs, they will utilise a scenario or a set of scenarios to determine what their needs would be. These scenarios are key to determine the volume of flexibility required into the future
Flexibility requirements	One of the main uses cases for the CEM is the evaluation of flexibility as a network option. There is specific functionality within the tool for DSOs to input their flexibility requirements into the evaluation of options. This can be tied to the load growth scenarios, or can be input manually
Point of view of economic assessment	
Ofgem CBA	The tool is built on the basis of the Ofgem CBA tool for network investment decisions [6], and as such there is consistency between the tool built and used by DSOs today. There are a number of inputs and values that will remain consistent with the Ofgem CBA, and some areas of the methodology that have been updated as a part of the scope of this project
Time horizon	The methodology sets out to analyse the discounted cash flow of each solution over the life time of an asset, or 45 years. The discounted cash flow starts at the beginning of the deferral period (given that an alternative solution would be used for the duration of the deferral period), and the discounted cash flow extends for 45 years from the end of the deferral period (given that the asset would be utilised fully from that point in time)

(continued)

Table 4.8 (continued)

Key area		Description
Totex treatment		The CEM is designed as a tool to help DSOs evaluate the costs and benefits of different strategies. As such, costs and benefits are represented from the DSO's perspective, which means applying the Totex treatment, consistent with Ofgem's CBA template
Assessment of network intervention options		
Costs	DSOs will input the appropriate costs across the baseline intervention and all alternative network intervention options for all scenarios	
Value of reinforcement deferral	A key element of value within the alternative assessment is the value of deferring network reinforcement into the future. When comparing two potential solutions (a baseline and an alternative network intervention), in many cases the alternative solution will involve the option to defer the decision to reinforce the network to some point in the future, and use flexibility in the meantime Through demonstrating the potential future value across a range of load growth scenarios, this methodology allows DSOs to explore the potential option value that is created in the future by decisions that they would make today. There is a facility within the tool to explore this option value further	
Wider network and societal impacts	The methodology considers some of the wider network and societal impacts of the different network interventions. This includes the impact of network losses, potential asset condition driven changes in CIs and CMLs, carbon emissions, and a range of other impacts measured in the original Ofgem CBA tool	
Outputs		
The outputs from CEM tool include: Table and charts showing, for each scenario and for a range of years, the benefit of flexibility at a specified price Additional insights and reporting: Least Worst Regret and Weighted Average analysis A table showing the maximum ('ceiling') flexibility price that could be justified given the benefits of deferral Results with and without uncertainty ('extrinsic' and 'intrinsic' value), demonstrating the potential option value of Flexibility Services Detailed CBA results for a given number of deferral years for a given scenario		

be a need to utilise multiple instances of the CEM tool to complete analysis across multiple network needs.

Load Growth Scenarios

As a part of network planning processes, DSOs will have individual approaches to define load growth scenarios, and assess network needs against alternative scenarios.

For all Flexibility Products that have network reinforcement as their baseline, these scenarios provide DSOs with a view of what the annual exceedance of the particular

4.4 Common Evaluation Methodology (CEM) Tool

asset that is under assessment (i.e. the amount by which electricity flows will exceed capacity), will be for a particular asset across a range of potential outcomes. There are a number of inputs that are required to determine the timeframe and windows for the decision being made. The "current year" is the year in which the decision to reinforce needs to be taken. Within the input section of the tool, DSOs will manually input the current maximum capacity for the asset (e.g. 30 MVA) and the forward-looking peak network load across the range of scenarios that are being considered within the tool. Peak load is then compared to the current asset capacity to determine the exceedance per year per scenario.

If the use case does not include reinforcement deferral (e.g. using flexibility to reduce Customer Minutes Lost (CML) and Customer Interruptions (CI) risk), the user can disable the model logic relating to network exceedance. The user then inputs the flexibility requirements, Incentives and Penalties, and Carbon impacts manually.

For all use cases where DSOs evaluate flexibility as a network intervention option, they would be required to input the annual flexibility requirements per year per scenario. The user should ensure that enough flexibility is procured to cover both the exceedance and any over-procurement required. The user can either specify the annual availability and utilisation volume directly, or can specify the days-per-year, hours-per-day and average hourly availability and utilisation requirement.

There is an empty Workings worksheet within the model for DSOs to include any justification and/or assumptions around the external calculations for availability and utilisation that are used within the model.

4.4.2 *Economic Assessment*

The tool that has been developed to replicate how costs and benefits are realised by DSOs through the price control framework. As such, it is largely based on the Ofgem CBA tool, and as this framework evolves, the CEM tool should evolve as well. There are a number of inputs and values that will remain consistent with the original Ofgem CBA, and a few key areas of the methodology that have been updated as a part of the scope of this project.

The standard inputs from the Ofgem CBA tool that the CEM methodology uses are listed in Table 4.9.

The methodology sets out to analyse the discounted cash flow of each solution over the life time of an asset, or 45 years. The discounted cash flow starts at the beginning of the deferral period (given that an alternative solution would be used for the duration of the deferral period), and the discounted cash flow extends for 45 years from the end of the deferral period (given that the asset would be utilised fully from that point in time). Within the Ofgem CBA, the Totex Incentive Mechanism (TIM) is applied to all costs. The CEM tool follows the Ofgem CBA template, applying the same Totex treatment.

Table 4.9 Standard inputs from the Ofgem CBA used for CEM

Input	Description
Customer Interruptions (CIs)	In order to evaluate certain asset condition related impacts of network interventions and also to evaluate the 'Restore' Flexibility Product there is a need to quantify and value CIs. The CEM tool utilises the Ofgem standardised value of £s per interruption. DSOs are able to manually insert the number of interruptions into the tool
Customer Minutes Lost (CMLs)	In order to evaluate certain asset condition related impacts on network interventions, there is a need to quantify and value CMLs. The CEM tool utilises the Ofgem standardised value of £s per minute lost. DSOs are able to manually insert the number of minutes lost into the tool
Weighted Average Cost of Capital (WACC)	This value will be unique to each DSO, and is used to convert capital costs into annual costs using each individual DSO's cost of capital
Discount rates	As defined by the Treasury's Green Book [7], this model uses the Social Time Preference Rate (STPR) of 3.5% (less than or equal to 30 years); 3% (greater than 30 years) to discount all costs and benefits, except safety where the Health Discount Rate (HDR) of 1.5% (less than or equal to 30 years); 1.2857% (greater than 30 years) should be used
Losses value	Where expenditures are justified using the reduction of electricity lost, the standardised value for £/MWh lost were utilised within the Ofgem CBA
Carbon prices	In order to calculate the cost of carbon associated with losses, the CEM tool utilises the BEIS traded carbon price [8] (in line with the Ofgem CBA). The CEM tool remains consistent with the Ofgem CBA to quantify carbon emissions that result from network losses
Cost per injury/fatality	In some use cases, DSOs may need to quantify benefits associated with reducing or preventing fatalities and injuries. The treatment in the CEM is consistent with the Ofgem CBA and requires DSOs to draw on guidance set out in HM Treasury Green Book and the HSE [10]. However, for the purpose of evaluating flexibility solutions there is no expectation that these sort of inputs would be required for the analysis

Costs and impact assessment

In order to evaluate the costs and benefits of different network options, the model requires DSOs to input the costs of the baseline [network] intervention. It is assumed that the baseline will usually involve asset reinforcement, but the user can specify other costs (e.g. those associated with losses, CI/CMLs or carbon emissions), provided they can be deferred (or avoided) through the use of flexibility.

4.4 Common Evaluation Methodology (CEM) Tool

In the assessment of the alternative interventions, input values should reflect the cost to the DSO of the alternative solution that is being assessed. In the case of flexibility, the user can either specify the volume and unit cost of flexibility being assumed, or can input the volume of flexibility required and allow the model to find the maximum price of the flexibility solution, beyond which it is no longer cost effective to defer the reinforcement (i.e. a net cost benefit of zero).

The value of the Flexibility Products is primarily derived from the time value of money from deferring large capex expenditure associated with network reinforcement. The CEM tool compares the Net Present Value (NPV) of discounted cash flows of the baseline (reinforcement scenario) with the alternative (flexibility solution) scenario. The CEM tool provides a view of the potential outcomes in terms of NPV for each set of forward-looking load growth scenarios.

Through demonstrating the potential outcomes across a range of scenarios, this methodology allows DSOs to explore the potential option value that is created in the future by decisions that they would make today. There is functionality within the CEM tool that enables DSOs to further explore this option value in two ways:

- Least Worst Regret: Identifying the strategy that minimises the worst regret outcome across the modelled load growth scenarios
- Weighted Average: by assigning probabilities to the each of the load growth scenarios, the user can identify the strategy that maximise the expected net benefit.

This analysis allows DSOs to test different flexibility procurement strategies, as well as understanding the option value (i.e. the value under load growth uncertainty) associated with flexibility.

Different network interventions will have an impact on the amount of electricity lost whilst transporting through the network. The tool accounts for this by utilising the value that is standardised and set by Ofgem in £/MWh, and allowing for DSOs to manually input the volume of losses that they would face with the specific network intervention that is being assessed. The Ofgem input for the £/MWh losses is included in the fixed inputs worksheet.

DSOs are required to input the expected reduction in losses for the baseline scenario as well as all alternative scenarios. The change in expected losses is therefore factored into the assessment of alternative flexibility solutions.

The CEM tool remains consistent with the Ofgem CBA to quantify carbon emissions that result from network losses. The option for DSOs to explicitly include the carbon value of different network solutions is also included in the tool. This includes the option to value the emissions associated with the energy used to release capacity under each option and embedded emissions in the baseline (reinforcement) option.

Where embedded emissions are deferred through the use of flexibility (e.g. delaying the date at which transformer reinforcement is required), the value of carbon (£/tonne) is kept aligned to the Baseline year, rather than reflecting the price in the year in which the deferred reinforcement occurs. The alternative would be to reflect the carbon price in the year in which reinforcement actually occurs. However, because the carbon price increases over time, this would mean that deferring the installation of carbon-intensive equipment would be seen as a negative. This approach is not believed to be appropriate in this context. However, whilst it is believed that the approach leads to sensible model behaviour, it is recognised that there may be alternative approaches that have not been considered. This area may, therefore, require further consideration in the future. In some use cases, there may be additional carbon emissions from alternative network interventions which can be incorporated in an 'other emissions' section of the model.

There is a range of other societal impacts that are included in the Ofgem CBA template, and captured in the CEM tool. These are unlikely to be affected by the choice of network solution, and hence are not expected to be used. However, there is an empty Workings worksheet within the model for DSOs to include any justification and/or assumptions around the external calculations for all societal impacts where appropriate.

The primary use case for this methodology and tool is for DSOs to compare traditional network investment to the use of Flexibility Products where network reinforcement would be the baseline scenario. Whilst, the methodology and report have been developed with this in mind, this methodology and tool can also be used to test alternative investment use cases, such as alternatives for managing export constraints/curtailment. The differences in the ways that these examples would be applied to the methodology have been explained in Table 4.10.

4.5 Illustrative Example

4.5.1 Background and Network Need Description

Over the last three years a GB DSO has identified in its Long Term Development Statement (LTDS), Distribution Future Electricity Scenarios (DFES) and Network Development Plan (NDP) that the demand on a primary substation, due to projected load growth, would exceed firm capacity of the substation in winter 2023/24 and beyond. The value of expected exceedance above firm capacity was small (but not negligible) relative to the firm capacity and the primary substation appeared, at first glance, a suitable candidate for exploring the flexibility options to mitigate the overload condition. The GB DSO's planning engineers have reviewed the primary substation both in isolation and then as one of several interconnected or grouped substations considering the lowest cost intervention options; for example, reviewing the utilisation of adjacent substations and looking into the option of reconfiguring the network to temporarily reallocate capacity across the network.

Having flagged the site as requiring intervention at some point, the DSO closely monitored connections activity in the surrounding area, as this may bring forward the expected date of intervention.

In its annual forecasting cycle for year 2022/23 the DSO confirmed an expected overload on the firm capacity at the local primary substation in winter 2023/24. Identification and declaration of this thermal capacity limitation initiated an optioneering

4.5 Illustrative Example

Table 4.10 Use cases for CEM methodology

Use Case	Key differences in application of methodology
Flexibility – Peak Reduction, Scheduled Utilisation, Operational Utilisation + Scheduled Availability, Operational Utilisation + Variable Availability (Using flexibility to defer network reinforcement)	Base case is reinforcement, triggered by, for example: Expected demand growth in an import-constrained area Expected net export growth (e.g. fall-off in local I&C demand) in an export-constrained area Model allows up to 10 network load growth scenarios to be tested Model shows the benefit of deferring that reinforcement by procuring flexibility for 1 or more years, along with associated benefits (e.g. losses, carbon, CI/CML) User specifies the flexibility that would need to be procured to achieve each year of deferral Output shown in two ways: • Net benefit of deferral by n years given a pre-specified flexibility price (availability and utilisation). User can see both the benefit of deferring by n years and the benefit of deferring by each additional year • Maximum flexibility price that can be justified by reinforcement and associated costs/benefits. Again, this can be seen as the maximum price for, say, a 3-year contract, or the maximum price that can be justified in the 3rd year of deferral
Flexibility – Operational Utilisation (Using flexibility to manage the re-energisation of the network, reducing the number and duration of customer interruptions)	The key difference for this product is that the counterfactual/baseline scenario is the cost of CIs/CMLs and/or the cost of stand-by generation, rather than the cost of network reinforcement Because this product does not relate to network reinforcement, there is no input required into the load growth scenarios Manual inputs would be required to determine the flexibility requirements, because the flexibility requirements are not driven by the network asset exceedance There would be zero capex for the baseline approach For CIs/CMLs inputs—there are two approaches the user could take (1) input zero for the baseline and the incremental change in CIs/CMLs in the alternative, or (2) input the absolute number of CIs and CMLs in the baseline and alternative

(continued)

Table 4.10 (continued)

Use Case	Key differences in application of methodology
Flexible connections[2]: (The DSO incurs some or all of the costs associated with new connections, including flexible connections. The assumption is that this would be facilitated through ANM, where ANM curtails export at network peak loads, allowing faster and cheaper connections)	The CEM tool should only be used to evaluate options against the DSO's share of reinforcement costs, and their expected contribution to the cost of curtailment The baseline is network reinforcement, driven by an export constraint and the connection of exporting assets (e.g. Distributed Generation or batteries) The user will need to enter the revised DSO-attributable reinforcement cost profile under the ANM scenario(s) All other inputs within the model would remain the same, assuming that the TIM would be applied in the same way
Future technology (e.g. dynamic network reconfiguration)	The CEM tool is able to accommodate any consideration of future technology applications, and provides options for users to input the appropriate costs into the CEM tool

process to identify the range of possible solutions, including the budget costing of the most probable network asset intervention that would mitigate the network need. In parallel the necessary information was gathered for tendering for a Flexibility Service to mitigate the network need.

In the Spring 2022 tender round the GB DSO sought to procure a 'Dynamic' Flexibility Services response for a single year in 2023/24 over the winter period to mitigate the capacity overload. Background knowledge of potential flexibility in this area of the network suggested there was limited availability, so the DSO chose a single year to test the market in terms of size and availability of Flexibility Services. The decisions on future years would be made later.

4.5.2 Application of the CEM Tool

Pre-tender stage

As previously described the DSO in the pre-tender stage will generate a ceiling price or price range for application in the tendering process. The DSOs have agreed sharing a ceiling/guide price or a price range that helps stimulate interest in the tender process and encourages participation in the provision of Flexibility Services. This approach aims to facilitate the development of the Flexibility Services market and ultimately drives liquidity and price competition.

---

[2] Further details about flexible connections can be found in Sect. 9.2.

4.5 Illustrative Example

In this example the DSO had only chosen to tender for the single year of 2023/24 but had specified that it is open to multi-year bids from Flexibility Services providers. In addition, this DSO had also confirmed that it is open to the bidder specifying their availability and/or utilisation prices and so the published guide price was the value available in one year for purchasing Flexibility Services.

To generate a guide price the DSO completes the CEM Tool with the information shown in Table 4.11.

The output from the application of the CEM Tool in the pre-tender stage is the guide price with a value of £20,000 per annum for winter 2023/24.

Tender stage

In the Spring 2022 tender round the DSO sought to procure a Flexibility Product in response for a single year in 2023/24 over the winter period with a guide price of £20,000 per annum. In its tender the DSO highlighted its requirement for 0.9 MVA for the winter months of 2023/24 everyday between the hours of 10:00 and 20:30 with availability and utilisation hours estimated as 108 and 48 respectively.

The DSO highlighted that existing or new distributed energy resource connected to the HV (High Voltage) and LV (Low Voltage) circuits supplied by the primary substation will be able to participate in the tender. The postcode areas supplied by the primary substation are highlighted in the tender details. Across the DSOs a range of facilities are employed for enabling any potential provider to check whether their assets in the area supplied by the primary substation. For example, a postcode checker is provided on the DSO's website/Data Portal, or the facility is provided by the hosting platform.

Table 4.11 Data and information for the CEM Tool for the ceiling/guide price or price range[a]

Data and information required	Inputs and source data	Reference
Reference year, start year and flexibility cost input type	INPUT the reference (2022/23) and start (2022) years and define how the flexibility costs will be input (Flex costs directly). Plus, name the scenario (Best View) and strategy (Flexibility)	Control Tab
Timing and costs of network reinforcement (i.e. the counterfactual)	INPUT the budget cost (£630 k) and intervention timing (2023 & 2024) for the network reinforcement. Plus determine the exceedance (0.9 MVA) in 2023	Baseline Reinforcement Tab
Flexibility volumes	INPUT the expected Availability (0.9 MVA) and Utilisation (45 MWh) volumes per year	Flex Volumes and Costs Input Tab
Operating parameters	INPUT initial contract years (1 year) and the scenario (Best View) for intrinsic value calculation. Plus set the probabilities for the weighted average approach (20% for all)	Additional inputs and control Tab

[a] To generate a price range instead of a ceiling/guide price the above steps will need to be undertaken twice using two sets of flex utilisation volumes

As the DSO has confirmed that it was open to the bidder specifying their own availability and/or utilisation price requirements, it has provided a cost calculator to aid the Flexibility Services providers generate appropriate combinations of availability and utilisation prices.

The DSO received multiple bids and evaluated each bid using the CEM Tool with the confirmed information shown in Table 4.12 (updating the parameters in Table 4.11) utilised to calculate the ceiling price. Those bids that were below the ceiling price were accepted.

The tender round was successful and the DSO has entered into a contractual arrangement with the service provider to procure the required Flexibility Product service.

Post-tender stage

All DSOs published the following information on its website (as a part of the reporting requirements of SLC 31E) [9]:

1. The counterparty to the contract;
2. The technology type of the counterparty;
3. The capacity and volume procured;
4. The length of the contractual agreement;
5. The payment structure of the contract; and
6. The price agreed for the provision of services.

Table 4.12 Confirmed data and information for the CEM Tool evaluation

Data and information required	Inputs and source data	Reference
Flexibility volumes and costs	INPUT the Availability (0.9 MVA and 108 h) and Utilisation (48 h) volumes per year; and the Availability (£/MW/h) and Utilisation (£/MWh) bid prices	Flex Volumes and Costs Input Tab
Losses and other penalties (optional)	INPUT the impact of losses (− 18.8 MWh) and any impact from the customer interruptions and customer minutes lost (zero). The associated Losses Tool should be used as the default method for calculating losses for input into the CEM Tool	Incentives and Penalties Tab
Carbon impact (optional)	INPUT the volume and types of assets for the reinforcement to calculate their embedded carbon impact (not utilised in this instance)	Embedded emissions input Tab
Operating parameters	INPUT initial contract years (1 year) and the scenario (Best View) for intrinsic value calculation	Additional inputs and control Tab

4.6 Suggested Further Reading

Listed below are the links to Open Networks publications relevant to this chapter in reverse chronological order. Please note, some of the documents/spreadsheets may now by outdated or superseded as the topics have evolved.

1. Network Development Plan (NDP) Form of Statement (FoS) 2022 Update v1.4 (2022).
2. FES & DFES Alignment Requirements and Justification (2022).
3. FES and DFES Building Blocks (2022).
4. Alignment of Grid Supply Point Definition (2022).
5. FES and DFES Purpose of Energy Scenarios (2022).
6. Network Development Plan (NDP) Form of Statement Template and Process (2021).
7. Best View Scenario Description and Justification Criteria (2021).
8. Network Development Plan webinar slide deck (2021).
9. Coordination of FES and DFES (2020).
10. Distribution Future Energy Scenario (DFES) standardisation (2020).
11. DSO Standard Network Capacity Report Publication (2020).
12. Common Evaluation Methodology Tool v3 & Supporting Materials (2024).
13. CEM Valuing Optionality Consultation Summary and Next Steps (2022).
14. Good Practice Guide for CEM Tool (2022).
15. Summary of CEM Tool Stakeholder Feedback (2022).
16. CEM and Whole System CBA Interactions Report (2022).
17. CEM Valuing Optionality Consultation Summary and Next Steps (2022).
18. Common Evaluation Methodology for Network Investment Decisions—Use Cases (2020).
19. Common Evaluation Methodology Tool for Network Investment Decisions—Background (2020).
20. Common Evaluation Methodology Tool Consultation (2022).
21. Common Evaluation Methodology and Tool Webinar Slide Pack (08 Apr 2021).
22. Common Evaluation Methodology (CEM) Tool V 2.2 (2022).
23. Common Evaluation Methodology (CEM) and Tool v2.1- User Guide (2022).
24. Statement for Common Evaluation Methodology for Network Investment Decisions v2.0 (2022).

References

1. RIIO-ED2 Business Plan Guidance, Ofgem (2021). [Online]. Available: https://www.ofgem.gov.uk/publications/riio-ed2-business-plan-guidance
2. "Clean energy for all Europeans package," Energy, Climate change, Environment—Europian Union (2019). [Online]. Available: https://wayback.archive-it.org/12090/20241209144917/https://energy.ec.europa.eu/topics/energy-strategy/clean-energy-all-europeans-package_en

3. Engineering Recommendation P2 Issue 8 2023," Energy Networks Association, 2023
4. Guidance on the domestic marketing licence condition Gas and/or Electricity—Standard Licence Condition 25, Ofgem (2010) [Online]. Available: https://www.ofgem.gov.uk/publications/guidance-domestic-marketing-licence-condition-gas-andor-electricity-standard-licence-condition-25
5. "EU energy policy," European Comission (2024) [Online]. Available: https://energy.ec.europa.eu/index_en
6. RIIO-ED2 Data Templates and Associated Instructions and Guidance, Ofgem (2021) [Online]. Available: https://www.ofgem.gov.uk/consultation/riio-ed2-data-templates-and-associated-instructions-and-guidance
7. The Green Book—Central Government Guidenceon Appraisal and Evaluation, HM Treasury (2022)
8. A brief guide to the carbon valuation methodology for UK policy appraisal, Department of Energy and Climate Change (DECC) (2011)
9. Electricity Distribution Licence Condition 31E: Procurement and use of Distribution Flexibility Services reporting requirements, Ofgem (2022)
10. "Appraisal values or 'unit costs'," Health and Safety Executive (HSE), 2022/23

Open Access This chapter is licensed under the terms of the Creative Commons Attribution 4.0 International License (http://creativecommons.org/licenses/by/4.0/), which permits use, sharing, adaptation, distribution and reproduction in any medium or format, as long as you give appropriate credit to the original author(s) and the source, provide a link to the Creative Commons license and indicate if changes were made.

The images or other third party material in this chapter are included in the chapter's Creative Commons license, unless indicated otherwise in a credit line to the material. If material is not included in the chapter's Creative Commons license and your intended use is not permitted by statutory regulation or exceeds the permitted use, you will need to obtain permission directly from the copyright holder.

Part III
Deployment of Flexibility Services

Chapter 5
Procurement Process

DSOs typically have had two structured procurement cycles per year. This balance ensures that as network flexibility requirements arise, they are communicated to the market as quickly as possible without overloading the marketplace with excessive procurement rounds. Each of the procurement cycles will have the following key milestones:

1. Signposting of requirements—DSOs publish network information highlighting areas where flexibility could arise in the near future. This is a good practice guideline as although every effort is made to provide network visibility as far in advance as possible, it may not be possible for all network constraints
2. Publish firm requirements—DSOs publish full network requirement, including the type of Flexibility Product that is set to be procured and the expected service delivery window
3. Publish tender/open procurement window—invite flexibility providers to formally bid in the procurement window/take part in auction
4. Close tender/procurement window—close opportunity for providers to bid/take part in auction
5. Contract execution—agreement of terms between contracting parties
6. Asset testing—DSO/flexibility provider test asset as per Flexibility Product/contractual requirements
7. Delivery window—flexibility provider is in a position to provide DSO with services.

5.1 Publication of Requirements

Constraints on networks are confined to specific geographical locations; DSOs define these locations as Constraint Management Zones (CMZ). Based on the common valuation assessment, DSOs may then seek to procure appropriate Flexibility Services (desired products of suitable volume) within these CMZs.

As on the date of publication, GB DSOs use independent market platforms to socialise the needs for flexibility.

The platforms presently in use (as of 2024/25) are

- *Piclo Flex*—Used by SPEN and NPg
- *Electron Connect*—Used by SSE and ENW
- *Local Flex*—Used by UKPN
- *Market Gateway* and *Piclo Flex* Used by NGED.

The platforms typically provide a map of CMZ locations and a postcode finder to allow potential flexibility providers to confirm their site is within the limits of the zone. The platforms also facilitate prequalification of assets.

5.2 Pre-qualification Process

Pre-qualification is the process to verify the compliance of a Flexibility Service provider (FSP) and/or the asset group with the requirements set by the DSOs. All assets looking to participate in the flex market are required to go through the prequalification processes.

There are several approaches to pre-qualify flexibility providers before the commencement of actual competition. In each of the methods some form of Pre-Qualification Questionnaire (PQQ) is used. PQQs typically consist of a list of organisational/technical/commercial requirements set forth by the DSO to ensure that flexibility providers meet the minimum eligibility criteria. (Details regarding the eligibility criteria are discussed further in the gap analysis below.)

Historically there was no prescribed approach to pre-qualification since the effectiveness and purpose is largely predicated upon the population of flexibility providers that are present in a licence area (or within a CMZ). All DSOs recognise the market benefits offered from standardising their pre-qualification processes. During 2022/23, through Open Networks, the DSOs sought to standardise procurement processes for market entry into Flexibility Services. This includes both commercial and technical criteria and will ultimately provide a standard pre-qualification data template which can be used by DSOs for adoption across their Flexibility Service procurement processes, whether they are facilitated through manual or system/platform procedures.

5.2 Pre-qualification Process

Table 5.1 Number of pre-qualification criteria questions per DSO (pre-alignment)

	Commercial	Technical	Total
Electricity North West (ENWL)	51	57	108
Northern Powergrid (NPg)	16	47	63
SP Energy Networks (SPEN)	25	30	55
Scottish and Southern Electricity Networks (SSEN)	45	32	77
UK Power Networks (UKPN)	44	80	124
National Grid Electricity Distribution (NGED)	17	16	33
Northern Ireland Electricity Networks (NIEN)	80	22	102
Grand Total	278	284	562

5.2.1 Review of Pre-qualification Process

A gap analysis was undertaken in 2022 that compared existing approaches to pre-qualification employed across DSOs and NESO to identify the best practices followed by a comprehensive stakeholder engagement to review and consolidate the views of the FSPs.

Table 5.1 shows the number of pre-qualification criteria questions asked by the DSOs in 2022.

The gap analysis concluded there was a significant amount of variation across DSOs with both number of criteria asked, and the complexity covered. One FSP would be required to answer 562 distinct questions in order to access markets across all the DSOs.

Criteria Consolidation

Throughout the process of agreeing a consolidated set of standardised pre-qualification criteria, Open Networks shared its proposals with stakeholders through various stakeholder engagement groups. Feedback received and actioned is detailed in Table 5.2.

5.2.2 Aligned Pre-qualification Criteria

Improvements in clarity was sought on the questions beings asked and remove unnecessary barriers to market entry, though

- Agreeing clear questions, reducing the number of 'in the case of' questions
- Significant reduction in the number of questions such that it is less onerous to complete
- Reduced insurance to only two requirements (Employers and Public Liability) and values lowered to minimum statutory limits (£5 m).

Table 5.2 Stakeholder Feedback on PQQ (pre-alignment)

Area	Feedback received	Our approach/ decision
Contracting Methods	DSOs should ensure that planned or un-recruited assets still have an opportunity to enter into procurement activities for longer term services	Retain the ability for planned and unrecruited assets to pass pre-qualification so that DSOs can consider them for the delivery of longer-term services
Technology Classification	The standardised Technology Classes should not restrict new emerging technologies from entering into flexibility opportunities	Aligned the technology classes with the criteria Ofgem require DSOs to report against. Currently Ofgem only has one broad bucket for 'Flexible Demand' additional technology classification were added within this bucket to increase our visibility of specific technology types. This is particularly important for the allocation of technology-specific baselines which are increasingly more in use
Insurance Requirements	Insurance requirements should reflect the nature of the services being delivered and respect at statutory limits	Undertook internal procurement and legal reviews to challenge instances where DSOs were asking for excessive insurance levels or for insurances that didn't reflect the service (i.e. motor insurance). The insurance requirements across all DSOs were agreed and standardised to the following minimum thresholds; Employers Liability £5 m Public liability £5 m
Due Diligence Questions	Current approaches have varying levels of complexity to complete dependent on the DSO. DSOs should all adopt the approach with the lowest barriers to entry	Feedback was greatly in support of all DSOs adopting the Due Diligence criteria already aligned across UKPN, ENWL and SPEN. This was agreed. With the addition of some further simplification of the criteria following wider DSO input

- Adopted a clear approach to accommodate DER at different stages of connection or recruitment.
- A standard set of groupings of technology type was adopted, aligned to DSOs reporting needs and ensured the DER parameters are adequately future proofed.

A consolidated and standardised prequalification criteria to now include;

- 31 standard questions for Commercial Qualification
- 35 standard questions for Technical Qualification.

This standardised criteria was formatted in a structure that allows its use as replicable data layer with common field labels, regardless of means of submission. Tables 5.3 and 5.4 provide examples of the commercial and technical submission formats respectively.

5.2 Pre-qualification Process

Table 5.3 Standardised commercial PQQ template (31 questions)

Area	Field NAME	Commercial qualification questions	Allowable Responses	Pass Criteria	Description
Reference	COMM_REF_ORG1	Assigned Reference for this Organisation, if known	Free text	Completed	If the Organisation has previously been submitted and assigned a reference, this can be detailed here
Company Information	COMM_CI_CNAME	Registered or legal name of the contracting party	Free text	Completed	Full registered/legal name of the party wishing to enter into the flexibility contract
	COMM_CI_REGNO	Company Registered Number [Or Charity/Trust]	Free text	Completed	Registered No. as shown on Companies House. Charity or Trust Registration No. is also acceptable
	COMM_CI_REGA1	Registered address 1	Free text	Completed	Registered Address line 1
	COMM_CI_REGA2	Registered address 2	Free text	completed	Registered Address line 2
	COMM_CI_REGA3	Registered address 3	Free text, blank	Completed, blank	Registered Address line 3
	COMM_CI_POSTC	Registered address postcode	Free text	Completed	Registered Address Postcode
	COMM_CI_FIRST	Key contact First Name	Free text	Completed	First name of main contact in respect of this submission and any subsequent contract

(continued)

Table 5.3 (continued)

Area	Field NAME	Commercial qualification questions	Allowable Responses	Pass Criteria	Description
	COMM_CI_CLAST	Key contact Last Name	Free text	Completed	Last name of main contact in respect of this submission and any subsequent contract
	COMM_CI_EMAIL	Key contact email	Free text	Completed	Email address of main contact in respect of this submission and any subsequent contract
	COMM_CI_TELNO	Key contact number	Free text	Completed	Contact telephone number of main contact in respect of this submission and any subsequent contract
	COMM_CI_WEBSI	Organisation website	Free text	Completed	Web address for contracting party
	COMM_CI_RELAT	Legal relationship with flexibility asset/s	Owner, Operator, Aggregator	One code completed	Declare relationship to asset/s, only select one
	COMM_CI_VATNO	VAT Registration Number	Free text	Completed	VRN of the contracting party
Terms and Conditions	COMM_TC_ACCEP	Confirm; You have read the applicable ENA_Standard Flexibility Services Agreement and understand it will be a requirement to accept this Agreement in order to form any contract for the delivery of Flexibility Services	Y, N	Y	Contracting party must confirm their acceptance of the ENA_Standard Flexibility Agreement applicable to the DSO they are pre-qualifying for

(continued)

Table 5.3 (continued)

Area	Field NAME	Commercial qualification questions	Allowable Responses	Pass Criteria	Description
	COMM_TC_DECLA	Do you declare that you have the authority to submit this application and by confirming you declare that to the best of your knowledge, the information in this form is accurate?	Y, N	Y	The individual completing this submission must have the authority of the contracting party to both accept the Agreement and provide the information listed in this application
Due Diligence	COMM_DD_FLEXA	Is the contracting party a member of Flex Assure Code of Conduct?	Y, N	Y, N	Being a member of flex assure is not required in order to pass commercial qualification, if the contracting party is member it will speed the required due diligence checks
	COMM_DD_ACHIL	Contracting parties Achilles UVDB Registered No. if applicable	Free text, blank	Completed, blank	Being Achilles UVDB (Utilities Vendor Database) registered is not required in order to pass commercial qualification, if the contracting party is member it will speed the required due diligence and credit checks

(continued)

Table 5.3 (continued)

Area	Field NAME	Commercial qualification questions	Allowable Responses	Pass Criteria	Description
	COMM_DD_CHECK	Where Achilles UVDB registration has not been advised, you understand that the DSO may access the contracting Parties most recent audited financial accounts via Companies House for the purpose of credit checks	Y, N, NA	Y, NA	Please confirm you understand that the DSO may perform credit checks as part of their due diligence checks relating to this application
	COMM_DD_RECEI	Is this contracting party currently, or has it ever been in receivership?	Y, N	N	This response will be assessed by the DSO as part of their due diligence checks
	COMM_DD_ADMIN	Is this contracting party currently, or has it ever been in administration?	Y, N	N	This response will be assessed by the DSO as part of their due diligence checks
	COMM_DD_LIQUI	Is this contracting party currently, or has it ever been in liquidation?	Y, N	N	This response will be assessed by the DSO as part of their due diligence checks
	COMM_DD_DEBTS	Is this contracting party currently, or has it ever been unable to pay its debts as they fall due (within the meaning of Sect. 268 Insolvency Act 1986)?	Y, N	N	This response will be assessed by the DSO as part of their due diligence checks
	COMM_DD_WINDI	Is this contracting party currently, or has it ever had, in the past 3 years, any petitions for winding up (other than vexatious petitions)?	Y, N	N	This response will be assessed by the DSO as part of their due diligence checks

(continued)

5.2 Pre-qualification Process

Table 5.3 (continued)

Area	Field NAME	Commercial qualification questions	Allowable Responses	Pass Criteria	Description
	COMM_DD_BANKR	Is this contracting party currently, or has it ever had any petitions for bankruptcy (or their equivalent in the country in which the Applicant is incorporated) within the last three years?	Y, N	N	This response will be assessed by the DSO as part of their due diligence checks
	COMM_DD_OFFEN	Is this contracting party currently, or has it ever been convicted of any of the offences or has any discretional exclusion occurred, as contained in Regulation 80 of the Utilities Contract Regulations 2016 (UCR), and listed in Regulation 57 (1) and 57 (8) of the Public Contracts Regulations 2015 (PCR)? [IF IN SCOTLAND, Is this contracting party currently, or has it ever been convicted of any of the offences or has any discretional exclusion occurred, as contained in Regulation 78 of the Utilities Contract (Scotland) Regulations 2016 (UC(S)R), and listed in Regulation 58 of the Public Contracts (Scotland) Regulations 2015 (PC(S)R)?	Y, N	N	This response will be assessed by the DSO as part of their due diligence checks
	COMM_DD_TERMI	Is this contracting party currently, or has it ever had, in the past 3 years, any similar contracts terminated prematurely and/or had damages claims or other comparable sanctions brought against the contracting party for any significant or persistent deficiencies in performance of a substantive requirement of the contract?	Y, N	N	This response will be assessed by the DSO as part of their due diligence checks
	COMM_DD_LITIG	Has the contracting party been subject to any material non-employment related litigation (pending, threatened or determined) or other legal proceedings against the contracting party within the last three years that may be relevant to your ability to deliver services	Y, N	N	This response will be assessed by the DSO as part of their due diligence checks

(continued)

Table 5.3 (continued)

Area	Field NAME	Commercial qualification questions	Allowable Responses	Pass Criteria	Description
Insurance	COMM_IN_EMPLO	Does the contracting party have or commit to have Employer's liability insurance with a minimum limit of £5 m	Y, N	Y	Confirm you meet the minimum insurance level required for Employers Liability
	COMM_IN_PUBLI	Does the contracting party have or commit to have Public liability insurance with a minimum limit of £5 m	Y, N	Y	Confirm you meet the minimum insurance level required for Public Liability
	COMM_IN_COPIE	Will the contracting party provide copies of such insurances upon request	Y, N	Y	Confirm you will be able to provide copies of insurances if requested by the DSO

5.2 Pre-qualification Process

Table 5.4 Standardised Technical PQQ Template (35 Questions)

Area	Field Name	Technical qualification questions	Allowable responses	Pass criteria	Description
Reference	TECH_REF_DER1	Assigned Reference for this DER, if known	Free text, NA	Completed	If the DER has previously been submitted and assigned a reference, this can be detailed here
Connection	TECH_CN_STATUS	DER Connection status	Archived, Energised, Awaiting Energisation, Planned, Speculative	Energised, Awaiting Energisation, Planned, Speculative	One eligible response per DER must be provided confirm the status of the assets connection to the distribution network; Energised means the DER is already connected to the network and is readily available for flexibility, Awaiting Energisation means the connection application and all installation work is complete but the connection is pending energisation therefore the DER is not yet available for flexibility, Planned means a connection application is in progress but no installation work has started and there is no known date for energisation therefore the DER is not yet available for flexibility, Speculative means that the DER is still being sought through recruitment and once recruited will form an aggregated group therefore the DER is not yet available for flexibility

(continued)

Table 5.4 (continued)

Area	Field Name	Technical qualification questions	Allowable responses	Pass criteria	Description
	TECH_CN_AWAI1	If awaiting energisation, firm date of energisation	DD/MM/YY, NA	Completed	If the DER is Awaiting Energisation the firm date of energisation must be provided, if energised, planned or speculative please respond NA
	TECH_CN_AWAI2	If awaiting energisation, connection reference number	Free text, NA	Completed	If the DER is Awaiting Energisation the connection reference number must be provided, if energised, planned or speculative please respond NA
	TECH_CN_PLAN1	If planned, connection voltage level	0.23, 0.40, 0.46, 3, 3.3, 6, 6.6, 7, 11, 13, 20, 22, 25, 33, 66, 132, NA	Completed	If the DER is Planned the voltage level at the point at which the DER is connecting onto the distribution network must be provided, if Energised, Awaiting Energisation or Speculative please respond NA
	TECH_CN_PLAN2	If planned, connection offer status	Not yet applied, applied awaiting offer, offer issued, offer accepted	Not yet applied, applied awaiting offer, offer issued, offer accepted	If the DER is Planned the status of the connection application must be stated, if Energised, Awaiting Energisation or Speculative please respond NA
	TECH_CN_PLAN3	If planned, connection reference number	Free text, NA	Completed	If the DER is Planned the status of the connection application must be stated, if Energised, Awaiting Energisation or Speculative please respond NA
	TECH_CN_PLAN4	If planned, what is the target delivery date?	DD/MM/YY, NA	Completed	If the DER is Planned, the target for delivery/energisation must be stated, if Energised, Awaiting Energisation or Speculative please respond NA

(continued)

5.2 Pre-qualification Process

Table 5.4 (continued)

Area	Field Name	Technical qualification questions	Allowable responses	Pass criteria	Description
	TECH_CN_SPEC1	If speculative, service readiness date	DD/MM/YY, NA	Completed	If the DER is Speculative, the anticipated dated the DER or aggregated group or DER will be available to deliver flexibility. If Energised, Awaiting Energisation or Planned please respond NA
	TECH_CN_SPEC2	If speculative, recruitment status	ASSET CONTRACTED, ASSET KNOWN, ASSET UNKNOWN, NA	Completed	If the DER is Speculative, the status of recruitment must be stated. If Energised, Awaiting Energisation or Planned please respond NA
	TECH_CN_SPEC3	CMZ Location, if known	Free text, NA	Completed	If known, the DSO Constraint Management Zone the DER or if speculative, the Aggregated Group of DER, are sited within. Speculative must complete this field with a valid DER
Site/Location	TECH_LN_POSTC	If Energised, Awaiting Energisation, Planned; Postcode	Free text	Completed	The full post code of the site where the DER is located
	TECH_LN_IMPAN	If Energised, Awaiting Energisation, Planned; Import MPAN (Meter Point Administration Number) If known	Free text (13 Characters). NA	Completed	The unique 13-digit identification of the meter associated with the DER's energy import from the network
	TECH_LN_EMPAN	If Energised, Awaiting Energisation, Planned; Export MPAN (Meter Point Administration Number) If known	Free text (13 Characters). NA	Completed	The unique 13-digit identification of the meter associated with the DER's energy export to the network

(continued)

Table 5.4 (continued)

Area	Field Name	Technical qualification questions	Allowable responses	Pass criteria	Description
	TECH_LN_MSID1	If Energised, Awaiting Energisation, Planned; MSID (where applicable)	Free text, NA	Completed	If applicable, the BSC unique identifier of the meter associated with the DER
	TECH_LN_ANAME	DER [If Speculative, then Aggregated Group] Name/Ref	Free text	Completed	The human readable name the contracting party wishes to name the DER
Technology	TECH_TG_GROU1	Asset Scale	DOMESTIC, I&C	Completed	Identifies whether the DER is connected at a site that is domestic, or industrial and commercial
	TECH_TG_GROU2	Metering Point	POINT OF CONNECTION, ASSET LEVEL	Completed	Describes where the DER is metered. 'Asset Level' means behind the meter and 'Point of Connection' means the connection at the boundary to the network
	TECH_TG_GROU3	DER Type; Generation &/OR Storage	Y, N	Y, N	Identifies if the purpose of the DER is for generation and/storage
	TECH_TG_GROU4	DER Type; Demand	Y, N	Y, N	Identifies if the purpose of the DER is for demand flexibility

(continued)

5.2 Pre-qualification Process

Table 5.4 (continued)

Area	Field Name	Technical qualification questions	Allowable responses	Pass criteria	Description
	TECH_TG_GSCL1	If Generation &/OR Storage, Energy Conversion Type	Battery, Compressed air system, Engine (combustion/reciprocating), Gas turbine (OCGT), Geothermal power plant, Hydro power system, Liquid air system, Offshore wind turbines, Onshore wind turbines, Photovoltaic, Steam turbine (thermal power plant), Steam-gas turbine (CCGT), Tidal lagoons, Tidal stream devices, Wave devices	Completed	Complete only if the DER is for generation and/storage
	TECH_TG_GSCL2	If Generation &/OR Storage, Energy Source	Advanced Fuel (produced via gasification or pyrolysis of biofuel or waste), Biofuel—Biogas from anaerobic digestion (excluding landfill & sewage), Biofuel - Landfill gas, Biofuel—Other, Biofuel—Sewage gas, Biomass, Fossil—Brown coal/lignite, Fossil—Coal gas, Fossil—Gas, Fossil—Hard coal, Fossil—Oil, Fossil—Oil shale, Fossil—Other, Fossil—Peat, Geothermal, Hydrogen, Nuclear, Solar, Waste, Water (flowing water or head of water), Wind, Stored Energy (all stored energy irrespective of the original energy source)	Completed	Complete only if the DER is for generation and/storage

(continued)

Table 5.4 (continued)

Area	Field Name	Technical qualification questions	Allowable responses	Pass criteria	Description
	TECH_TG_DDCLS	If Demand, Technology Type	Air source heat pump, Ground source heat pump, Water source heat pump, Hybrid Heat pump, EV Charger DSR, EV Charger V2G, On site Battery, Flexible Site Demand	Completed	Complete only if the DER is for demand flexibility
DER parameters	TECH_PS_INCAP	DER [If Speculative, then Aggregated Group] Installed capacity (MW)	Free text	Completed	The full-load output of the DER
	TECH_PS_FCDTU	DER [If Speculative, then Aggregated Group] Flexible Active Capacity—Demand Turn-up (MW)	Free text	Completed	The capacity, in MW, of your DER to increase its generation output. If Zero, please respond '0'
	TECH_PS_FCDTD	DER [If Speculative, then Aggregated Group] Flexible Active Capacity—Demand Turn-down (MW)	Free text	Completed	The capacity, in MW, of your DER to reduce its generation output. If Zero, please respond '0'
	TECH_PS_FCGTU	DER [If Speculative, then Aggregated Group] Flexible Active Capacity—Generation Turn-up (MW)	Free text	Completed	The capacity, in MW, of your DER to increase its demand on the network. If Zero, please respond '0'

(continued)

5.2 Pre-qualification Process

Table 5.4 (continued)

Area	Field Name	Technical qualification questions	Allowable responses	Pass criteria	Description
	TECH_PS_FCGTD	DER [If Speculative, then Aggregated Group] Flexible Active Capacity—Generation Turn-down (MW)	Free text	Completed	The capacity, in MW, of your DER to reduce its demand on the network. If Zero, please respond '0'
	TECH_PS_REACE	DER [If Speculative, then Aggregated Group] Reactive Export Capacity—Generation Turn-up (MW)	Free text	Completed	The capacity, in MVar, that an DER has to generate reactive power: providing an inductive load and lagging effect on current with respect to the network voltage. If Zero, please respond '0'
	TECH_PS_REACE	DER [If Speculative, then Aggregated Group] Reactive Import Capacity—Generation Turn-down (MW)	Free text	Completed	The capacity, in MVar, that a DER has to absorb reactive power: providing a capacitive load and leading effect on current with respect to the network voltage. If Zero, please respond '0'
	TECH_PF_MINOD	DER [If Speculative, then Aggregated Group] Min Operating Duration (HH: MM)	HH: MM	HH: MM	Minimum amount of time the DER is able to provide a flexibility response. Where a DER is able to run continuously beyond 24 h, respond 'Unlimited'
	TECH_PF_MAXOD	DER [If Speculative, then Aggregated Group] Max Operating Duration (HH: MM)	HH: MM, Unlimited	HH: MM	Maximum amount of time the DER is able to provide a flexibility response. Where a DER is able to run continuously beyond 24 h, respond 'Unlimited'
	TECH_PS_RESPO	DER [If Speculative, then Aggregated Group] Response Time (minutes)	HH: MM	HH: MM	The minimum time required for the DER to respond to a utilisation instruction

(continued)

Table 5.4 (continued)

Area	Field Name	Technical qualification questions	Allowable responses	Pass criteria	Description
	TECH_PS_RECOV	DER [If Speculative, then Aggregated Group] Recovery Time	HH: MM	HH: MM	The time required by the DER [If Speculative, then Aggregated Group] to recover from one instruction until the next instruction can be actioned
Metering	TECH_PS_METER	Metering Granularity (Second by Second, Minute by Minute or Half Hourly)	SS, MIN, HH	MIN, HH	The metering frequency available for the DER, or if speculative then for the Aggregated Group. Entry should be relevant to the service applicable, for example many DSOs do not accept second x second metering

5.3 Tendering and Contracting 99

Standardised Commercial Template

See Table 5.3.

Standardised Technical Template

See Table 5.4.

5.2.3 Accepted Deviations to the Templates

The following deviations to the template, noted in Table 5.5, were accepted and understood by stakeholders. It is expected that these aspects will evolve towards alignment in the near future as DSOs align further on contractual processes post pre-qualification.

5.3 Tendering and Contracting

Registered and/or pre-qualified parties will be notified of all flexibility requirements that open for tender. These tenders are typically split between CMZs. Flexibility providers with eligible assets within a CMZ that the DSO are looking to procure for should respond to the ITT within the timescales. The competition that is organised shall seek to assess the bids to meet the volume requirement, at a cost that is within budget, as economically as possible. DSOs therefore have to take into account technical and pricing considerations when designing their competition mechanism. This section explores the factors and split between the technical and pricing considerations and the bidding rules employed by DSOs.

Historically the network operators used different criteria when assessing and awarding flexibility contracts as listed in Table 5.6.

It was then agreed that all DSOs will undertake 4 key stages when assessing flexibility operators and the asset, as shown in Table 5.7,

1. the PQQ commercial assessment,
2. the PQQ technical assessment,
3. the procurement commercial score and
4. the procurement technical score.

5.3.1 Procurement Timelines

An indicative timelines of the key procurement milestones are shown in the Fig. 5.1. Note that this timeline is for the procurement of flexibility applicable to the deferral

Table 5.5 Noted deviations from the PQQ template

Area	Deviation	Means of collection
Terms and Conditions	COMM_TC_ACCEP: NGED will replace wording here to allow digital acceptance of the Standard Agreement at the point of completing Commercial Qualification. This is because NGED have implemented the Standard Agreement as an Overarching Contract, requiring all FSPs to accept the T&Cs ahead of entering competitions for the delivery of Flexibility Services	NGED will collect this information through its Procurement Portal, Market Gateway
Billing	As NGED have implemented an Overarching Contract, billing information and compliance with the self-billing system NGED have adopted through its Operational Portal, Flexible Power, will also be collected from FSPs at the point of completing Commercial Qualification	NGED will collect billing info through a separate secure system. Agreement to self-billing arrangements will be collected through NGEDs Procurement Portal, Market Gateway
Information Security	NPg will collect additional information to verify compliance with its corporate information security requirements that are mandated by its parent company	These will be collected through its employed procurement platform, Piclo Flex
Sub-assets	NGED will collect an additional granularity of data where multiple DER of varying technology types are metered at the same point of connection	NGED will collect this information through its Procurement Portal, Market Gateway
MPANS	At present SSEN do not use MPAN data in the process of validating DERs at the pre-qualification or contract award stages of Flexibility Service procurement. MPANs have been assessed to be Personal Data across industry (reference: Data Protection Impact Assessment v4 030,718 (Ofgem.gov.uk). As MPANs are not used for a specified, explicit purpose for Flexibility Service procurement by SSEN then it would not be considered to be a valid reason for collecting such data under the applicable data protection regulations (Data Protection Act 2018). This is why, in the immediate term at least, SSEN will deviate on this one point in the industry standard pre-qualification questionnaire for Flexibility Services	SSEN will make this change clear within whichever platform it employs to facilitate procurement processes

(continued)

5.3 Tendering and Contracting

Table 5.5 (continued)

Area	Deviation	Means of collection
Other	Additional Platform specific information, where used, may be required (i.e. long/lat for mapping features)	
	Where DSOs require evidence (i.e. insurance, delivery plans etc., these maybe requested via platforms where doc upload capability exists)	

Table 5.6 Historic assessment weightage

DSO	NPg (%)	NGED (%)	ENWL (%)	UKPN (%)	SSEN (%)	SPEN (%)
Financial	100		60	100	30	30
Technical		100	40		70	70

Table 5.7 Scoring criteria and recommended weightage

Stage		Scoring criteria	Parameters (visible to the market prior to tender round)
1	PQQ commercial assessment	Pass/fail	Meet the minimum requirements regarding including corporate regulatory obligations, legal offences, creditworthiness, conflicts of interest etc. (to be agreed)
2	PQQ technical assessment	Pass/fail	Meet the criteria of flex product parameters (min capacity, location (geographic and voltage), G59 [1], G99 [2] compliance, run times, ramp up times, etc.)
3	Procurement commercial score	70%	Assessment of price of bid
4	Procurement technical score (relevant to assess value add above minimum product criteria set out in product parameters)	30%	Does not cause with any other issues on the network
			Conflicts with other services being provided
			Effectiveness in alleviating the constraint
			Ramp rates of the asset
			Energised status of asset
			Type of connection (flexible/firm)
			Type of metering (assuming that minimum is met)

of network reinforcement only. The industry responses to the Open Network consultation questions reflected a general agreement that the timing of the PQQ technical assessment should be prior to the opening of the bidding window. There was also broad agreement with the timing and number of procurement cycles within the year which was an approach that reflected the level of development of DSO flexibility markets at the time. As the network flexibility market in GB evolve the procurement methods and timescales are regularly adjusted and vary across DSOs to reflect their needs. It is noted, however, that DSOs continue to develop both real-time and long-term procurement capabilities

The gap analysis undertaken in 2020 indicated that all DSOs operated in similar timeframes from identifying requirements to the delivery of contracts, in line with best practice noted above and Procurement obligations, such as providing at least 3 months from Invitation to tender (ITT) opening to bidding closing. However, the point in the year in which they are issued do not align across all networks: some will align with the Load Index reporting undertaken in September–October, with sites that require further analysis being issued in March in line with the new Standard Licence Condition (SLC) 31E submissions; other DSOs issue their competitions in January and July each year, however both timelines cater for summer and winter requirement windows following stakeholder feedback from previous consultations.

A further review was undertaken to investigate the options and types of procurement alignment to identify opportunities for alignment for DSO and NESO procurement timelines.

The findings together with the stakeholder feedback, as shown in Table 5.8, did not support or provide any real evidence to indicate benefit to market participants from procurement timeline alignment, but rather that the staggered timescales should be continued whilst improving the visibility of procurement timescales.

5.4 Standard Agreement (Contract)

Since 2019 the Open Networks has led the formation and adoption of a common agreement for use within Flexibility Services market. The development has been marked by staged evolution and deployment of updated versions of the agreement.

In 2019, Ver 1.0 was focussed purely on DSO contract alignment, Ver 2.0 develop in 2021/22 was a version of the common agreement which included some NESO services, migrating towards a 'Framework' based contractual process.

It must be noted that the speed in which some of the developmental steps were undertaken were and will remain dependent on engagement with DSO and NESO stakeholders as the changes will impact the process of securing Flexibility Services as well as potentially impacting legacy contracts. It's also critical to note that throughout the evolution the services were secured competitively based on network scenarios, geography and market maturity.

A standard contract developed by Open Networks is presently used by all DSOs and for applicable NESO products. The standard contract consists of:

5.4 Standard Agreement (Contract)

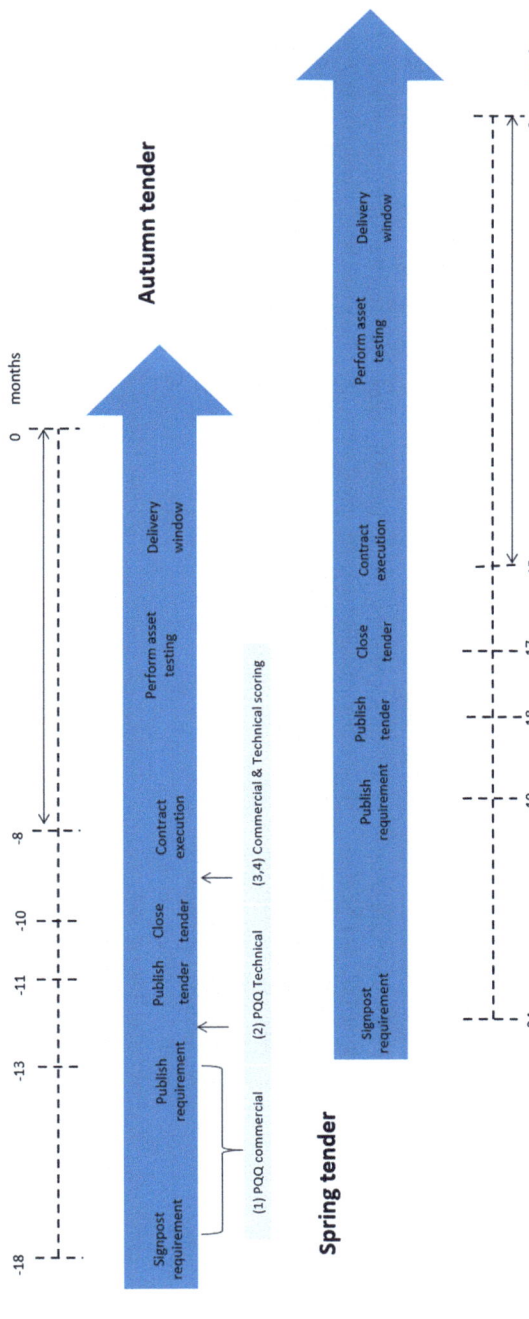

Fig. 5.1 Typical procurement milestones timelines (indicative)

Table 5.8 Pros and cons of procurement alignment between DSO and NESO

Type of procurement alignment	Pros	Cons	Findings
Scheduled aligned timescales	FSPs who participate in multiple DSO and NESO procurement processes can prepare bids at the same time FSPs weighing whether to apply for both DSO and NESO services would be better informed	FSPs may not have resources to participate in multiple bids at the same time especially if being procured across different platforms. Strong stakeholder support in favour of avoiding clashes with the Capacity Market & other NESO competitions that prevent revenue stacking. Internal processes and regulatory submissions govern the time of the year. DSO markets may suffer as FSP's opt for the most rewarding offer due to time constraints	There is no real evidence whether this would be of overall benefit or not to FSPs, feedback from stakeholders has been both for and against
Scheduled staggered timescales and Improved visibility	Any FSPs who participate in multiple DSO and NESO procurement processes can spread out and plan their bid preparation. This allows them to stack assets into multiple markets, and to focus on recruitment of aggregated assets in specific locations at different times of year. Strong stakeholder support in favour of avoiding clashes with the Capacity Market & other NESO competitions	FSPs will not be able to prepare bids at the same time as administrative tasks are spread across the year. However, if all operators used the same platforms this may not be an issue	As is but there is no real evidence whether this would be of overall benefit or not to FSP, as feedback from Stakeholders both for and against. First step towards greater visibility of Flexibility Services procurement processes and timescales. Alignment of pre-qualification reduce administrative burdens and simplify the procurement process
Timelines	Allows FSPs to better plan their bid resources	Time pressure for FSPs bidding for multiple services under a single procurement round Requires greater co-ordination Doesn't fit with the drive towards real time procurement and operating dynamic systems	Timeline alignment as part of wider process alignment could be seen as beneficial overall

5.4 Standard Agreement (Contract)

- The 'front end'—sets out the background and purpose of the specific agreement
- The schedules—standard templates, that can be adapted if needed, where specific details of the agreement are captured
- The conditions—standard terms and conditions to remain common across all Flexibility Services agreements

The agreement presently being used is Standard Agreement Ver 3.0, which is the latest version at the time of publication of this book.

5.4.1 Bilateral Versus Framework Approach

DSO's have historically utilised a bilateral contract approach for Flexibility Service delivery (2018–2021), which meant contracts are implemented per service to each successful provider and remain explicit to that service or services detailed within the contract. For each new service being procured a new contract complete with service specific schedules and agreed prices is required to be agreed between the DSO and the provider through a full procurement process. With the first BAU DSO Flexibility Services placed in 2018, the markets have not reached the same level of fluidity as those open to NESO driven services (although this is expected to occur in a more rapid fashion as the UK moves towards a zero-carbon future). The bilateral approach utilised historically worked well for the lower frequency of procurements required, smaller service values and a smaller base of engaged providers as well as the relative infancy of the approach, where risk aversion is understandably a core concern and the assurance of more defined, specific contracts is required.

The framework approach also relies upon a contract between the DSO and a provider, however the contract covers the key expectations, obligations between the two parties, with separate services being independently procured under the 'framework' agreement. This means a provider can respond to multiple tenders across a defined period without needing to sign specific contracts each time, the new services are added as schedules or sub-agreements under the core contract with the agreed price for that service being stipulated. Figure 5.2 shows the timeline of evolution from a bilateral to framework approach

It is important to stress that different product or service 'types', for example Fast Frequency Response (FFR) and Dynamic services may need separate overarching contracts, but the individual 'auctions' or requirements within those types will be awarded underneath that parent contract, even though the parent contract would remain 'common' across the NESO and DSO services. Both cases still require specific procurement runs for new services to ensure these are awarded fairly and transparently, however the framework approach is more efficient within mature markets where new services or changing requirements are more frequently needed from a population of engaged market participants.

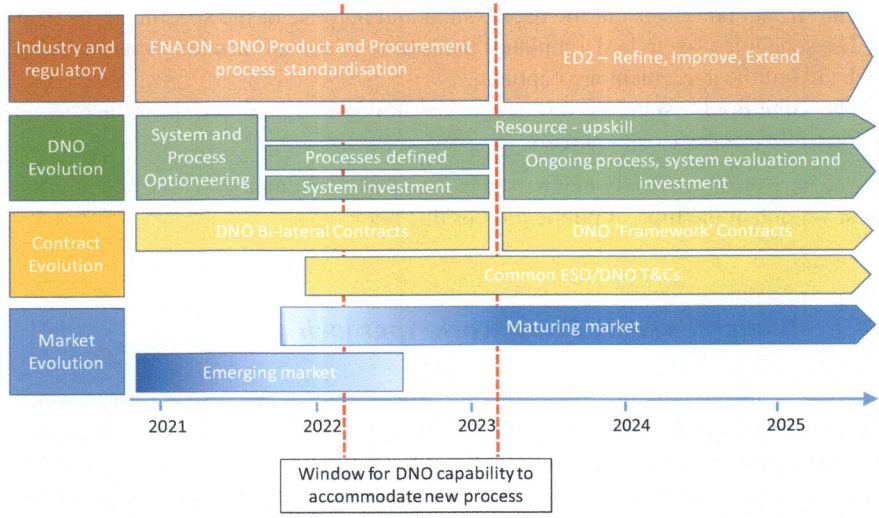

Fig. 5.2 Timeline of evolution from bilateral to framework approach

As local flexibility markets evolve, more providers will seek to engage in those markets, more services will become available and as both DSO's and the NESO move towards day ahead procurement, greater ease in the contract process will be required. In addition, the risks associated with managing Flexibility Services become better understood, mitigated and alleviated by increased market fluidity, automation and data, a framework approach moves from being a desired (but not essential) solution to be a critical development for the increased growth of the DSO and NESO markets.

The NESO utilises a Framework approach to securing Flexibility Services. It's essential to note that the NESO has been securing 'Flexibility Services' through network stability and balancing products for several years and their approach has evolved to enable greater accessibility and uptake as more providers have become available to the system.

Considerations while moving from bilateral to framework approach

- Technical capabilities: All DSOs utilise some form of Dynamic Purchasing Systems (DPS) which can support a more framework-oriented approach while retaining the bilateral contract requirement traditionally utilised. In the past few years the experience in DPS utilisation has grown across DSOs and these systems have mostly become embedded in BaU use, enabling procurement within a framework environment.

Beyond this initial procurement requirement there were also contract/service management systems which needed to be developed/procured which reduced the resource impact of managing multiple contractual relationships across a wider range of providers, with providers also having multiple contracts within the 'framework'.

Different DSOs have developed systems to facilitate visualisation of constraints, procurement processes and dispatch of Flexibility Services. These DER Management Systems (DERMS) offer slightly different functionalities as well as other potential developing systems. To ensure overarching alignment remains deliverable, it is vital that steps are taken to ensure alignment in the contract management elements of any development and utilisation of these systems. Given the recent adoption of these contractual and service management systems, it must also be acknowledged that DSO are continuing to build experience in these systems and the wider systems required to support 'closer to real-time' tendering, or 'trading' of services (month, week, day ahead etc). Reliance on both internal and external support from a cyber security, IT and real time systems perspective would be required as well as significant support from organisations commercial, legal and procurement teams.

Since the beginning of RIIO-ED2 (2023), all DSO's have the required system functionality to support framework contracts and service management

- Regulatory and Legal: Some DSOs consider Flexibility Services as subject to Utilities Contracts Regulations (UCR) compliance, however others feel they're exempt as Flexibility Services are seen as trading/buying energy. The procurement and legal teams of different DSOs still display variance in the interpretation of regulatory compliance. Given these variances, it must be considered that in some cases, any change to the procurement process could result in a breach of licence if the new process deviates from those regulations. DSOs could need to release Official Journal of the European Union (OJUE) notices on individual service requirements or Prior Information Notices (PIN) notices covering a group/yearly requirement across services, with variations driven by the host DSO's appreciation of the regulations.

The types of regulatory reporting processes undertaken by the DSO's and the NESO, will require a uniform approach to application and compliance to regulations, which in turn will enable alignment in the procurement approach applied. From this point and from a contractual adaptation point, it is important to note that DSOs and the NESO also display different approaches to legal support, with some having internal Legal teams capable of sustained support, whereas other organisations contract the majority of work out to 3rd party, albeit regularly used suppliers. This does create some variance in how legal definitions are interpreted, the length of time it can take to shift contractual processes and can accentuate the risk appetite of the host organisation in terms of mitigating the contractual risk of undertaking new processes.

5.4.2 Adoption of Service-Based Schedules

In order to standardise the format of schedules across organisations, it was important to adopt a structure that can accommodate a range of contract forms, including traditional bi-lateral based contracts utilised by DSOs and auction-style framework agreements through which the NESO procures balancing services. A gap analysis carried out highlighted key differences in contracting journeys and associated procurement processes across the DSOs and NESO. Whilst contracting and procurement processes evidently differed across the DSOs and NESO. It is important to equally acknowledge that subtle variations exist even amongst DSOs in areas like contract duration, contract award processes, price determination and contract award procedures. Furthermore, even within the respective organisations, each service is procured differently, with a distinct set of requirements and technical parameters. Service-based schedules allow service specific details to accommodate these service specific differences whilst still enabling uniformity across organisations in the structure and format of each schedule.

The NESO's service-based schedules were used as a blueprint in developing the structure for the standard agreement. The NESO experience has been that service-based schedules make for an easier provider onboarding experience, as all legal and technical documents relating to the service are in the same schedule. It also enables any service/contractual documents to be isolated, in the event of industry consultation. Service-based schedules are ideal where requirements differ across services. The structure enables uniformity to be retained in the look and high-level content of the schedules.

Therefore, the service-based structure adopted was the first step in the alignment of the schedules that sit alongside the standard agreement. As shown in Fig. 5.3, it was a key steppingstone that sets the foundation for future work, as the industry moves forward. This initial structure allowed for the differences across the organisations in regard to product design, procurement process and contract award, whilst providing providers consistent structure and high-level content. Each service-specific schedule would have the same naming convention and follow a similar order of arrangement, giving a familiar look and feel to FSPs.

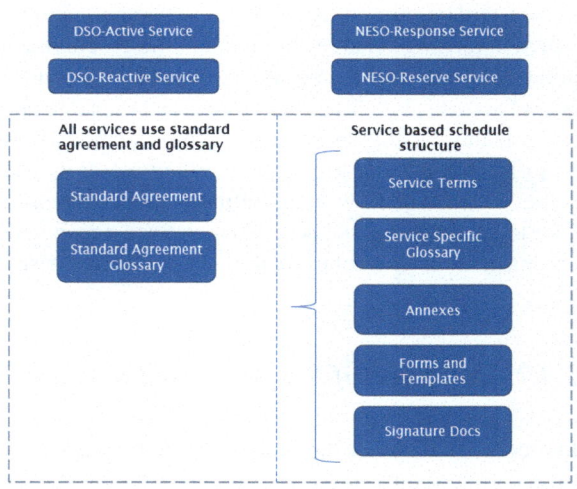

Fig. 5.3 Schedule structure (in Ver 2.0 and above)

5.4 Standard Agreement (Contract)

The first schedule (Service Terms) would contain the description of the Service as well the applicable payment terms. The next schedule titled "Annexes" would cover information on communications, performance monitoring, system/ technical requirements, special requirements, participation guidance, tender conditions, and testing. The third schedule would serve as a repository for the applicable forms and templates, including unavailability/ remedy forms, performance reporting templates, any forms detailing sites and DER, provider data templates and tender templates. The last schedule (Signatures) would contain the DSO and NESO signature page, post-award contract notices and any registration or prequalification forms. These schedules would be used for:

- DSO Active Service
- DSO Reactive Service (if and when procured)
- NESO Response Service
- NESO Reserve Service.

5.4.3 Standard Agreement Version Tracker

Figure 5.4 tracks the evolutionary versions of the Standard Agreement for Flexibility Services over the years.

2020: Standard Agreement Ver 1.2

- Introduced Standard Agreement that were used by DSOs, with standard DSO terms

2022/23: Standard Agreement Ver 2.0 (and Ver 2.1)

- Ver 2.0 successfully secured alignment between the NESO and DSO by introducing a framework contractual structure, which allowed contracts to be awarded, ranging from day ahead (auction-style) agreements to bilateral contracts.
- Ver 2.0 utilised service-based schedules, which defined how each service is to be procured and provided. The structure made it easier for providers to follow what is required per service and enables any service/contractual documents to be isolated, in the event of industry consultation. Please see diagram on next page.
- Ver 2.1 modified relevant clauses to allow for the implementation of Primacy Rules (See Chap. 10)

2024: Standard Agreement Ver 3.0

- Refined Standard Agreement contractual terms to address barriers to participation within the flexibility markets, provide clarity, and further align wording within the service terms schedules, creating an improved contract for DSOs, the NESO and FSPs alike.

Key topics of change were:

- Duplicate registration of assets

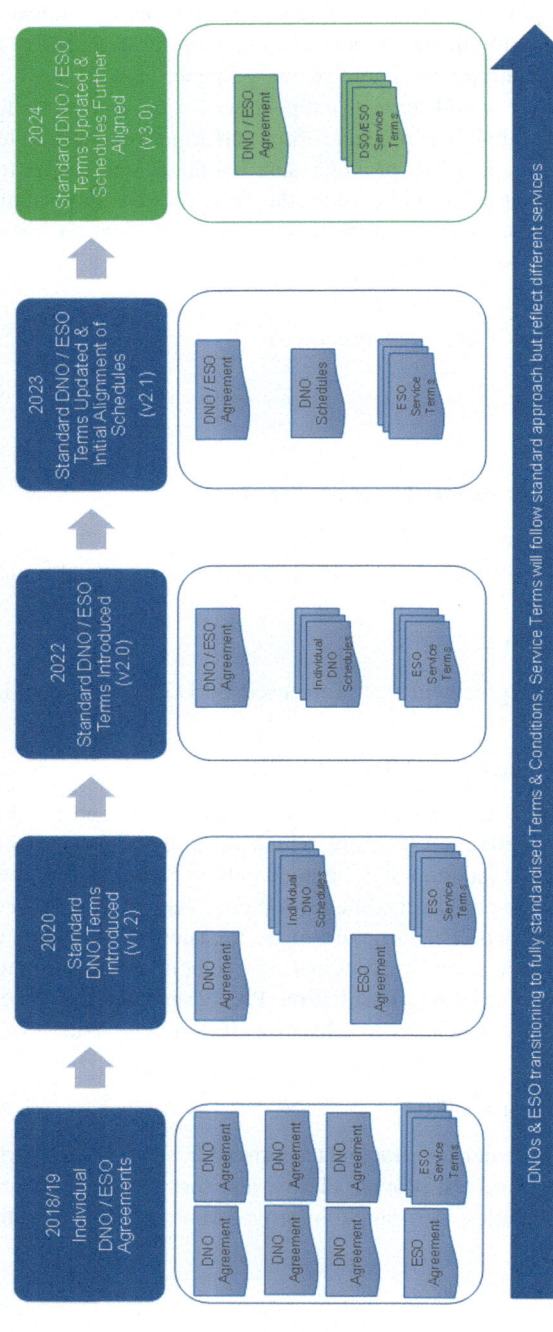

Fig. 5.4 Published versions of Standard Agreement for Flexibility Services

5.4 Standard Agreement (Contract) 111

- Cyber Security
- Voluntary participation
- Liabilities
- Definition of services(Metering; Payment calculations; Performance monitoring; and System comms)
- Re-organisation/Alignment of Schedules

Enduring Differences in the Standard Agreement

- Contract duration: For both the NESO and DSOs, award duration is based on several factors such as operational requirements, economics and compliance with legislation, which are documented in the procurement processes. Currently, variations exist amongst the DSOs, with some DSOs adopting one year rolling contracts, with a total duration of 5 years and others awarding contracts for longer timescales. For most of the NESO services, there is a move towards day ahead procurement. Newer services like Dynamic Containment, STOR, Dynamic Moderation and Dynamic Regulation are procured daily, whilst the Static and Dynamic FFR Products (which are being retired imminently) are procured monthly. However, services like Constraint and System Stability are procured on an ad hoc basis.
- Contract Award Process: The NESO's newer services are procured at EFA block granularity (i.e., every four hours). Hence, each provider can have up to 6 contracts for each unit per day. This results in a significant number of contracts being processed daily, for which manual award processes are unsuitable. For these services, the contract is formed once the auction results are published online. Contracts for some of the NESO's older services like Constraint and Static and Dynamic FFR are awarded via Signed Contract and Award Letter respectively. Whilst the System Stability service which is typically procured on an ad hoc basis, is awarded by instruction.

Amongst the DSOs, the award process is either through signed contracts or, where contracts have been signed ahead of the ITT, by issuing a contract award notice.

- Price Determination (Pay-As-Bid versus Pay-As-Clear): The mechanism for market pricing varies between DSOs and the NESO, either adopting fixed pricing, pay-as-bid, pay-as-clear, or a combination depending on a liquidity test. Whilst there is a move towards pay-as-clear pricing for the NESO's newer services, pricing for existing services is determined via the pay-as-bid methodology.

Methodologies for calculating value adopted by the DSOs include net present value (NPV) of deferred reinforcement to determine ceiling cost (using the CEM Tool (Sect. 4.4)), cost of alternative generation/ avoided Customer Interruption (CI) and Customer Minutes Lost (CMLs), and assessment of tendered pricing against investment case.

- Contract Award Volumes: The procurement granularity of the NESO's newer services, contract volumes per award are in the region of 100–150 per service.

Although the older services are procured less frequently, and tend to cover longer periods, approximately 20–50 contracts per award respectively.

The contract volumes per award for each DSO currently range between 1 and 25 per procurement, with higher contract volumes expected in future.

- Mix of Manual and Online Platforms for Tender Assessment: Amongst the DSOs and NESO, there is a mix of manual and automated processes for bid submission, whilst the bid assessment process is largely manual. The NESO runs automated bid submission and assessment processes for its new range of services. Some of the older services utilise automated bid submission processes alongside manual assessment processes.

5.4.4 Future Considerations

It is recognised that greater alignment of procurement processes and practices across the DSOs and NESO is required to support full contractual alignment. The NESO's operational requirement for closer to real time procurement has benefitted from a longer period of implementation, enabling greater evolution and catering towards the more NESO service provision. With both the DSOs and NESO now using the Standard Agreement, incremental steps are certainly being taken towards the adoption of a common approach for the procurement of Flexibility Services.

- Whilst DSOs and the NESO generally agree on the need to move closer towards real-time procurement (as this is commercially expedient and helps meet operational needs), the wider systems required to support 'closer to real-time' tendering, or 'trading' of services (month ahead, week ahead and day ahead) are still being considered.
- As the distribution flexibility markets become more fluid and the volume of awarded contracts increase, DSOs recognise the need to implement online contact award processes, to support increased market activity. The NESO will continue with online contract award with the high volume current and future services.
- Some DSOs and the NESO currently utilise manual bid assessment platforms for some or all of their auctions. Manual processes are time-consuming, have risks associated with manual data handling and are generally unsuitable for mature, high-volume markets. Therefore, a move towards the adoption of automated bid submission and assessment platforms by DSOs may be required to support this.

5.5 Suggested Further Reading

Listed below are the links to Open Networks publications relevant to this chapter in reverse chronological order. Please note, some of the documents/spreadsheets may now by outdated or superseded as the topics have evolved.

References

1. Procurement Process High-level Summary Paper (2021).
2. Review of DSO-NESO Flexibility Procurement Timelines (2021).
3. Procurement Coordination Implementation (2020).
4. Flexibility Service Pre-qualification Standard Template (2024).
5. Pre-qualification Standardisation Report (2024).
6. Pre-Qualification Alignment Recommendations (2022).
7. Standard Agreement for Flexibility Services Ver 3.0 (2024).
8. Consultation on Standard Agreement for Flexibility Services Ver 3.0 (2024).
9. Standard Agreement Ver 3.0 Consultation Feedback and Response Publication (2024).
10. Standard Agreement for Flexibility Services draft Ver 3.0 (2024 Consultation draft).
11. Standard Agreement for Flexibility Services Ver 2.1 (2023).
12. Standard Agreement Ver 2.0 Service Based Schedule Alignment (2022).
13. Standard Agreement for Flexibility Services Ver 2.0 (2021).
14. Flexibility Services Standard Agreement Ver 2.0 Consultation (2021).
15. Standard Agreement Ver 2.0 Consultation Webinar Slide Deck (2021).

References

1. "Engineering Recommendation G59, Issue 3 Amendment 7," Energy Network Association, 2019.
2. "Engineering Recommendation G99, Issue 1 Amendment 9," Energy Networks Association, 2022.

Open Access This chapter is licensed under the terms of the Creative Commons Attribution 4.0 International License (http://creativecommons.org/licenses/by/4.0/), which permits use, sharing, adaptation, distribution and reproduction in any medium or format, as long as you give appropriate credit to the original author(s) and the source, provide a link to the Creative Commons license and indicate if changes were made.

The images or other third party material in this chapter are included in the chapter's Creative Commons license, unless indicated otherwise in a credit line to the material. If material is not included in the chapter's Creative Commons license and your intended use is not permitted by statutory regulation or exceeds the permitted use, you will need to obtain permission directly from the copyright holder.

Chapter 6
Dispatch of Flexibility Services

Dispatch of Flexibility Services is defined as the process through which the DSO informs a flexibility provider of the required level of service within operational timescales. It is recognised that in order to provide confidence in DSO Market, DSOs must provide Flexibility Service providers (FSPs) with clear and transparent guidance on any criteria used to inform its dispatch decisions.

Applicability and timing of dispatch decisions depends on product type. All DSOs carry out initial forecasting to establish a level of requirement (as detailed in Chap. 4), but some products need further forecasting as requirements move closer to real time. It is at this later forecasting stage that decisions around the scheduling of dispatch are applicable.

Dispatch decisions fall into two categories: capacity availability and energy utilisation.

- Capacity Availability: the point at which the flexible capability has been scheduled, at which point it is firm and for which an availability fee may be paid. This can occur at the procurement stage or some time in advance to real time where certainty of availability is required.
- Energy utilisation: where the energy delivery has been scheduled (when done in advance) or dispatched (when done near to real time), and a utilisation fee would be be paid for the energy delivered. This schedule may be firm at the procurement stage for some products, or more commonly for Flexibility Services, close to or in real time.

Ahead of making any decisions on both availability and utilisation of Flexibility Services DSOs consider the following guiding principles to inform dispatch decisions. No principles would be considered ahead of another, all must be considered to ensure network security is delivered for the most cost effective outcome.

1. Security—the needs of the system will be met using flexibility in such a way that security of supply is maintained. DSOs will conform with applicable standards with an appropriate management of risk.
2. Cost—flexibility will be operated, without undue discrimination against any provider, to meet system need at the minimum level of cost.
3. Operability—DSOs will seek to dispatch services that offer compatible levels of operability. Accepted offers need to match/partially match product requirements.
4. Competition—DSOs will provide transparency of their dispatch decisions and activities. By sharing this methodology in advance, flexibility providers may be able to align their flexibility offering to best meet requirements
5. Fairness—DSOs will operate a fair dispatch methodology and provide equal opportunities to participate.

How a DSO chooses to operate each product will inform when relevant dispatch decisions are made. Open Networks has therefore endeavoured to minimise the work required by a FSPs who is successfully providing services to one System Operator (DSO or NESO) to provide comparable services to other System Operator(s). To ensure that any proposed standardisation does not limit the development of future Flexibility Products, the process of dispatching services has been decoupled from individual products as much as possible and, instead, discussions have been focused around a number of key phases of dispatch that are common across most existing products. Depending on the nature of individual products some of these phases may be completed as part of the procurement process or may be combined into another phase.

6.1 Phases of Dispatch

1. Declaration of availability by Service Provider
2. Acceptance of offered services by System Operator
3. Scheduling of services to run by System Operator
4. Instruction of services to run by System Operator
5. Cease instruction to stop operation of services
6. Variation of dispatched service
7. Monitoring of services
8. Post-action reporting
9. Cancelation of dispatch

Declaration of availability by Service Provider: The Service Provider informs the System Operator of the relevant technical and commercial parameters of the services that they are offering to provide. Depending on how the System Operator runs their particular services some of these parameters may be specified by the System Operator as part of the procurement process, in which case the Service Provider would have confirmed their ability to meet or acceptance of these parameters as part of the procurement process.

6.1 Phases of Dispatch

Historically there has been a fair degree of variation between different System Operators. Most notably some System Operators assume that contracted Service Providers will be available unless they declare otherwise, while others require a regular declaration from Service Providers that they will be available and one System Operator currently proactively contacts Service Providers to confirm their availability. There is also variation regarding the timescales that System Operators operate services in, which can also vary between different products operated by the same System Operator.

Further more significant variation exist between how System Operators currently receive declarations of availability from Service Providers, with dedicated electronic systems, email and phone calls being used to communicate availability. Longer term most System Operators agree that the best way to communicate this information at scale is through the use of APIs (Application Programming Interface: A standard way of communication between computer systems utilising predefined messages), but there is acknowledgement that this will require an overhead for Service Providers to integrate with that may put off smaller providers, especially during early stages of market maturity.

Acceptance of offered services by System Operator: The System Operator reviews the services that have been offered and decides which services to utilise. Depending on the service this may result in services being scheduled for delivery or it may secure availability of these services to operate if called on closer to real time by the System Operator.

Historically variations in acceptance practices were generally seem to be the result of differences between the types of products being run by different System Operators. For example, Scheduled services (e.g. Sustain or Scheduled Utilisation) have acceptance as part of the contract while offers of operational services (Restore or Operational Utilisation) tend to be automatically accepted by System Operators. The major difference between how System Operators process acceptance is the method of communication in use. The method of communication used by System Operators is generally the same method as is used by the Service Provider to submit their availability. As with declarations there is a general consensus that in the long term this process would be best managed at scale via APIs

Scheduling of services to run by System Operator: Depending on the service, the System Operator may schedule the usage of services with the Service Provider in advance.

Most System Operators have a service that is scheduled in advance, although there is variation between System Operators as to how far in advance the schedules are decided and the products that have advance scheduling. Differences method of communication is used for scheduling of services, although the longer-term view is that APIs would be the best way to manage this at scale.

Instruction of services to run by System Operator: The System Operator provides a real time instruction to the service operator to start delivering a service. This instruction will normally be issued a pre-agreed ramp up time period ahead of the time the System Operator requires the service to start being delivered. For service types that are scheduled with the Service Provider in advance the System Operator may provide a dispatch instruction at the start time or may require the Service Provider to start delivering the service at the scheduled start time.

The method of sending dispatch instructions varies between System Operators and products offered. For products that are scheduled with a Service Provider in advance, some System Operators still provide a dispatch signal at the start of the scheduled period, while others will instead require the Service Provider to self-dispatch at the start of the scheduled time. For System Operators that have a product dispatched in real time there is generally a ramp up notice time period, which is the time period the Service Provider has from the issue of the dispatch instruction to start delivering the dispatched service in full. This time period may be fixed for the product or agreed with each Service Provider on an individual basis. Currently System Operators send instructions by a range of communication methods including APIs, phone, email and SCADA. The general consensus is that APIs will be the longer-term method for sending instructions. Given the likely long-term importance of services it is likely that back up communication methods will be necessary for critical services, to ensure that these can be dispatched in the event of any failures in an API based system.

Where the availability of contracted flexibility assets exceeds the network's needs at any given time or place, fair and transparent decision-making criteria must be followed to determine the dispatch order. In most existing network flexibility approaches, cost order is sufficient determine the order of dispatch. However, in a fixed price approach to flexibility or in a market clearing approach, additional guidance will be required. A real-time price based flexibility market is an aspirational goal for DSOs, and may come about in the short or even medium-term.

Cease instruction to stop operation of services: The System Operator provides a real time instruction to the Service Provider to stop delivering a running service. For service types that are scheduled with the Service Provider in advance the System Operator may provide a cease instruction or may require the Service Provider to stop delivering the service at the scheduled end time.

For services that are scheduled in advance the stop time of operation is normally sent along with the start time as part of the scheduling process. For services that are instructed in real time, the end time may be sent as part of the start instruction or may be sent separately based on current network conditions. Generally the cease instruction will be sent by the same method as the instruction to start delivering the service.

Variation of dispatched service: The System Operator provides a real time instruction to a Service Provider to vary their delivery of a running service. Use of this sort of instruction is predominantly limited to applicable NESO services and there is currently no DSO currently utilising these.

Monitoring of services: Depending on the nature of the service being delivered the System Operator may require the ability to monitor the status and delivery of services in real time.

Generally amongst DSOs, there is limited real time monitoring of individual services at this stage, as existing wider system monitoring processes are used to ensure the network remains within limits. For some services this means that for real time operational purposes service delivery is assumed unless the System Operator is informed otherwise. Real time monitoring of individual services is generally via SCADA, where there is SCADA already fitted at the connection point of the service. Where Service Operators submit metering data for billing in real time this can be used to monitor delivery. Where APIs are used for dispatching, the structure of HTTP APIs means that there is an acknowledgement that dispatch messages have been received by the API server through HTTP status messages. Longer term there is likely to be an increase in the level of real time monitoring required by System Operators, both in the form of enhanced status messages from Service Providers and real time data of the volume of service currently being delivered.

Post-action reporting: Following delivery of a service, the System Operator may carry out analysis on the delivery of the service and provide various reporting.

There is currently significant variation between DSO and NESO System Operators as to the methods they use for settlement of services and the level of other post-action reporting / analysis they carry out. NESO use a number of established systems, to process event data and provide settlement, depending on the type of service. Some DSOs use dispatch Platform (e.g. Flexible Power) whilst other DSOs have more manual processes to review delivery post event and provide settlement. This area is likely to have further development in future as DSOs are likely to move to more automated reporting and settlement solutions as the volume of services used within their networks increases.

The Electricity Distribution Standard Licence Condition Procurement and use of Distribution Flexibility Services (C31E) [1], was implemented in December 2020 and transposes Article 32 of the Clean Energy Package (CEP) [2] (Incentives for the use of flexibility in distribution networks) into the GB regulatory framework. This sets out three distinct reporting requirements for DSO.

- The Distribution Flexibility Services Procurement Statement ("the Procurement Statement")
- The Distribution Flexibility Services Procurement Report ("the Procurement Report")
- Ongoing Reporting

Cancelation of dispatch: Stopping of planned or running services either as the result of the Service Provider no longer being able to deliver the service or because the System Operator needs to stop a planned service from running.

With the exception of NESO, System Operators do not have particularly established methods for cancelation of services. This may in part be due to the availability declarations and dispatch decision processes meaning that cancelation of services should be a relatively rare event, either because Service Providers would not offer availability if they knew their plant would be unavailable, or because System Operators would not schedule excessive services in advance, such that they would need to cancel services to keep their network within limits. As the volume and criticality of services within distribution networks increases it is likely that there will be further development around the process and communication of cancelation of services.

6.2 Areas of Development

Longer term the consensus is that the communication of dispatch requirements is best handled at scale via Application Programming Interfaces (APIs) as this will enable the use of automated systems to process dispatch requirements. It is recognised that this approach will require an overhead for Service Providers to integrate with APIs and that this could be off putting for smaller providers. In addition, as the use of services becomes more critical it is likely to be necessary to provide backup methods for dispatching services to ensure services can be dispatched in the event of a failure of systems associated with an API. As a result of these it may be that System Operators still maintain other methods of dispatching alongside APIs.

Given the rapid pace of development of Flexibility Services it will be essential to ensure that a common API has suitable flexibility such that it allows further innovation and development of new products. A poorly designed API that only addresses the current needs of Flexibility Services may bring some short term alignment, but runs a significant risk of becoming quickly outdated and ultimately hindering the future development of flexibility markets.

Furter developments and approach to adopt a common dispatch API is discussed in Chap. 8.

6.3 Suggested Further Reading

Listed below are the links to Open Networks publications relevant to this chapter in reverse chronological order. Please note, some of the documents/spreadsheets may now by outdated or superseded as the topics have evolved.

1. Dispatch interoperability and settlement review of existing practices and gap analysis (2022).
2. Dispatch Interoperability and Settlement Key Service Parameters (2022).
3. Dispatch Recommendations for Alignment (2022).
4. Initial Dispatch Process Comparison Spreadsheet (All DSO and NESO) (2022).
5. Flexibility Services–Dispatch and Settlement Processes (2020).

References

1. Electricity distribution licence condition 31E: Procurement and use of distribution flexibility services reporting requirements, Ofgem, 2022
2. Clean energy for all Europeans package, Energy, Climate change, Environment–European Union (2019). [Online]. Available: https://wayback.archive-it.org/12090/20241209144917/https://energy.ec.europa.eu/topics/energy-strategy/clean-energy-all-europeans-package_en

Open Access This chapter is licensed under the terms of the Creative Commons Attribution 4.0 International License (http://creativecommons.org/licenses/by/4.0/), which permits use, sharing, adaptation, distribution and reproduction in any medium or format, as long as you give appropriate credit to the original author(s) and the source, provide a link to the Creative Commons license and indicate if changes were made.

The images or other third party material in this chapter are included in the chapter's Creative Commons license, unless indicated otherwise in a credit line to the material. If material is not included in the chapter's Creative Commons license and your intended use is not permitted by statutory regulation or exceeds the permitted use, you will need to obtain permission directly from the copyright holder.

Chapter 7
Settlement

Settlement is the payment for the supply of Flexibility Services from a Flexibility Service Provider.

Individual service design can vary, dependent on the need of a system or network operator, but in general there can be two payment components:

1. Availability payments
2. Utilisation payments.

An availability payment may be made to an FSP who is ready and available to deliver a Flexibility Service with a utilisation payment made to a Service Provider for the actual provision of a service. The utilisation payment may or may not be based on the energy delivered. Services can include an availability only payment (with no additional payment when utilised), utilisation only or both availability and utilisation payments.

While service design can vary widely and not all the items below will be applicable to all services provided by FSPs, in general the items below will be required to calculate settlement. The first fours items in the list have cross over and commonality with dispatch.

1. Service Provider availability for service and payment
2. Metering/other data to determine Service Providers baseline, pre any dispatch
3. Dispatch instructions (if any) and corresponding cease instructions
4. Service Provider metering/other data during instruction period (either real-time metering or provided ex-post)
5. Calculation of payments due according to pre-defined methodology
6. Information exchange between System or Network Operator and Service Provider
7. Pre-payment dispute process
8. Payment
9. Post payment dispute process.

Table 7.1 Payment structures for new flexibility products

Product name	Payment structure
Peak reduction	Utilisation payment only
Scheduled utilisation	Utilisation payment only
Operational utilisation	Utilisation payment only
Operational utilisation + scheduled availability	Availability and utilisation payment
Operational utilisation + variable availability	Availability and utilisation payment

As the use of Flexibility Services increases across system operators there would be increased volumes of settlement, whereby FSPs are paid for the services they deliver. Historically, between 2018 and 2023 the process for settlement has varied between DSOs which created a fragmented and complex approach whereby an FSP could be paid differently between one DSO and another despite providing the same service type primarily due to the difference in data, performance calculations and processes.

Open Networks has endeavoured to standardise and align settlement processes to ensure the settlement process is consistent for FSPs. The focus has been to create standard equations for Availability and Utilisation payments. The concept being, equations can be used across all the Flexibility Products as defined in Sect. 3.3 and summarised in Table 7.1. This standardisation will streamline administrative procedures, reduce discrepancies, and enhance efficiency in delivering Flexibility Services. As a result, service providers will encounter fewer barriers to entry and benefit from a predictable, transparent process, which will foster greater participation in flexibility markets.

Following a gap analysis and stakeholder engagement, The scope of the following key areas were identified as a priority for standardisation; calculation methodologies, service terms, requested metering data and granularity, defined process steps, performance metrics and penalties, settlement procedures, dispute resolution processes, and data transmission methods (API, manual, or other). These are discussed in further detail, in the following sections.

7.1 Performance/Payment Calculations

This section sets out the calculations to allow DSOs to make payments to FSPs for the delivery of Flexibility Services. Two types of Flexibility Services were considered during the development of this methodology: Turn-up/Turn-down and Peak Reduction. The signage used in calculations differs for the different asset types with the Table 7.2 showing a summary.

7.1 Performance/Payment Calculations

Table 7.2 Signage used in calculations for different asset types

Response type	Direction	Baseline value	Meter readings	Dispatched capacity
Demand	Turn down	Negative	Negative	Positive
Demand	Turn up	Negative	Negative	Negative
Generation	Turn down	Positive	Positive	Negative
Generation	Turn up	Positive	Positive	Positive

7.1.1 Availability Payments (for Turn-Up/Turn-Down Services)

Where the contracted Flexibility Service includes Availability, payments are applicable to the FSP in respect of Accepted Availability Windows, whereby the FSP is ready and available to supply the Utilised MW in the event of a Utilisation Instruction. When applicable, Availability is calculated, based on performance on a minute or 30-minute granularity. The Availability Payment due, is subject to Utilisation Performance and a Grace Factor can be applied at the DSO's discretion to calculate the total Availability Percentage due.

Utilisation Performance is calculated as an average across the Utilisation Events within month. Where a grace factor is applied to this average, the DSO may pay up to 100% of availability for the month where average delivery falls within the grace factor band. For example, if the Grace Factor is defined as 5%, and Utilisation Performance is determined to be 95% then 100% of the Availability Payment is due. If there are no utilisation events in the month then full availability is paid.

The parameters to calculate the Availability Payments are detailed in Table 7.3. There are two examples shown, minute by minute and 30-minute granularity. The calculations are the same in both instances however minute by minute granularity would allow the MW or price to change (if needed) on a greater granularity level than 30-minute granularity.

Mathematical Formula

The Availability Payment (AP_{sm}), to be made by DSO to the FSP in respect of the Accepted Availability Window(s) in calendar month, **m**, for Flexible Unit, **s**, can be calculated using the following formula:

$$AP_{sm} = \sum_{j=1}^{j=t} AC_{sj} \times AVP_{sj} \times CC_{sj} \times SA_{sj} \times MPF_{sm}$$

t	represents the number of Metered Time Periods
AC_{sj}	in respect of each Flexible Unit, **s**, and each Metered Time Period, **j**, means the Availability Price in £/MW/h.
AVP_{sj}	in respect of each Flexible Unit, **s**, and each Metered Time Period, **j**, means the Availability Period in minutes/60.

Table 7.3 Parameters used in availability calculations

ID	Parameter	Description	Example 1, 1 minute granularity	Example 2, 30-minute granularity
A	Start datetime	Start date and time of the Accepted Availability Window	01/07/23 00:00	01/07/23 00:00
B	End datetime	End date and time of the Accepted Availability Window	01/07/23 00:01	01/07/23 00:30
C	Metered time period (minutes/60)	The Metered Time Period divided by sixty. In the datetime in example one the period is 1 minute so 1/60, while in example 2 the period is 30 minutes.	0.016666667	0.5
D	Availability price £/MW/h	The contracted fee for the availability component of the service in £/MW/h	£2	£2
E	Contracted capacity (MW)	The Available MW the service provider has been accepted to supply	5	5
F	Availability (1,0)	1 if the service was available, 0 if not available for the period	1	1
G	Grace factor	Availability Grace Factor	5%	5%
H	Utilisation performance	The calculated Utilisation Performance taking into account the Grace Factor in G.	85.33%	100%[1]
I	Pre-utilisation performance payment £	The payment for the availability period without any performance applied Using the data above the value shown in the next columns is calculated by (C*D*E*F)	£0.17	£5.00
J	Post-utilisation performance payment	The actual Availability payment value using the applied performance %, values calculated by (C*D*E*F*H)	£0.14	£5.00

CC_{sj} in respect of each Flexible Unit, s, and each Metered Time Period, j, means the Contracted Capacity in MW

SA_{sj} in respect of each Flexible Unit, **s**, and each Metered Time Period, **j**, is 0 where the Flexible Unit is declared (or redeclared) unavailable or the DSO deem unavailable, otherwise 1.

MPF_{sm} in respect of each Flexible Unit, **s**, and each calendar month, **m**, means the Monthly Utilisation Performance Factor and is 1 where:

(a) there are no Dispatch Events in a month.
(b) the Monthly Utilisation Performance Factor is not being applied by the DSO.

---

[1] The Utilisation performance is greater than 95% hence 100% of the Availability is being paid.

(c) the calculated Monthly Utilisation Performance Factor is 100% or within the Grace Factor applied by the DSO.

otherwise, is calculated using the formula below:
For a Flexible Unit, **s**, for each calendar month, **m**:

$$MPF_{sm} = \frac{1}{p}\sum_{e=1}^{e=p}\frac{1}{n}\sum_{k=1}^{k=n}\text{Max}(\text{Min}((\frac{(MM_{ek} - BM_{ek})}{DC_{ek}}), 1), 0)$$

BM_{ek} in respect of each Dispatch Event, e, and each Minute, **k**, **BM_{ek}** represents the Baseline MW with the signage negative for a demand unit and positive for a generation unit.

MM_{ek} in respect of each Dispatch Event, e, and each Minute, **k**, **MM_{ek}** represents the Metered MW with the signage negative for a demand unit and positive for a generation unit.

DC_{ek} in respect of each Dispatch Event, e, and each Minute, **k**, **DC_{ek}** represents the Dispatched MW with the signage positive for a demand reducing asset or a generation increase asset and the signage negative for a demand increasing asset or generation reducing asset.

n represents the number of minutes for each Dispatch Event

p represents the number of Dispatch Events in a calendar month for a Flexible Unit.

7.1.2 Utilisation Payments (for Turn-Up/Turn-Down Services)

Utilisation payments are paid to the FSP for the energy delivered during a Utilisation Event. Utilisation is calculated, if applicable, on a minute granularity, the payment is subject to a performance metric known as a Performance Multiplier which applies after, if applicable, a Grace Factor.

The parameters to calculate the Utilisation Payments are detailed in Table 7.4. Two examples are detailed a Demand turn-down asset and a Generation turn-up asset, note the signage used differs dependent on asset type and is shown in Table 7.2.

Mathematical Formula

The Utilisation Payment (UP_{sm}), made by DSO to the FSP in respect of the Utilisation Events(s) in calendar month, m, for Flexible Unit, s, can be calculated using the following formula:

$$UP_{sm} = \sum_{j=1}^{j=t} UF_{sj} \times UVP_{sj} \times UC_{sj} \times PM_{sj}$$

Table 7.4 Parameters used in utilisation calculations

ID	Parameter	Description	Example 1, 1 minute granularity Demand Reducer	Example 2, 1 minute granularity Generation Increase
A	Start datetime	Start date and time of the Utilisation Event	01/07/23 00:00	01/07/23 00:00
B	End datetime	End date and time of the Utilisation Event	01/07/23 00:01	01/07/23 00:01
C	Dispatched capacity (MW)	The Utilisation MW instructed.	5	5
D	Utilisation price £/MWh	The contracted fee for the utilisation component of the service in £/MWh	£25	£25
E	Utilisation period (minutes/60)	The duration of the Utilisation Event divided by sixty.	0.016666667	0.016666667
F	Baseline (MW)	The baseline of the asset in the minute in question.	−5	10
G	Metered (MW)	The actual Metered MW of the asset in the minute in question.	−0.712	14
H	Delivered (MW)	The actual delivered MW of the asset in the minute in question. Calculated by G-F	4.288	4
I	Delivery percentage[*]	The actual delivery percentage in the minute in question, calculated by H/C	85.76%	80%
J	Grace factor[**]	A utilisation Grace Factor applied to the MW delivered, if the delivered % is within this tolerance then full payment is made for the minute in question.	5%	5%
K	Performance multiplier[***]	If the delivery % is outside the Grace Factor, then a Performance Multiplier is applied to reduce the % paid.	3	3
L	Payment percentage	The % calculated for payment considering the Grace Factor and penalisation multiplier for the minute in question	67.28%	50%

(continued)

7.1 Performance/Payment Calculations

Table 7.4 (continued)

ID	Parameter	Description	Example 1, 1 minute granularity Demand Reducer	Example 2, 1 minute granularity Generation Increase
M	Utilisation payment (£)	The actual utilisation payment value for the minute in question considering the applied performance % calculated by (D*E*C*L)	£1.20	£0.83

*This percentage is collared at 0% and capped at the POD (payable over delivery) in the formula specified in Detailed Utilisation Formula section.
**In these two examples if either asset had delivered >=95% (taking the grace factor into account) then the full utilisation payment would have been made.
***A performance Multiplier of 3 is used in these examples with a grace factor of 5%. Example 1 has a actual delivery of 85.76% this falls 9.24% (95%-85.76%) outside of the grace factor level, a multiple of 3 is applied (3 * 9.24% = 27.72%) and a payment 67.28% (95–27.72%) of the max amount for that minute is paid.

t   represents the number of Metered Time Periods

UF_{sj} in respect of each Flexible Unit, s, and each Metered Time Period, **j**, means the Utilisation Price in £/MWh.

UVP_{sj} in respect of each Flexible Unit, s, and each Metered Time Period, **j**, means the Utilisation Period in minutes/60.

UC_{sj} in respect of each Flexible Unit, s, for each Metered Time Period, **j**, means the delivered MW calculated as below:

$$MAX\left\{MAX\left(MIN\left(\left(\frac{(MM_{sj} - BM_{sj})}{DC_{sj}}\right), POD_{sj}\right), 0\right) * |DC_{sj}|, |DC_{sj}|\right\}$$

BM_{sj} in respect of each Flexible Unit, s, and each Metered Time Period, **j**, BM_{sj} represents the Baseline MW with the signage negative for a demand unit and positive for a generation unit.

MM_{sj} in respect of each Flexible Unit, s, and each Metered Time Period, **j**, MM_{sj} represents the Metered MW with the signage negative for a demand unit and positive for a generation unit.

DC_{sj} in respect of each Flexible Unit, s, and each Metered Time Period, **j**, DC_{sj} represents the Utilised MW with the signage positive for a demand reducing asset or a generation increase asset and the signage negative for a demand increasing asset or generation reducing asset.

POD_{sj} in respect of each Flexible Unit, s, and each Metered Time Period, **j**, POD_{sj} represents Payable Over Delivery, the percentage over delivery beyond 100% for which remuneration would be allowed. A POD value of 1 will not allow payment for any over delivery whilst a POD value of 1.1 will allow payment for up to 10% over delivery.

PM_{sj} in respect of each Flexible Unit, s, for each Metered Time Period, **j**, means Performance Multiplier calculated using the formula below:

$$PM_{sj} = IF\begin{pmatrix} Delivery\% \geq (1 - GraceFactor) \\ , 1, MAX\begin{pmatrix} 0, (1 - GraceFactor) \\ -\begin{pmatrix}(1 - GraceFactor - Delivery\%) \\ *(Mutiplier\% *100)\end{pmatrix}\end{pmatrix}\end{pmatrix}$$

Delivery% in respect of each Flexible Unit, s, for each Metered Time Period, **j**, means the actual MW delivered divided by the dispatched MW expressed as a percentage.

Grace Factor: in respect of each Flexible Unit, s, for each Metered Time Period, **j**, represents the percentage of under delivery below 100% for which renumeration of 100% would be applied.

Multiplier%: in respect of each Flexible Unit, s, for each Metered Time Period, **j**, represents a multiplier applied to under delivery expressed as a percentage.

Figure 7.1 shows the impact on Utilisation Payments when there is less than 100% delivery. The figure shows a Grace Factor of 5% and a Performance Multiplier of 3 and as can be seen the payable delivery tapers away to zero once the delivery percentage hits 63%.

7.1.3 *Utilisation Calculations (for Peak Reduction Services)*

Flexibility providers commit to reduce their highest demand peaks during pre-contracted windows. Payments are for utilisation only.

Mathematical Formula

The Utilisation Payment (SP_{sm}), made by DSO to the FSP in respect of a calendar month, **m**, for Flexible Unit, **s**, can be calculated using the following formula:

$$SP_{sm} = CC_s \times UF_s \times H_{sm} \times PM_{sm}$$

CC_s in respect of each Flexible Unit, s, means the Contracted Capacity in MW.
H_{sm} in respect of each Flexible Unit, s, and month, m, means the number of service hours awarded in the month in hours.
UF_s in respect of each Flexible Unit, s, means the Utilisation Fee in £/MW/h
PM_{sm} in respect of each Flexible Unit, s, for each Month, m, means Performance Multiplier calculated using the formula below:

7.1 Performance/Payment Calculations

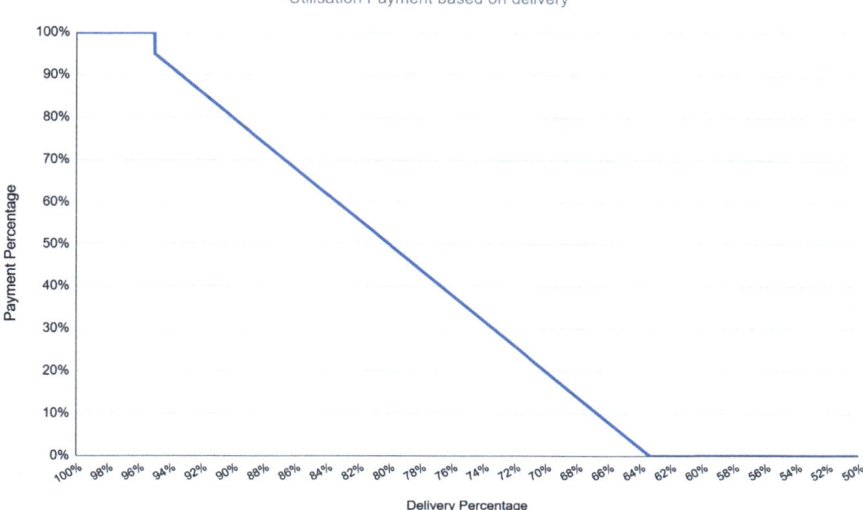

Fig. 7.1 Indicative visual representation impact on payment with changing delivery percentage (Turn-up/Turn-down)

$$PM_{sm} = IF\begin{pmatrix} Delivery\% \geq (1 - GraceFactor) \\ , 1, MAX \begin{pmatrix} 0, (1 - GraceFactor) \\ -\begin{pmatrix}(1 - GraceFactor - Delivery\%) \\ *(Multiplier\%*100)\end{pmatrix}\end{pmatrix}\end{pmatrix}$$

GraceFactor: in respect of each Flexible Unit, s, for each month, represents the percentage of under delivery below 100% for which renumeration of 100% would be applied.

Multiplier%: in respect of each Flexible Unit, s, for each month, **j**, represents a multiplier applied to under delivery expressed as a percentage.

Delivery% in respect of each Flexible Unit, s, and month, m, is calculated as below:

$$Delivery\% = \frac{((Min(MM_{sj}, MM_{s,j+1}, \ldots, MM_{s,j+n})) - (Min(BM_{sj}, BM_{s,j+1}, \ldots, BM_{s,j+n})))}{CC_{sm}}$$

BM_{sj} in respect of each Flexible Unit, s, and each Settlement Period, **j**, BM_{sj} represents the Baseline MW with the signage negative for a demand unit and positive for a generation unit.

MM_{sj} in respect of each Flexible Unit, s, and each Settlement Period, **j**, MM_{sj} represents the Metered MW with the signage negative for a demand unit and positive for a generation unit.

CC_{sm} in respect of each Flexible Unit, s, and each Month, **m**, CC_{sm} represents the Contracted MW with the signage positive for a demand reducing asset or a generation increase asset and the signage negative for a demand increasing asset or generation reducing asset.

n represents the number of dispatched settlement periods.

7.2 Metering Specifications

7.2.1 Metering Granularity

Agreed Alignment: Minute-by-minute and half hourly data, would be accepted for settlement purposes. Certain products rely on minute-by-minute metering granularity for accurate performance monitoring and settlement. Where an alternative to minute-by-minute granularity is provided the data may be disaggregated. As such, this could result in performance monitoring and calculation inaccuracies. [Prompt: Option to insert table to specify granularity preference for products]

Historically, DSO's have taken various approaches to Metering granularity. Aligning on both minute-by-minute and half hourly gives FSPs consistency and clarity on what is required to be provided to settle a service. This was deployed through the adoption of Open Networks' "Standard Agreement" (Ver 2.1 and beyond) for Flexibility Services.

7.2.2 Metering Data Requested

Agreed Alignment: A common set of API parameters was defined shown below in a 'CSV template', in the Table 7.5. These parameters are applicable to the measurement data. Parameters of baselining are not included in this template

7.2.3 Metering Accuracy Standards

Agreed Alignment: For Asset Point Metering (the metering measured directly from the DER and is downstream of the Boundary Point Metering.), the FSP will ensure compliance with the following metering standards set out within the most recent published relevant Balancing and Settlement Code of Practice Eleven: code of practice for the metering of balancing services assets for settlement purposes [1]:

7.2 Metering Specifications

Table 7.5 Aligned metering parameters

Settlement phase	Message	Parameter	Definition	Example (1MW Generator)
Measuring	Operational metering/ delivery readings	Unit ID	ID for the specific asset, discrete or aggregated, which has been dispatched	Aggregate_Alde
		start_time_ timestamp_ utc	Timestamp in UTC ISO-8601 format. If the meterable unit is marked as Half-Hourly metering then the timestamp must start exactly on the hour or half-past the hour (ie, 00 or 30 minutes) If the meterable unit is marked as minute-by-minute metering then the timestamp must start exactly on the minute (ie, 01:00)	2023-07-01T00:01:00Z
		end_time_ timestamp_ utc	Timestamp in UTC ISO-8601 format. If the meterable unit is marked as Half-Hourly metering then the timestamp must start exactly on the hour or half-past the hour (ie, 00 or 30 minutes). If the meterable unit is marked as minute-by-minute metering then the timestamp must start exactly on the minute (ie, 01:00)	2023-07-01T00:01:00Z
		Power	Instant active power for the minute-by-minute period, in MW. Positive values indicate power export to grid and negative values indicate power import from grid	516
		Energy	Metered energy, in MWh for the granularity period. Positive values indicate energy export to grid and negative values indicate energy import from grid	425

- Metering accuracy requirements
- Asset meter calibration test certification
- Limits of error
- Sealing requirements.

For Boundary Point Metering (the metering measured at the point of supply from the DSO network), the Provider should be compliant with Balancing and Settlement Codes of Practice 1, 2, 3, 4, 5 and 10 as applicable [2].

If requested by the DSOs, the FSP shall provide evidence of compliance with the above standards. This may be in the form of certification, photo, or written confirmation.

The FSP can develop their asset and boundary metering based on a standardised code of practice adopted by the wider industry. By adopting these industry codes of practice, there will be potential alignment with future NESO services and markets. Wording has been drafted and a new section added to the Standard Flexibility Agreement (Ver 2.1 and beyond)

7.2.4 Site Meter Location

Agreed Alignment: DSO's have agreed to accept both boundary and asset metering locations and appropriate wording has already been added to the Flexibility Agreement. FSPs have an option based on their capabilities to provide boundary or metering data which will deliver consistency when providing services to different DSO's.

7.3 Billing and Invoicing

7.3.1 Process Steps

Agreed Alignment: Figure 7.2 shows the steps in the settlement process, starting with the entering of baseline data and finishing with payment. It is not possible to fully align on the timescales for each step due to the various systems used by each DSO, in some cases there are multiple systems that are also used for wider business functions. However, the timescales do not diverge significantly, in most cases it is a few days difference.

Performance Data Feedback

Agreed Alignment: The Performance report/Initial Statement should be accompanied by the following details, ensuring all FSP's receive the minimum data outlined below from all the DSO's:

7.3 Billing and Invoicing

Fig. 7.2 Processes steps in the settlement process across all DSOs

1. The period(s) during which the Flexibility Services were made available to the DSO and, if applicable, where utilised;
2. Availability charges (if any), reflecting any reduction for periods of where the service was unavailable or had reduced capacity;
3. Utilisation charges (if any) reflecting any reductions for periods of where the service was unavailable or reduced, including any utilisation payment cap;
4. Summary of the relevant Charges (if any); and
5. The event and zone/service the performance data relates to.

Regularity of payments

Agreed Alignment: Payments would be made monthly.

Whilst the date of payment may be different due to DSO internal processes, by agreeing to make payments on a monthly basis, FSPs will benefit from having a consistent approach across the DSO's allowing them to build a single process rather than multiple processes which need to accommodate different payment dates.

Treatment of over and non-delivery

Agreed Alignment: DSO's have agreed there would be no additional payment for over-delivery and no payment for non-delivery. However, the payment mechanics developed allow for over-payment to be considered in the future.

Grace Factors

Agreed Alignment: The following Grace Factors, listed in Table 7.6 were agreed by all DSOs as a recommendation for alignment. It was noted that further analysis and review is necessary to understand these in operation. This is following feedback from stakeholders and their concerns around fixing the grace factor. There is a view that by fixing the grace factors it could stagnate the market if they are too onerous or could prevent DSOs trying new methods to support FSPs and markets. Therefore, DSOs are adopting their grace factors to 80% of their Flexibility Services, which allows DSOs to be innovative when launching new products or when trying to increase liquidity.

There were also proposals to adopt a grace factor of 0-5% for new products such as the Peak Reduction service estimated to be deployed later in 2025.

Table 7.6 Grace factors and performance multipliers

Product	Grace factors (%)	Performance multiplier
Peak reduction	0–0 5	3
Scheduled utilisation	5	3
Operational utilisation	10	2
Operational utilisation + scheduled availability	5	3
Operational utilisation + variable availability	5	3

7.4 Baselining

Baselining is a key consideration of the commercial aspect of Flexibility Services. The baseline is defined as the volume of electricity demand or generation which is expected to have come from a FSP if they were not providing flexibility; the watermark from which a change in demand or generation is measured. The difference between the baseline and the actual metered volume during the period of service provision is the volume of flexibility that the FSP has provided and would be compensated for.

Determining this volume correctly is important to DSOs, the NESO and FSPs to ensure both the correct amount of flexibility is procured and is properly rewarded. Equally, effective and transparent baselining is necessary for FSP to judge how it can stack Flexibility Services.

Baselining is therefore the 'Established level of Distributed Energy Resources base load from which a delta is measured to calculate level of service delivered' this is the measurement on the basis of which the FSP is paid for the dispatched service.

All DSO Flexibility products are technology-agnostic, meaning that they are open for all technologies to participate and do not include explicit requirements (or) favouring certain technologies. It is important to understand, however, that certain technologies have a better match with certain products than others. For example, products with long utilisation periods will favour (dispatchable) generation assets to storage facilities. Yet this is not driven by the technical product definition, but by the economics of the assets looking to participate. Historically a limited set of technologies are / have been participating, However this is set to change with many new (DER) technologies entering the market in large numbers, e.g. EV chargers and heat pumps.

Flexibility Product parameters have a large impact on the choice of the right baseline methodologies, which is a theme that stakeholders feedback. The parameters with the largest impact on baseline assessment are:

- Metering configuration
- Type of remuneration
- Utilisation instruction notification period
- Utilisation period
- Frequency of use of the Flexibility Product
- User and technology segment
- Pool or asset level activation and
- Controller of the assets (i.e. DSO or the FSP).

7.4.1 Baseline Methodology Groups

Meter Before—Meter After (MBMA) Baseline

MBMA is a flat baseline set at a pre-activation level determined by one or a small number of intervals (settlements) which is consistent with no delays. The baseline can be the single meter reading or average/median/min/max of meter readings before the activation window. Metering readings of the activation window are compared against the meter readings prior to the activation to calculate the delivered flexibility. The Meter Before/Meter After is a widely used methodology for accurately estimating the level of service delivered under real-time dispatch conditions and short utilisation periods. It is also a preferred baseline for frequent activations as it is not dependent on historical data. As such, MBMAs are very common in balancing products which have these characteristics. Meter Before/Meter After requires DERs with relatively flat load profiles during the utilisation period. If a resource has periods of ramping up or down or general variability, the meter Before/Meter After approach can over- or underestimate the actual level of load reduction even for the shorten period. MBMA requires sub-metering at DER asset level.

Data: MBMA baselines have extremely limited data selection rules, essentially a single interval.

Estimation method: No estimation method is used for MBMA baselines, as they are a flat line at the prescribed load level. Single meter reading is sufficient.

Window before and after (WBA) Baseline

This is similar to the MBMA. In this case the window after is also taken into account to compensate for direct rebound effects. It has same characteristics and applications to the MBMA.

Data: This baseline is not dependent on historical data or external data (e.g. weather data)

Estimation method: No estimation method is used.

Nomination Baseline

The baseline is the forecast of the generation or demand profile of the DER asset, as if no flexibility utilisation would take place. The forecast is sent by the FSP to the DSO before gate closure or at another predefined deadline. For example, the physical notifications which are used in the Balancing Mechanism belong to the Nomination baseline types. The DSO can then use this profile to calculate the deviation of the metered data from the planned profile. The FSPs can use different methods to forecast their profile. The nomination is a technology agnostic baseline for real-time utilisation instruction. In addition, as per stakeholders' response it works well when submetering is available, but it creates challenges and inaccurate forecasts when sub-metering is not available.

Data: There are no data requirements. The FSP can select if they want to use any historic data in forecasting their demand/ generation.

7.4 Baselining

Estimation method: The FSP can choose their forecasting approach. The estimation method is not applicable to this baseline.

Historical Baseline (a.k.a. Rolling baselines)

This baseline methodology uses historical data to calculate the baseline, usually based on recent data prior to the utilisation day. For example, the 'average interval load of last 5 business days' can be the baseline from which the deviation is calculated.

Data: Two important aspects of the historical baselines are the data selection and exclusion rules which determine what data will be used to produce the baseline. Historical baselines require usually 5 to 10 days of data. In addition, historical baselines exclude prior-event days and non-similar days (e.g. weekends and holidays). In addition to the exclusion of prior dispatch and non-similar days, many historical baselines have an additional set of data exclusions based on load characteristics. For example, a baseline may drop low load days if they fall below a threshold related to the mean of the selected days (e.g. the low load day is less than 20% of the mean load during the selected days during dispatch window hours). Other baselines rank the included days based on load and exclude a subset of those chosen days, either extreme days (e.g. mid 8 of 10) or just low-load (high 4 of 5). For example, mid 8 of 10 baselines excludes the highest and lowest demand days of the 10 selected and high 4 of 5 excludes the lowest demand day. These exclusions are designed to target the recent days that are most likely to predict the dispatch day load.

Estimation method: Historical baselines are primarily calculated using a simple mean for each interval across the final set of chosen days. A median approach is also used by some historical baselines. DSOs usually provide the baseline which is based on historical data to the FSPs. DSOs use metered historical data and FSPs can replicate the baseline using their own data.

Regression-based Baseline

A regression model is used to calculate the baseline. Regression-based baselines use schedule, weather and other variables to explain customer load variability. The regression summarises how load interacts with the variables allowing for the prediction of load levels at any combination of the variables. The baseline is the regression-based predicted load for the dispatch day based on that day's weather and schedule characteristics.

Data: Regression-based baselines require substantially more data than historical baselines. The make up for the lack of recent data by using advance methods to identify the relevant load characteristics from a large pool of data. Most regression baselines required at last a full year of data. Regression approaches use all of the available data but control for excluded days in the regression structure. In the regression context, controlling for weekends days removed their effect on the weekday baselines while informing the weekend baseline.

Estimation method: Regression estimation methods are substantially more complicated than the mean used by historical baseline approaches. For example, a regression baseline may include calendar, weather and daylight variables in multiple forms. The specification controls for heat build-up over days, heat gain within the day, hour of light and fraction of dark as well as a range of temperature-time interactions. The regression uses this specification to develop a baseline that reflects the calendar, weather and daylight characteristics of the dispatch day. The regression produces both weekday and weekend baselines. Regression-based methodologies have the potential to be the most effective baseline without same-day adjustment (SDA) (see below for explanation on same-day adjustments). The difficulty of developing an accurate baseline without the SDA is a primary potential justification of the additional technical challenge of the regression approach.

Calculated Baseline

This method involves calculation based on external parameters (e.g. weather based), without relying on historical data. These baselines are not very common as they are applicable to technologies and assets with a demand or generation profile which can be calculated based on formulas and inputs which do not use historical data. Some examples are as follows:

- Generation of a wind turbine base on wind speed and capacity.
- Generation of a solar PV based on solar panels characteristics, radiation, sunlight weather information etc.

Control group Baseline (a.k.a. peer group)

The baseline is calculated using as inputs measurements of similar customers who do not participate in the Flexibility Service.

Data: This baseline methodology uses data of assets/ technologies/ customers which are of the same type as the participating DER technologies. The main requirement is that these assets do not participate in Flexibility Products. The methodology uses data of the same utilisation period as the flexibility event.

Estimation method: The baseline is calculated based on a simple estimation method, using a simple mean or median.

Adjustments

1. Same-Adjustment (SDA) Method and Period

In addition to baseline methodologies above, there is a variant of historical and regression-based baselines which includes a same-day adjustment of the baseline. Adjustment method and period are additional components of historical and regression baselines. Adjustment method refers to adjustments that can be made to the initial baseline to make it a better fit for the dispatch day load/generation. Adjustments use the most up-to-date information to inform the final position of the baseline. The adjustments bring the baseline into line with the pre-dispatch intervals on which the adjustment is based. This means that the baseline starts the dispatch period relatively close to actual load and will only diverge if the load shape from the baseline is different from the actual profile of the day.

Adjustment period refers to the specific intervals that are used to make the adjustment. Figure 7.3 visualises how this mechanism works in practice. A customer with a weather-sensitive load profile is shown, with the meter data displayed in green. A historical baseline calculation (in this example a mid 3 of 10, displayed in blue/dotted trace) without same-day adjustment is constantly below the measured values (probably the current day is much colder than the preceding days). The adjustment window is set to 8:00–12:00AM, during which the average (relative or absolute) difference between the baseline without SDA and measurements is determined. During the DR event (utilization period), the baseline is adjusted according to this difference. This results in the adjusted baseline (in pink), which, based on a visual inspection, approaches the counterfactual more closely. This has direct influence on product validation, as shown by the difference in calculated load drop, which corresponds to the activated energy.

Whilst there is evidence to show that same-day load-based adjustments improve the accuracy of baselines, load-based adjustments also carry a degree of risk because the adjustment process relies on a handful of load intervals to adjust the whole baseline. SDAs can be more susceptible to gaming as few intervals can have a big effect on the baseline. Beyond gaming, there are reasonable pre-dispatch load characteristic that may also affect the baseline in ways that would make the baseline less accurate for both system and customers. For these reasons adjustment approaches that do not use pre-dispatch load are also available.

2. Additive and Scalar same-day load adjustments

Adjustments are made either using an additive or scalar approach. Both approaches have the effect of bringing the baseline in line with actual load on average, during the pre-dispatch adjustment period. During the dispatch period the adjustments differ:

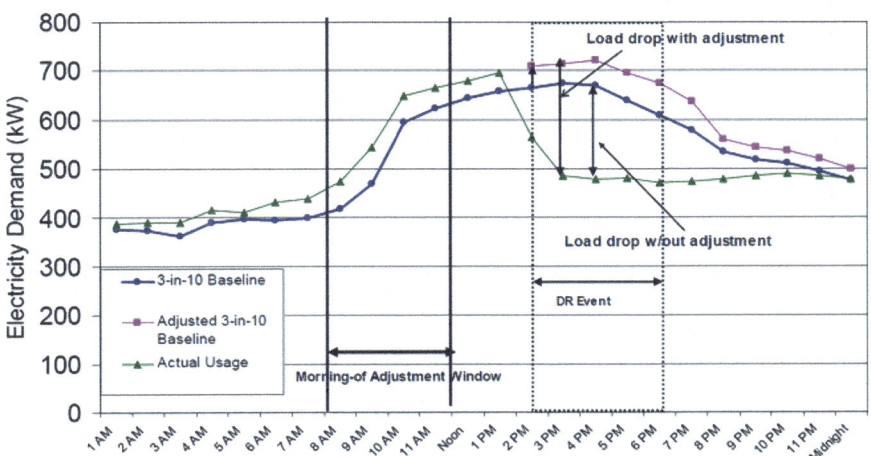

Fig. 7.3 Illustration of morning of adjustment for a weather-sensitive asset

- For the additive adjustment, the baseline is shifted by this same amount for each interval. For example, the difference between the measurement and the baseline value in the adjustment period is added to the baseline value.
- In contrast, for the scalar approach the magnitude of the adjustment will vary depending on the level of baseline load to which the percentage adjustment is applied. For example, the baseline is multiplied by the ratio d/b, where d is the measurement in the adjustment period and b the baseline value for the adjustment period.

3. Drop-to mechanism

Although baseline methodologies are suitable for most of the Flexibility Products, other mechanisms have also been used, such as the drop-to mechanism which is used to validate that a flexibility asset has provided the contracted capacity. The drop-to mechanism also known as firm load in case of demand response, identifies load levels below which an asset will stay during an activation period. Rather than attempting to explicitly measure load reduction from a counterfactual baseline, the drop-to mechanism assures that an asset is not contributing to system load beyond the specified amount. The key characteristics of a drop-to mechanism are:

- Drop-to typically is used for large customers to address local emergency conditions.
- Drop-to is suitable for capacity-only products. When energy is remunerated, the activated energy needs to be calculated for which the drop-to mechanism is not suitable.
- Drop-to mechanisms generally do not, on their own, support a quantified settlement process. Ex-post application of drop-from baselines or regression techniques can provide estimates of the actual load reduction supplied at the time of activation. Similarly, a deemed load reduction value can be calculated as the average of customer load net of the drop to value for relevant historical intervals.

7.4.2 Use of Baselining Methodology

Historically the use of baseline methodologies by DSO shows that priorities vary across Flexibility Products, markets, organisations. Whilst accuracy of the baseline is a key principle for mature markets, simplicity and inclusivity may be prioritised at markets which are at their infancy. The below list of principles guide DSO decisions

- FSP implementation simplicity (Very high Priority): Are the costs for implementing and operating the administrative processes proportionate for the FSP?
- Inclusivity (Very high Priority): Is the Baseline Methodology technology agnostic and not biased to a particular type of solution, technology and provider?
- DSO implementation simplicity (High priority): Are the costs for implementing and operating the administrative processes proportionate for the DSO?

7.4 Baselining

- Replicability (Medium priority): Is the baseline reproducible by the DSO, FSP, and third-party validator for settlement (verification) purposes?
- Accuracy (Medium priority): Does the Baseline Methodology provide an accurate estimate of the flexibility load impact at a level expected by DSO and FSPs, or does it show a relatively high variance?
- Bias (Medium priority): Does the Baseline Methodology provide an unbiased estimate of the flexibility load impact at a level expected, or does it show a relatively high bias?
- Integrity (Medium priority): Does the Baseline Methodology avoid or minimize the risk of gaming and strategic behaviour?
- Stackability (Medium priority): Does the Balancing Methodology allow the FSP to combine the delivery (Availability and/or Utilisation) of DSO products with other markets?
- Design fit (Low Priority): Are there high requirements on data to calculate the baseline? Do data quality issues undermine the baseline quality? Can specific parameters of the service design be met?

Table 7.7 shows the common DSO baseline methodology descriptions.

Choice of which baselining methodology is used is the prerogative of the DSOs. These methodologies are not designed to be paired with specific products, instead

Table 7.7 Common DSO baseline methodology descriptions

Methodology	Description	Example use(s)
Mid 8-in-10	A rolling historical baseline which uses data from the "middle" of the last 8 of 10 days	FSPs with a regular weekly profile
Mid 8-in-10 with Same Day Adjustment	A rolling historical baseline which uses data from the "middle" of the last 8 of 10 days, but also applies a "same day adjustment"	FSPs with a regular weekly profile but with adjustment to allow for provision of other Flexibility Services
Mid X-in-Y	A custom rolling historical baseline, where the user can choose how many days to consider and what length of same day adjustment to use	FSPs providing a long-term regular Flexibility Service (e.g. Peak Reduction)
Nominated	A nominated baseline, which allows the user to input the self-declared baseline of the asset in advance of the flexibility dispatch event	Bespoke methodology for FSPs with highly irregular profiles
Zero	A baseline which assumes that the asset is not operating except for when providing a flexible service	FSPs who have zero baseline

allowing optionality for a range of FSPs to be accurately baselined by DSOs based on the information that is best available.

Across NESO services (as well as the wholesale market), there are a wide range of baselining approaches that are used (Table 7.8).

Table 7.8 NESO baselining methodologies for balancing services

Market/service	Baseline method
Wholesale market	Effectively a zero baseline. Asset measured against contracted position–Energy Contract Volume Notifications
Balancing mechanism	Physical notifications–Final PNs submitted at gate closure (1-hour before delivery)
NIV chasing/imbalance	Effectively baselined against wholesale contracted position
Capacity market	For generation, effectively a zero baseline is used. For demand–six-week baseline of historical demand data used. 16 data points are used in calculating the DSR baseline. Baseline is adjusted for provision of balancing services
Short term operating reserve	BM STOR–physical notification with a zero baseline required
Firm frequency response–static only	Effectively real time baseline–output or demand measured immediately prior to the Relevant Frequency Incident
Local constraint market	A nominated baseline is used for LCM (providing immediately after bidding (day-ahead or intra-day as appropriate)
MW dispatch service	Delivery is measured from the point of instruction (effectively real time baseline)
Demand flexibility service	Historical baseline (unadjusted), using 60-day window with a selection of most recent days used
Slow reserve	TBC. At the time of publication, the Quick Reserve Technical and Procurement Service Design has proposed the use of Final Physical Notifications for baselining. Slow Reserve is expected to be the same
Quick reserve	
Balancing reserve	Final physical notifications submitted at gate closure (1-hour before delivery)
Dynamic containment	Physical notifications for BMUs–Submission of Operational Baseline. Final PNs submitted at gate closure (1-hour before delivery). Non-BMUs to provide equivalent via a Non-BM Data Submission
Dynamic moderation	
Dynamic regulation	

7.5 Suggested Further Reading

Listed below are the links to Open Networks publications relevant to this chapter in reverse chronological order. Please note, some of the documents/spreadsheets may now by outdated or superseded as the topics have evolved.

1. Standardised DSO Settlement Methodology (2024).
2. Aligned DSO Settlement Processes (2024).
3. Dispatch interoperability and settlement review of existing practices and gap analysis (2022).
4. Dispatch Interoperability and Settlement Key Service Parameters (2022).
5. Flexibility Services–Dispatch and Settlement Processes (2019).
6. Baseline Methodologies Review (2022).
7. Baseline Principles for measuring delivery of Flexibility Services (2020).
8. Flexibility Baselining Tool—Mathematical Specification (2022)-Tool no longer in use.
9. Flexibility Baselining Tool–Functional Specification (2022)-Tool no longer in use.
10. Baseline Tool Dissemination Webinar Slides (2022) -Tool no longer in use.
11. Flexibility Baselining Tool User Guide (2022)–Tool no longer in use.

References

1. "Code of Practice Eleven. For the Metering of Balancing Services Assets for Settlement Purposes; Issue 1.0, Version 1.0," Balancing and Settlement Code, 2022
2. BSC digital code, Elexon (2024) [Online]. Available: https://bscdocs.elexon.co.uk/bsc

Open Access This chapter is licensed under the terms of the Creative Commons Attribution 4.0 International License (http://creativecommons.org/licenses/by/4.0/), which permits use, sharing, adaptation, distribution and reproduction in any medium or format, as long as you give appropriate credit to the original author(s) and the source, provide a link to the Creative Commons license and indicate if changes were made.

The images or other third party material in this chapter are included in the chapter's Creative Commons license, unless indicated otherwise in a credit line to the material. If material is not included in the chapter's Creative Commons license and your intended use is not permitted by statutory regulation or exceeds the permitted use, you will need to obtain permission directly from the copyright holder.

Part IV
Interactions Across Markets

Chapter 8
Interoperability of Flexibility Dispatch Systems

Interoperability of Flexibility Systems refers to the idea that the process and experience for an Flexibility Service Provider (FSP) being dispatched will be the same, irrespective of which network operator they are providing services to. This means that, from an FSP perspective, the systems, protocols, communications requirements, and expectations should be aligned, to avoid ambiguity or different interpretations of message semantics.

With more intelligent utilisation and distribution of existing resources, it was determined that a common interface for all Flexibility Service providers to provide flexible power in local constrained areas to electrical network operators using a common interface was essential to prevent eco-system fragmentation, and reduce barriers to entry for FSPs. If FSPs were to implement discrete solutions for each regional network operator would reduce the commercial incentive to provide solutions and increase costs for bill payers.

8.1 Application Program Interface (API)

An Application Program Interface (API) is a specific technical interface definition used for applications or systems to communicate across. An API is fundamentally a way for two or more different pieces of computer software to communicate. An API may be defined through documentation or a specification, which is a technical document setting out the parameters required to interoperate with other systems implementing that API. An API can provide a practical way to allow independent systems with different architectures to communicate, by creating a well-defined abstraction. The term API is often used to refer both to the technical specification of communications, as well as a specific implementation of that technical specification. It is important to note that for an API to be interoperable (i.e. allowing others to implement it), there must be a technical specification and documentation around it. Best practice

© The Author(s) 2025
A. Aithal et al., *Distribution System Operation: Flexibility Services*, Power Systems,
https://doi.org/10.1007/978-3-031-92905-2_8

is for this specification and documentation to be versioned, and subject to change control and governance, to ensure that appropriate integration and interoperability tests are defined correctly for each version.

It is important to note the technical distinction between a standard and an API. A standard, as generally considered by standards development organisations (SDOs) to be technical specifications and procedures, which enable interoperability across consistent protocols which can be universally understood and adopted [1]. A standard could, for example, include the technical definition and specification of an API, but the standard would not cover the implementation itself—the standard would define how implementations should act and behave.

The process of technical standards development is one of broad consensus and mutual agreement. This means that the standards development process moves relatively slowly, in order to ensure that all stakeholders have an opportunity to have their views heard, and input to the process. Rapid evolution would likely hinder this. To ensure that a standard gains adoption and uptake (since there is generally no specific obligation to adopt or implement a standard) over other alternative proprietary solutions, the process of developing standards needs to be inclusive and open to be credible.

Standards Development Organisations (such as the IEEE, IEC, ETSI/3GPP etc.) have significant rules and procedures designed to ensure that standards are developed transparently and openly, in order to avoid negative effects. Decisions such as this are likely to lead to a need for formal defensible decision-making and governance around such decisions, in order to ensure that all stakeholders' needs and interests are given due consideration.

8.1.1 Stakeholder Views

Rigorous stakeholder engagement was undertaken to gain input from a variety of stakeholders, for understanding of the challenges, requirements and considerations for FSPs their requirements/considerations/concerns in the flexibility market. This was undertaken through workshops, a series of direct interview style engagements In addition to DSO and NESO the stakeholders included distribution, SDOs, dispatch service providers, subject matter experts, aggregators, industry groups, government, consumer representatives, equipment vendors, established market providers and new entrants.

From these engagements, a series of key insights were gained

- They agreed that developing consistency in the market and wider eco-system would be extremely beneficial to grow liquidity.
- They had a strong preference for deploying a solution, which can then be iterated on, rather than waiting to develop "the perfect solution".
- They preferred a common digital life-cycle engagement between all DSOs, including tendering, PQQ, contracts, dispatch and settlement.

8.1 Application Program Interface (API)

- They had a preference to employing modern technologies, (HTTP REST as opposed to XML SOAP), given the more established ecosystem of developers.
- FSPs were keen to have stable APIs for automation but also email notifications for information.
- FSPs recognised the importance of Cyber Security, but mostly considered this a platform issue.
- FSPs generally sought iteration on a design, albeit without breaking backwards compatibility—discussions tended to focus on examples of versioned systems, and the potential to support different versions of an API for a longer period of time, with implementers able to upgrade when they felt it was worthwhile.
- FSPs generally didn't mind which dispatch platform was employed, provided it was consistent, had longevity, and relatively simple to deploy.

The most pronounced gap noticed during the stakeholder engagement was the varying understanding and interpretation of the system wide requirements. Notably, the distinction between a standard and an implementation, while we were focused on exploring the requirements for a standard.

8.1.2 Review of Architectural Requirements

The section first explores the required minimum architectural decisions which need to be made for a communication standard, minimum technical requirements a Flexibility Service dispatch interface would require in order to provide a minimum viable product (MVP).This determines the required components, and the associated communication parameters, to share data around the different system components and different potential architectures for a dispatch system to demonstrate how the requirements vary depending on the architecture.

The first stage in developing any system is understanding and agreeing on a common view of the architecture—that is to say, understanding the different components which have to communicate, and at fundamental level, how they will communicate. With a focus on dispatch, the simplest version of this communication was explored, as shown in Fig. 8.1. The operator will send dispatch messages to the Flexibility Service Provider (FSP), and the FSP will respond with a physical change in electrical generation or consumption to match the operator's request. Operator in this context is defined as either a, DSO or NESO. The content of these messages can then be defined, to include things such as start time, duration and requested capacity. The format, units and interpretation of these messages then needs to be agreed between the FSPs and the Operators to enable interoperable communication between both parties.

At this fundamental level, there is an immediate fundamental technical implementation decision required—should the FSP or the operator's dispatch and management system be the "server" in a traditional REST-based client/server Application Programming Interface (API) architecture?

Fig. 8.1 Simple dispatch architectural diagram

Answering this robustly will be a key API requirement before other architecture decisions take place, and will impact on the evaluation of potential architectures, APIs and standards. This will drive other technical and architectural requirements, but indicates the kind of fundamental decisions which need to be taken before selecting protocols, standards or APIs. Conventionally in API design, the requester of a service would be the client in a client/server architecture, with the provider of the service being the server—this means that a long-running server service can listen proactively for an incoming request, and trigger an interrupt to service the request when it is received, act upon the request, and send a response back to the initiator of the request in real-time to acknowledge and confirm the status of the request. This is how an HTTP client works—the user's browser makes a request of the web server—the browser is the client, making a request of the server. Following this logic, an FSP would run a server, and the DSO would be the client in this architecture.

This raises several design and security considerations which should be explored as part of making this decision. If the network operator were to act as the server, then an FSP needs to either poll the server periodically to look for a new dispatch request, carry out TCP long-polling, regular short-lived heartbeat requests to look for pending dispatches, or establish, through a protocol such as WebSockets or similar, a route for the server (DSO) to initiate communication with the client (FSP). This would be counter-intuitive compared with conventional logic.

Conversely, if the FSP was to act as the server, they would need to take responsibility for implementation and enforcement of a number of security measures (i.e. TLS verification of communications and similar), and would need to operate an exposed attack surface over the internet. There would be a need for DSOs to enrol the appropriate FSP URLs into their dispatch clients, and handle error states at the DSO end where a dispatch request failed due to a DNS or IP routing failure. The FSP would also need to ensure that their external attack surface (including their dispatch service API) was patched and maintained, since it would be exposed outwardly, and to ensure that they only respond to correct requests from the authorised network operator. This would place significant business logic responsibility for validation and authorization on the FSP to implement correctly, and would be likely to present effective or practical barriers in FSPs entering the market, with this level of responsibility on them to host and operate a resilient API like this, in the face of denial of service attacks and similar, which may be directed towards them. Many FSPs may also struggle to access and afford the skilled personnel needed to maintain and secure such a service on an ongoing basis.

8.1 Application Program Interface (API)

It is therefore a requirement to determine the directionality of any API, taking into account security architecture and design, as well as the means through which an FSP receives dispatch instructions. This should take into account the wider architecture of a flexibility system, in order to create a cohesive and consistent approach to system architecture. As demonstrated by API directionality, in this simple example of two components communicating can change considerably depending on the architectural decisions. Accordingly, three potential operating models of this simple FSP to operator interaction are detailed below in the three architectural design patterns

Common Dispatch Management Systems

The first of these architectural design patterns, as shown in Figure 8.2, would create a single centralised national common dispatch and management system. Every FSP and electrical operator wishing to engage in the flexibility market would connect to this common system and use it as an intermediary. The advantage of this design is that it creates a common system for all operators and FSPs, overcoming the currently fragmented nature of DSO regions, and reducing the barrier to entry for FSPs to grow their portfolios new regions. It would also be easier for new entrants as commonality would grow the overall ecosystem and enable a new network of supporting companies to assist smaller FSPs to engage. This would reduce integration and interoperability testing requirements, as every market participant knows the (one) implementation they need to be able to work with.

The governance, security and accountably of this shared system are the disadvantages of this model. A shared governance model would allow for joint advancement of functionality and features; however, this would likely be constrained by requiring

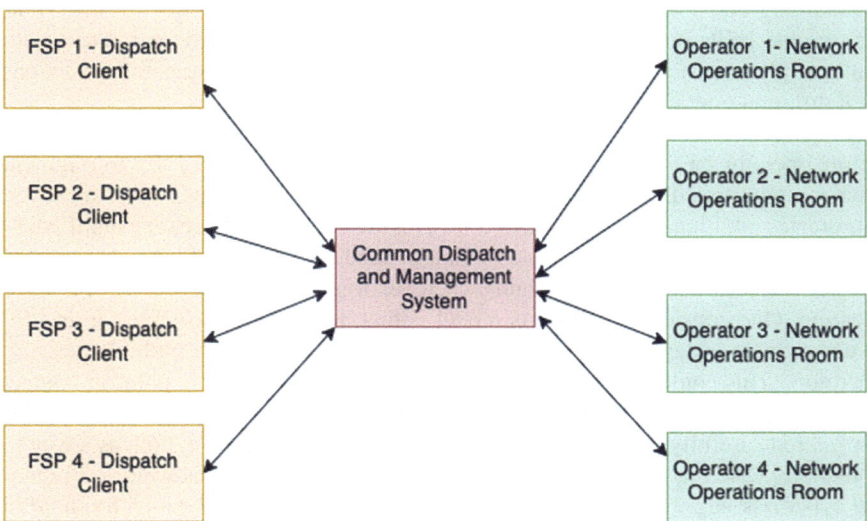

Fig. 8.2 Common dispatch management systems

group consensus. From an accountably perspective, clear legal agreements would have to be agreed between operators and providers of this common platform as to determine who is responsible if operational failure resulted in loss of power for consumers. Operational uptime also poses some consideration as in the event of an operational outage all regions would be affected, thereby, creating a disproportionate issue and additional operational reserves would have to be held than if this was segmented. This would also create a single dispatch platform across all DSOs, which would be a significant and attractive target for foreign state actors. Such a platform would, from a security perspective, need to be designed to resist attack by the most determined and capable foreign intelligence services and state-aligned attackers, given its centralised nature across all distribution network operators, and thus the potential for use to cause cascaded impact across the wider UK energy system as a means of a high-profile central attack on the UK's energy networks, without the adversary crossing into directly attacking OT systems.

Region Specific Dispatch Systems

In this second potential model, as shown in Fig. 8.3, each operator would control and deploy their own instance of a dispatch system. This system would follow a common standard so that all the functionality and capabilities, would be common throughout the industry to ensure interoperability and prevent isolation. Through a competitive tendering process these systems could be created by a common vendor or disperse vendors. Provided all the platforms followed a common standard operators would be able to switch systems and migrate FSPs data over to a new system to prevent vendor lock-In, with minimal interruption for FSPs. The advantage of this model is the operators would have additional oversight and control over the dispatch platform, including maintaining and monitoring based on their individual constraints. The disadvantage is that FSPs operating in multiple DSO regions would have to interact with multiple platforms, and potentially support and maintain multiple parallel versions with different operators.

For example, operator 1 could update to the latest standard version/security patch, before operator 2 FSPs. Accordingly, it would be best practice for FSPs to fragment their systems based on operator region to avoid incompatibility issues, however, this creates additional overhead for FSPs. This would remain the case even if backwards & forwards compatibility were implemented, as there would remain the need to integration test and validate against each version and implementation of a specification. One consideration of this model is that, it would be possible for an FSP to use a single instance of a dispatch client to communicate with multiple network operators. This could create an aggregation risk, whereby a single point of contact in the FSP network is interacting with multiple network operators. This should be considered carefully from an architecture and security perspective, both as a way to potentially ease the technical and operational burden on FSPs, as well as a potential cyber risk aggregation point, where a successful attack could gain a foothold in multiple network operators' flexibility supply chain.

8.1 Application Program Interface (API)

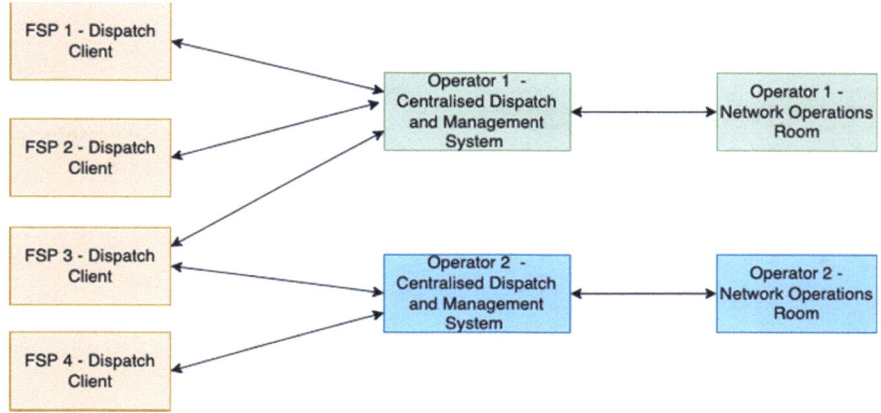

Fig. 8.3 Region specific dispatch systems

Single Operator's Shared Dispatch System

Finally, each operator could deploy a provider-sharded dispatch management system, as shown in Fig. 8.4. This model would provide a 1:1 relationship between each FSP and a Dispatch Frontend Instance (DFI). Each DFI would map to a Centralized Dispatch and Management System (CDMS) which would manage all the dispatch operations. Once a FSP is onboarded as a flexibility provider to a network operator, they would be granted access to a unique dispatch front-end instance, which itself is connected to the respective network operator's CDMS. In essence, in this architecture, there is a "factory" concept, where the Frontend Instance Factory is instructed to create a new instance of a DFI for each new FSP. The FSP would then connect to this instance of the DFI, and the CDMS will be able to issue instructions through the DFI to the FSP. The Network Operations Room (with oversight over the overall electrical network) would detect a requirement for local generation and instruct the CDMS to meet these requirements. The CDMS would then decide on the optimal combination of FSPs, and issue dispatch instructions to each of the required FSPs through the matching DFI. In such an architecture, the level of isolation and segmentation between DFIs could be determined by an operator's security posture—some may be software-isolated slices, while other more critical ones could be deployed in hardware-isolated environments.

This segmentation of different providers increases cyber security protection by isolating potential exposure to operators, and limiting the potential for a compromised component to cause wider impact on the system. The separation of the frontend and the CDMS into separate components would also enable decoupled version updates to take place, as this split moves the update requirements from FSP <-> CDMS, to FSP <-> Frontend <-> CDMS, therefore, each of these links can be updated independently. This reduces the barrier to entry for smaller FSPs and would improve longevity, provided the Frontend <-> CDMS link maintained backwards compatibility. Early

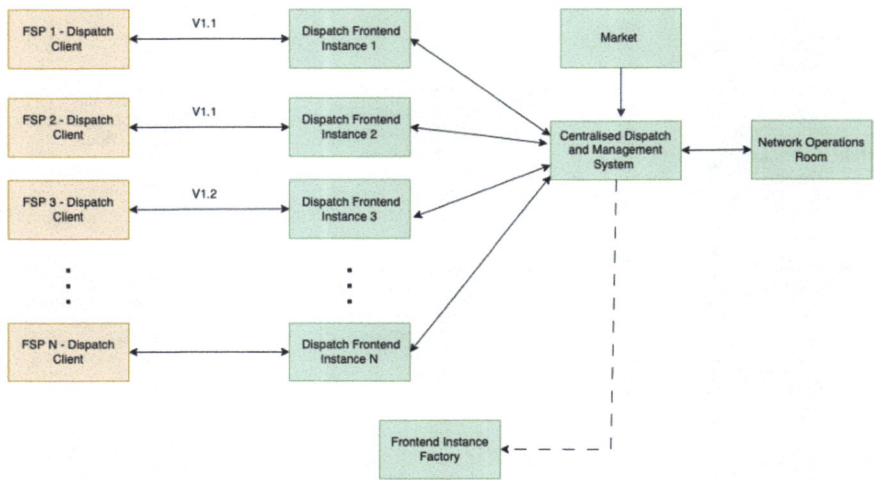

Fig. 8.4 A Single operator's sharded dispatch system

FSP adopters could continue to engage on Version 1.1, and remain on that version without any changes while the rest of the eco-system grows around them to V2, V3, V4 etc. Additionally, as the Frontend <-> CDMS link is entirely within the control of the operator it would be easier to update as required, as FSPs wouldn't need to be involved. As new FSPs come online they would be required to connect in at the latest version, and when FSPs require a new function, they would be able to update in their own time, but this would provide a slow but steady progression forward without limiting innovation or leaving parties behind. The disadvantage of this design it is more complex for operators to implement and maintain.

Provider-Sharded Architecture with Multiple Operators

The above concept can be extended to include multiple network operators, as shown in Fig. 8.5. The FSPs in the middle of the diagram each connect to different DFIs using different version numbers to enable FSPs to engage in the most markets possible while limiting "re-inventing the wheel". There are various models demonstrated in Fig. 8.5, where FSPs may, without any other rules, choose to implement. Firstly, FSP 1 operates two client systems each of which connect into a different operator. FSP 2 on the other hand uses a single client to connect to two different systems as they are both operating on the same version number. FSP 3, operates two identical client systems to engage with operators' systems but have employed segmentation between the systems for additional security. FSP N, operates three segmented clients, however, uses commonality between the two V1.2 systems to link to both the NESO and an operator, enabling them to engage in multiple markets seamlessly.

While the minimum required message content between the FSP and the Dispatch system remains consistent within all the architectures, each of these designs choices would result in a unique set of minimum architectural requirements, and accordingly

8.1 Application Program Interface (API)

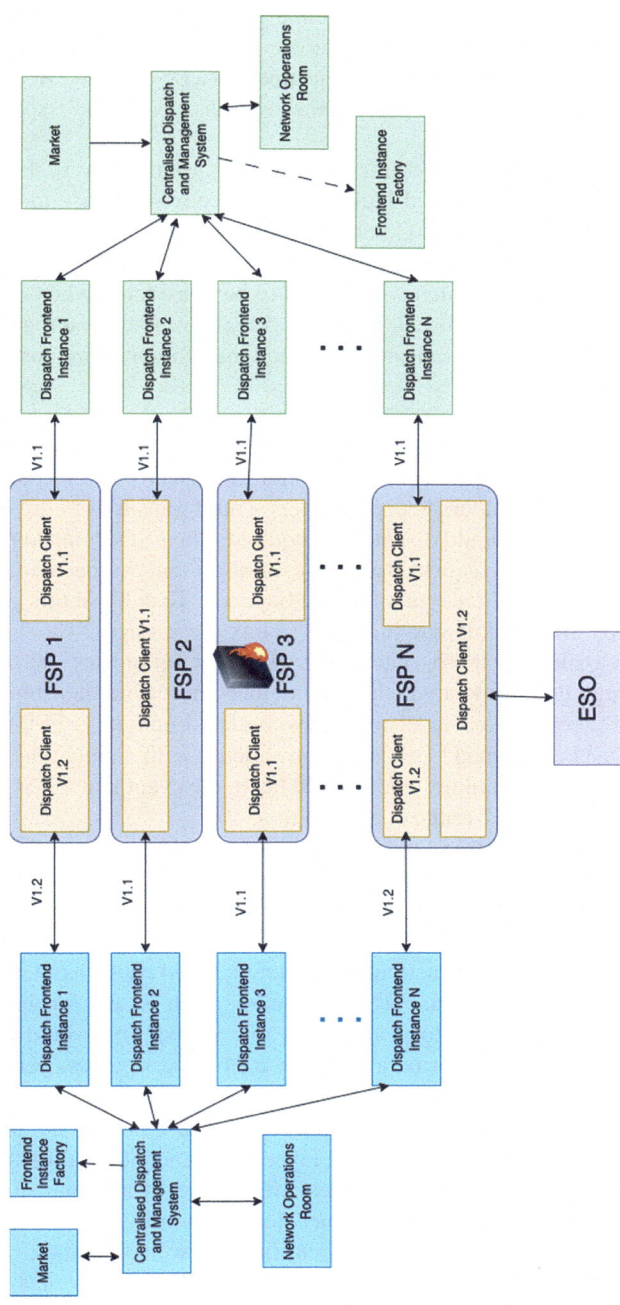

Fig. 8.5 Provider-sharded architecture with multiple operators

would affect the overall design of the architecture. Additionally, as set out previously, the wider technical architectural decisions around REST API implementation also need to be considered—such as whether FSPs are indeed dispatch clients, or whether FSPs implement the REST API server to be consumed by a DSO client. Moreover, as additional external systems are added the minimum message content requirements may vary to provide linkages.

For example, Fig. 8.6 outlines a fuller whole eco-system view of Flexibility Services including Ofgem's proposed Flexibility Service marketplace module [1], settlement and monitoring modules. Combined these modules present an architecture including the three core functions of Flexibility Services—market engagement, dispatch and settlement, all of which would need to be considered when designing the minimum requirements for a dispatch system. As each of these external systems interact with dispatch components in some way, and therefore, require passing data effectively.

The most common way of linking these systems together would be employing globally unique message identifiers which would link to an entire lifecycle for each event. For example, when an operator would tender for capacity in a local area, the FSP would respond to this tender. Accepted offers would transition to the dispatch management system, which would dispatch as required. The FSPs demand response would be measured against the requested capacity variation, and compared to the original request and the appropriate payment would be made. To achieve this smoothly it would be required to have a unique identifier to link these stages. There is therefore a requirement to identify appropriate primary key and foreign key "references", and establish through the protocol which entity defines these, and their uniqueness constraints (global energy-system level uniqueness, vs DSO-unique, vs FSP-unique). In addition, this must be designed to trade off convenience with maintaining a single source of truth, and what the failure mode would be, in the event that a DSO or FSP is unable to access their full IT estate at a given time.

For example, it is proposed that a dispatch request should be executed with reference to a specific contract identifier (and that this contract identifier should be namespaced by the DSO issuing it, with the DSO setting the identifier). From a practical perspective, however, dispatching against a contract requires a lookup process at both ends, in order to reconcile the action–for the DSO, based on the asset or service they require, what the correct contract reference to dispatch is, and for the FSP, to translate from a contract reference to an asset to be dispatched. This would therefore suggest that, while from a normative data structure and business process perspective, a dispatch operation is carried out against a contract (as the subject of the API operation), the information communicated should also contain the DSO-defined asset identifier being dispatched. This introduces a layer of redundancy and resilience—the FSP could dispatch without access to their own record of all contracts, in the event they had an outage affecting their contract store. Similarly, a DSO could issue a manual dispatch through knowledge of the identifier they were seeking to dispatch, without access to their contract records. This introduces a second-order layer of complexity around reconciliation and validation—in the event that a dispatch against a contract does not match the contracted flexibility asset, the standard or protocol will

8.1 Application Program Interface (API)

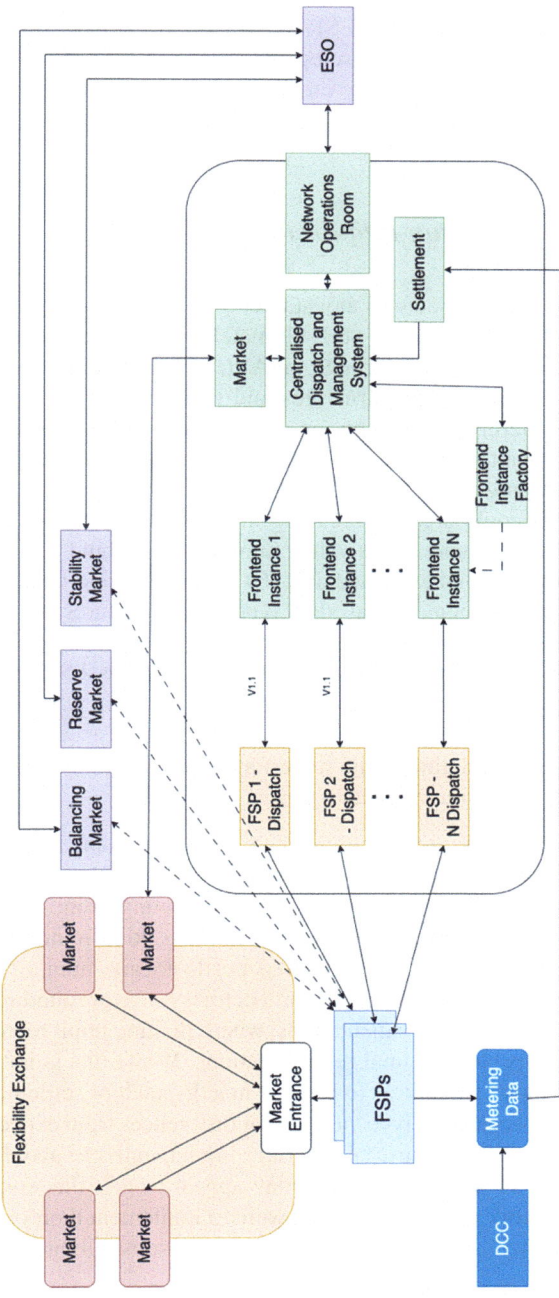

Fig. 8.6 Market, dispatch and settlement functionary combined

need to define how this situation should be handled, and whether the dispatch should be rejected. There would also be a third-order layer of complexity around handling reconciliation and settlement where a FSP was unable to validate the asset being dispatched against the contract, and a discrepancy is later identified—such a process is likely to be relatively manual, or rely on after-the-fact manual reconciliation by an independent system on past dispatches and contracts.

8.1.3 *Data Representation Requirements*

One of the most important aspects of any standard or API is the proper definition of data structures and representations—this ultimately defines what information can be represented and communicated through an API, between system operators and Flexibility Service providers. Data structures and representations must be agreed in advance, so that different implementations can understand messages from each other, in whatever format they are encoded or encapsulated in. For instance, an asset identification number might be considered an integer by one party, and a string by another party. This could create a breaking interoperability failure if one party added a non-numerical character to the value. Similarly, this could cause a failure if the integer number represented exceeded a 32–64 bit integer field value. Therefore, it is a fundamental requirement that all data types and structures are defined and documented unambiguously and, in an implementation-neutral way, at wire-format level (rather than programming language level). This enables FSPs and network operators to understand the expected format and implement it in whatever language they choose.

Two specific examples relating to the representation of a decimal number, and a date/time field highlight the importance of this.

1. Where decimal numbers are communicated, it is important to consider how they are encoded and represented—a common means of representation of a decimal number is IEEE754-2019 for decimal floating-point numbers. This standard defines 32, 64 and 128-bit decimal floating point numbers [2]. IEEE floating point representations are widely used, but do not give numerically precise decimal representations—while the standard defines rules for rounding, there are a number of edge cases in implementations, where floating point representations may differ from expected decimal representations. While this is not a material difference under most units likely to be used in a dispatch or settlement system, floating point numbers are likely to yield minor differences from expected values, and comparison operations must be carefully implemented to avoid failures in reconciliation of values [3]. While this may appear trivial, this could become problematic if network operators or FSPs were to implement their own business logic in different ways, resulting in Boolean comparisons failing on edge cases.

For example, a dispatch request for 0.1 MW by a DSO might be erroneously rejected by a FSP which was working in floating point numbers, and was only able to

8.1 Application Program Interface (API) 161

offer 0.099999997 MW—this is less than the dispatch request, so the request cannot be fulfilled by strict interpretation of these numbers. This highlights the requirement for robust definition of the correct format and encoding scheme for all values communicated through an API or specification of an API, in order that there is no ambiguity. This also extends to the correct format and encoding scheme being used for internal representations of such values, or a robust testing regime, such that any relevant internal business logic that would impact on the reliability of the flexibility dispatch or response system also uses a suitable representation, to prevent unintended downstream consequences of format selection. It is important to note that, generally, standards and APIs do not define requirements on external functions and how they operate—the correct functionality should be tested using adversarially-selected field values through end-to-end interoperability testing (as described later), to ensure that inter-connected systems behave correctly when presented with data fields.

2. In various places where dates/times are specified, it is essential to specify the exact format to be used for dates and times, including time zones. This is critical as date and time inconsistencies and misinterpretations have historically caused numerous interoperability issues. This is due to the various different date time formats employed, e.g. "10/09/23" (as a string) could represent an arbitrary string, or be decoded to represent the 10th of September 2023, or the 9th of October 2023, or the 10th of September 2123, depending on regional settings and assumptions around this ambiguous date. These issues permeate into programming languages and supporting libraries. Similarly, operating across time zones or seasonal time changes, e.g., British Summer Time (BST) and Greenwich Mean Time (GMT) can cause coordination issues. For example, if a dispatch order is setup within GMT, but executed in BST the dispatch could be an hour out of sync. A non-time zone aware implementation may naively assume that all times are represented in local time or UTC, which would be incompatible with a system correctly indicating a time zone.

APIs have traditionally solved this issue by employing either Unix/Epoch Time or ISO 8601 time. Unix/Epoch is a 32-bit integer representation of the number of seconds since January 1, 1970 (midnight UTC/GMT), and is displayed as a large integer number, for example, 1694698011. This gives second level precision but is not easily human readable. ISO-8601 on the other hand follows a human readable format of year-month-day, followed by the time and any time zone offset, for example, 2023-09-14T13:21:52Z. A disadvantage of this it is slightly more verbose, but it is human-readable. This verbosity often does not benefit inter-machine communication as being human readable is not usually a requirement. It may however provide a useful redundancy for dates and times to be readable in messages, especially where they are relied on in contractual or other disputes. Many APIs would implement Unix/Epoch Time. However, a constraint of the most common 32-bit implementations of the current Unix/Epoch system is that it does not support dates past 19th January 2038, thereby, potentially limiting longevity for a long-term standard. It would be possible to use a larger integer value, but this would need to be carefully tested for compatibility in implementations, to ensure interoperability in future.

Another consideration is that certain programming languages do not natively support the parsing of ISO 8601, and external libraries may be required. Once a standard format is agreed, dates and times should always be stored and transmitted in an agreed format (including time-zone representation), to ensure temporal coordination and non-ambiguity. From a future-proofing perspective, it may be wise to consider a requirement for all times to be expressed in UTC, in order to reduce complexity of integrating with assets in another time-zone in future.

8.1.4 Security Requirements

This section captures some of the initial requirements and considerations which have arisen during the course of the technical review and stakeholder engagement. It also sets out some best practice principles for the design of a security architecture, and some of the rationale for importance of this process. (Note: this is not a complete requirements specification or security model, but rather a summary of security-related outputs that have been noted thus far.)

Security systems and architectures are best not the product of "design by committee", since they require trade-offs to be made according to design principles, which holistically must come together to deliver the desired security properties. In a multi-stakeholder environment, good governance around burden sharing and clarity of responsibility and accountability is essential. The National Cyber Security Centre (NCSC) has some public cloud security guidance setting out the cloud security shared responsibility model. At architecture level, there is a need for a cohesive and comprehensive view of the security properties required, in order to ensure a robust solution is developed. Similarly, the concept of the shared responsibility model sets out the kind of governance view that may be required to ensure that FSPs and network operators are clear on security roles, responsibilities and expectations.

In addition, there is a recent move by most Western cyber security agencies to drive a move towards "Security by-Design and -Default" [4]. This is supported by the NCSC. At this point of consideration of a dispatch and flexibility API, the principles of security by design and default should be given serious consideration from the outset, rather than applying security as an overlay to another solution.

Information Provenance

Some stakeholders expressed a view that there would be provenance requirements in the information transmitted through any dispatch API. Security properties of non-repudiation (i.e. the ability to prevent a party claiming they did not authorise/send a message) and replay resistance (i.e. the ability to prevent a recipient presenting a previous authorisation and claiming it was new) were suggested as requirements. Following specific requirements were initially noted:

- Timestamping of messages
- Relative ordering of messages

8.1 Application Program Interface (API)

- Replay resistance (potentially achieved via reliable/trusted timestamping or message sequence numbering)
- Enduring message digital signatures (encompassing the timestamp/sequence number and contents of message)
- Transport layer security to prevent tampering with messages in transit
- Clear and robust security design to ensure unambiguous definitions of correctness, avoiding undefined or ambiguous states and values in protocol messages
- Versioned APIs and messages (potentially down to versioning of fields), in order to facilitate rejection of unsupported versions of messages
- Use of fields such as HTTP user agent to provide DSOs with visibility of security risk exposure by client software consuming the API
- A concept of criticality (or mandatory comprehension) per-field, such that an implementer of an API can evaluate, from a message, whether it has the capabilities to properly parse the message with its API version understanding
- A general principle, where possible, of idempotency in the API, to avoid duplicated requests causing changes in state, and to attempt to avoid creation of "state machines" in client/server implementation logic where possible
- Clear and robust definition of responsibilities and accountability in API documentation, to make clear which party (or parties) are responsible for carrying validating and verifying fields, and how to respond in the event of a validation logic failure.
- Robust definitions of types for every field, to prevent type-confusion attacks or comparison failures
- Use of a secure and well-reviewed serialisation format for encapsulation of API messages, in order to minimise risk of buffer overflows in parsing of API messages through having clear and well-defined field lengths, rather than relying on variable field lengths and complex serialisation formats

A complete security architecture and design will be required, in order to develop specific technical requirements that could validate specific solutions' applicability here. It is believe that at this point no "off the shelf" protocols will deliver these capabilities, since the industry standard (HTTPS) is focused on protection of information "in-transit", rather than longer-term information provenance at-rest. While there are technical solutions and standards available for the signing of data, these signatures generally do not extend to timestamps or message sequencing. Given the potential for a bad actor to abuse flexibility for commercial gain, by attempting to seek settlement for a false dispatch, there is likely to be a requirement for linkage of settlement requests against dispatches and contracts, to facilitate robust and unambiguous reconciliation processes.

General Security and Resilience Posture

The concept of thresholds for security requirements have come up during discussions with stakeholders, including thresholds at which traditional generation would face additional security and communications requirements, based on their connected capacity. From a network perspective, 3 key areas of concern were noted during the discussions:

1. Risk of cyber-attack on DSO dispatch systems (which would sit as a point of common interconnection), originating from an API participant. A DSO's dispatch system is likely to be relied upon for critical dispatch, and is also (depending on architecture) likely to present a common attack surface that is linked to multiple FSPs. This presents a potentially interesting target for an attacker, since a single point in a DSO's system may present an opportunity to cause disruption or outages.
2. Risk of being unable to dispatch when required, due to API outages, technical failures or communications issues. This could include wider telecoms outages, as well as cloud provider or datacentre outages (making the FSP's connection to the DSO inoperable), or upstream supply chain outages. There will be times where the inability to dispatch an FSP could be detected (i.e. the FSP is no longer sending heartbeat signals), and times where this will not be detectable. As an example for the latter case, if an aggregator responds to dispatch by issuing a push message or SMS to end customers, and their SMS provider fails to deliver messages, or Firebase Cloud Messaging (FCM) had an outage, they may be unable to dispatch when required, due to wider technical issues. Similarly, if they rely on other third party APIs (such as to electric vehicle charging points), and those APIs are unavailable, they may be unable to deliver services on dispatch.
3. Risk of misunderstanding nature or diversity of Flexibility Service offered, and dispatch proving ineffective for other reasons. This could include unintended correlation or overlap of Flexibility Services, as well as a lack of visibility of the level of criticality (or connected capacity) of dispatchable Flexibility Services. This may not always be visible to the DSO, since an FSP could potentially have control over a significant number of geographically dispersed (or proximate) assets, and the full extent of this may not be visible to the DSO.

There could be situations where multiple FSPs or aggregators are unwittingly re-selling the same underlying asset flexibility to the DSO, resulting in risks of non-delivery when dispatched in a period of network constraint – for example, where Flexibility Services are offered via an OEM, as well as an installer, as well as an operator, all relying on control of the one asset. There may also be downstream aggregation of risk, where an FSP is highly dependent on a single provider (such as one EV manufacturer's API), and indeed multiple different FSPs may themselves rely on that provider, or infrastructure shared by that provider.

In general, new-entrants and smaller providers are less ready for complex and high-touch security practices or formal audit mechanisms like ISO27001 or SOC-2 etc. Smaller providers often have confidence in their technical security practices, but are unlikely to have formal certifications in place. Smaller providers generally favour moving at pace, and iterative development and change. Similarly, it is important to recognise that even the largest organisations (with significant cyber security resources and certifications) can make security-fatal errors in their implementation of critical security functions, and therefore the presence of certifications and governance should

8.1 Application Program Interface (API)

not be replacements for demonstrable technical security measures. There could be risks to the success of wider flexibility if the requirements for dispatch API interoperability were unduly onerous, or this may make it more difficult for DSOs to interact with smaller providers. Higher security requirements may well be justifiable (and could lead to aggregators who do adhere to higher security practices adding value into the chain), but could also reduce the diversity of Flexibility Services available to DSOs.

Communications Protocols

Before selecting technology or protocol options, it is important to design a communications system with an accepted and agreed lifespan. It is also critical that, before selecting or specifying requirements for communications security, a wider security architecture is developed—protocol requirements cannot be set without a wider architecture, as the overall security of the system will rely on the protocol used, as well as wider design parameters. Attempting to do one without the other will result in a piecemeal approach to security that does not deliver a secure system. The longer a time period that a system is designed to be secure for, the more consideration must be given to making use of the most modern equipment, as well as the non-technical drivers and levers to ensure that participants in the dispatch API are adopting updated standards and technology.

Standard approaches such as PKI (public key infrastructure) [5] are likely to provide effective and viable methods for identity management, but these will require robust governance and security measures around creation of certificates and the business processes around issuance and revocation of certificates, to ensure only authorised organisations and individuals at those organisations can request certificates. Security design patterns for the whole system surrounding communication protocols should be considered, including avoiding any "single line of contact" to the internet, and other pieces of best practice guidance from the NCSC.

Given the inherent links between a flexibility/dispatch API, and the UK's critical national infrastructure, it is likely that the infrastructure around such an API would be an attractive and visible target to nation-state attackers, and that they would attempt to gain a foot-hold in it for future exploitation, or exertion of influence. There are also potential economic benefits to market participants, to the detriment of others, were they able to exploit functionality of such an API in a difficult-to-detect manner. The likelihood and incentives/motivations for attack would make such a system at high risk of attack.

The Network and Information Systems (NIS) Regulations Implications [6]

1. Designation of Operator of Essential Service Status: At present, Flexibility Services would not be covered by the current criteria and thresholds as set out by the Secretary of State for BEIS (since replaced by DESNZ) and Ofgem as NIS joint competent authorities for the energy sector, since Flexibility Services would generally not fall under the criteria unless they were also a large generation site or similar. The NCSC is a national technical authority in cyber-security, and does

not hold specific regulatory responsibilities, or the ability to endorse any OES' specific activities.

2. NIS Regulation Obligations: The NIS Regulations require operators of essential services (if designated as an OES) to take appropriate and proportionate technical and organisation measures to manage the risks posed to the security of the network and information systems on which their essential service relies, per regulation 10(1) of NIS; and to take appropriate and proportionate measures to prevent and minimise the impact of incidents affecting the security of the network and information systems used for the provision of an essential service, with a view to ensuring the continuity of those services. This means from a NIS perspective that DSO covered by NIS should consider their NIS compliance when incorporating flexibility into their network, and in particular consider the security of the network and information system that their essential service relies on, which is likely to include their Flexibility Service interface, as well as any hosted provider or aggregator whose systems they rely on to deliver an essential service.

Given the threshold for electricity distribution covering a loss of supply incident to 50,000 customers for more than 3 minutes, DSOs would need to evaluate their current and potential future use and reliance on FSPs, and the extent to which they may be exposed to incidents exceeding NIS reportable thresholds in the event of a flexibility failure, or a need to take emergency actions which could cause a similar impact.

3. NIS Digital Service Providers: There is an additional route through which a flexibility platform could be deemed a NIS-covered service, regulated by the Information Commissioner's Office (ICO) as competent authority, were it to be deemed to be an online marketplace. Per ICO guidance (A9) [6], an online marketplace provided to external customers (individuals or organisations) by an entity that is not a small or micro-business is a "Relevant Digital Service Provider" (RDSP) under NIS. There is a requirement for RDSPs to register proactively with the ICO. Online marketplaces are considered to be:

"digital services that allow individuals or traders to conclude sales or service contracts with traders, either on their own website or by means of providing services to traders' websites. Online retailers that sell directly to individuals on their own behalf are not covered..."

It is likely that one or more aspects of the properties of a flexibility marketplace or platform could be construed to be an RDSP, on the grounds that a digital service (i.e. the flexibility marketplace, platform, or system) allows traders (FSPs and aggregators) to conclude sales or service contracts (i.e. flexibility contracts) with traders (DSOs and Flexibility Service buyers), on their own website (i.e. the flexibility platform). RDSPs are covered by regular NIS regulations (Part 4, and Regulation 12) [7] as well as a separate "DSP Regulation" [8]. ICO is clear in their guidance that when measures are implemented to manage risks, implementers are allowed to consider the state of the art when evaluating available measures, however they are not allowed to consider costs of implementation in this evaluation. This means that security measures implemented for NIS compliance must be taken without regard to costs of implementation.

8.1.5 Testing Requirements

Another significant consideration for the implementation and successful growth of a standard/common interface is the ability to reliably ensure when two parties are unable to correctly interoperate understanding where the issue lies. This can occur where two parties have both implemented their interpretation of a standard, but the interpretations differ. A specification document is harder to robustly test against, especially where interpretations of wording are required in order to validate behaviours against the specification. Many technical standards use "test vectors" of pre-defined inputs and outputs to provide ways to validate behaviours at algorithm level, but for a system or interoperability level test, a reference implementation is generally required. Therefore, to enable testing, it is a necessary requirement for there to be, at the point of selection of a standard or development of an API/specification, a plan to create an agreed "reference implementation" of both client and server.

From a network operator's perspective, there will also be a requirement for any API or standard to be sufficiently well-defined to enable exhaustive informed testing to take place. This means that, for each value or function, the permitted and disallowed inputs and outputs should be suitable well-defined to enable a compliance test suite to be run for inputs and outputs. In addition, there will be a requirement for a sufficiently well-defined functional test suite, which tests business logic flows and message sequences (i.e. to ensure an invalid dispatch message is rejected, to ensure that an implementation acknowledges the correct dispatch).

To specify good practice for testing, this should encompass (for each of the areas identified above, i.e. for data structures, for functional invocations, and for the overall system):

- Positive testing—testing of an implementation under normal circumstances, to ensure that correct inputs and stimulus result in expected behaviours, and that all valid inputs are accepted.
- Negative testing—testing of an implementation under abnormal and boundary circumstances, to ensure that invalid inputs are appropriately rejected, and do not result in incorrect actions being triggered.
- Fuzzing-based testing—a kind of testing often associated with security testing, where randomly generated values are injected into fields, and valid or invalid messages are replayed out-of-sequence, in order to ensure that implementations are robust to unpredicted errors.
- Interoperability/integration testing—a set of testing carried out pairwise between two implementations as they would be deployed. This would be between a given client implementation, and a given network operator's implementation, in order to validate that, once an implementation passes the other tests, it also interoperates correctly at system-level with a network operator's sandbox implementation.

Integration test suites will validate that the implementation under-test correctly interoperates with simulated versions of the rest of the eco-system. This will include reliability tests to ensure the correct response occurs, even when the system is under high load or experiencing unexpected input. Integration testing is important to be carried out in representative environments, with the same security measures and protocols in place as would be in place in a deployed environment. Prior to a large-scale deployment these test suites would have to be developed or acquired. Such a process is likely to need to take place through a standards development approach, working with network operators and Flexibility Service providers, in order to appropriately define such tests. To avoid "false starts" and insufficient tests being defined, this process should seek input from those with experience in standards and test development. As part of this process, there will be a requirement for the wider design of any API/protocol/standard to be fully-defined, including any stateful behaviours and state machine design, messages, error states, and expected responses. As such, there is a requirement to establish appropriate governance around test suite development, evolution, and implementation, since implementation and definition of tests will effectively set entry criteria to the flexibility market.

The following technical requirements, would be the minimum required in order to enable verifiable and demonstrable interoperability, and permit for a governed process to facilitate system interconnectivity:

- A set of "positive test vectors", provided as stimulus to the "system under test" (which could be either a DSO server, or a FSP client), to provide a set of
- Sufficient definition of the necessary outputs or exposed state of the "system under test" in order to verify the outputs or behaviours in response to the test vectors provided as stimulus for the test.
- A set of "negative test vectors" (which is likely to evolve over time), to provide a set of invalid inputs, and the expected error state and output behaviour to be triggered in the event that the "system under test" receives such a message.
- A defined "interoperability test suite", which can be automated to ensure that different client and server implementations correctly correspond and interact, with the correct business process and high-level logic flows implemented—if any messages or API endpoints are stateful rather than idempotent, such state should be exhaustively tested here (i.e. by attempting to stop services not yet dispatched, to dispatch after stopping a given service, etc.)

8.2 Delivering Interoperable Dispatch

A range of delivering interoperable dispatch were identified, and carefully evaluated. This was informed by the previous stakeholder engagement, and a detailed technical evaluation of standards options. The main conclusion was that there is no immediate "off the shelf" solution that can be adopted in the immediate term to deliver a fully interoperable solution. Existing solutions either do not meet the requirements to dispatch each of the five DSO Flexibility Services, or there would be other very significant trade-offs to their adoption (i.e. major commercial downsides to, for example, selecting a single vendor solution for every network operator to adopt).

At the time of publication Open Networks have endorsed the proposal to continue the technical delivery and implementation of OpenADR 3.0 as a specification and implementable means of providing an interoperable dispatch solution. It was noted that OpenADR 3.0 is based around a modern REST-based API, which appears to be sufficiently versatile to support the communication of both the current and future data parameters needed to enable the dispatch of Flexibility Services.

8.2.1   OpenADR 2.0 and 3.0

OpenADR is a standard for communication of information to trigger a demand response action from energy using devices. It features functionality which could be used for dispatch of flexibility resources. The 2.0 version of the standard is based around an HTTP/SOAP architecture or XMPP communications, and the newer 3.0 version of the standard is based around HTTP/REST. In stakeholder engagement with the OpenADR Alliance, It was noted that SOAP was perceived by many FSPs to be a barrier to adoption, hence a transition towards REST-based APIs with version 3.0. OpenADR 3.0 was formally launched in late November 2023 [14]. OpenADR 3.0 was not designed to replace the previous 2.0 version of the standard, and 3.0 was designed to provide an extra, simplified way to add OpenADR functionality to devices. It is important for context to note that OpenADR is intended to be used to send demand response signals to consumer equipment in the household (for example, electric vehicle chargers or water heaters). OpenADR 3.0 opens up more opportunities for expansion and extension within the confines of the standard (which is not particularly relevant to this comparison), through extensibility in API fields, as well as in adding new values to existing enumerated-type fields. Taking advantage of such extensibility features is effectively creating a new bespoke standard, as the expectation of, and use of, such additional fields will require standardisation and adoption by the industry.

One important learning from OpenADR was the importance of having an agreed single testing specification, rather than allowing for a proliferation of different test specifications by different test houses—resulting in products passing their respective tests but not being truly interoperable. They also reflected on the lesson of offering too much flexibility (such as supporting XMPP-based transports in addition to HTTPS-based transports in OpenADR 2.0), and how this ultimately resulted in multiple parallel implementations of the same logic being required to deliver a usable system. From a standards selection perspective, it is worth noting that OpenADR 3.0 is explicitly not backwards compatible with 2.0 in terms of protocol layer data, due to the changes in communications layers and architectures (i.e. removal of XMPP, introduction of HTTP-based PUSH via webhooks, removal of whole message signatures of the XML payloads, and removal of mandatory client certificates for TLS sessions, introduction of an OAuth2 client credential flow for authentication).

From a technical perspective, either version of OpenADR appears capable of supporting dispatch communications, although there would need to be standardisation and agreement of which API messages are used to communicate which messages, since OpenADR 3.0 is extensible, and there are a number of different "events" which can be announced to FSPs through OpenADR, and these would need to be common across network operators to deliver dispatch API interoperability.

It is important to note the areas of a dispatch process which are not covered by specifications and APIs, and which OpenADR has called out as being excluded from scope, as these are a helpful reference in evaluating other options and considering additional work that would be required to support these. For example, OpenADR does not define various requirements for HTTP client and server capabilities, such as compression algorithms supported and TLS cipher options which must be supported. OpenADR does not define how a network operator-side implementation should validate content, and this is left as an implementation detail for those implementing OpenADR. Similarly, content validation of messages received by clients is considered outside the scope of the specification, and the service provider hosting the OpenADR platform needs to maintain a secure web platform, including updating TLS ciphers as required. Significantly, OpenADR also points out the need for out-of-band business processes to onboard FSPs, and agree specific details of "programs" or "tariffs", and then issue credentials to a FSP so they can participate. This will be a requirement for any option considered, and therefore becomes a common requirement for any solution. On balance, OpenADR 3.0 appears to be more appropriate than 2.0, based on the previous stakeholder feedback indicating a preference for more modern use of web technologies, rather than legacy XML/SOAP oriented protocols. It is worth noting that significant time and effort would be needed to turn the standard into a specification that GB network operators could adopt and then implement. At the time of publication, effort to implement a GB wide standard is underway.

8.2.2 Other Standards and Approaches Considered

While OpenADR 3.0 was selected for further development and implementation, a range of other international standards and solutions were included in the analysis. The following section presents a detailed review.

Universal Smart Energy Framework (USEF) & Universal Market Enabling Interface (UMEI)

Based on the discovery work looking for standards, as well as stakeholder feedback, Universal Smart Energy Framework (USEF) [9] and Universal Market Enabling Interface (UMEI) [10] were also reviewed as potential standards for dispatch.

The UMEI project was part of the EU-funded Horizon 2020 EUniversal project, aiming to develop a universal approach to the use of flexibility by network operators. Similarly, the USEF project was developed as part of the USEF Foundation, and has

8.2 Delivering Interoperable Dispatch

since transitioned to part of the Linux Foundation Energy project, and been rebranded as ShapeShifter.

The USEF standard (version 3.0, latest available as of December 2023) [9] sets out a flexibility trading protocol, which is focused around the delivery of a "market structure, roles, rules and tools for the commoditization and trading of flexible energy usage work with existing energy markets." In the USEF phased paradigm, the "Operate" phase is the most relevant one to flexibility dispatch, as this is the phase in which actual flexibility is delivered. The USEF protocol however for this is complex, and does not directly expose a straightforward paradigm for flexibility dispatch of the standard defines the Operate phase, where *"The actual assets and appliances are dispatched and the AGR adheres to its D-prognoses. When required, network operators can invoke additional flexibility from AGRs to resolve unexpected congestion."*

The USEF standard itself does not describe a semantic through which dispatching is carried out. The documentation describes the Operate phase as exchanges of updated D-prognoses, the revocation of flexibility offers, and the exchange of flexibility orders. However there isn't a mechanism through which Flexibility Services can be dispatched independently of other complex processes. USEF, therefore would not be a viable tactical dispatch API solution, as it does not define a method to dispatch an FSP. The USEF standard appears to make an implicit assumption that acceptance of a FlexOrder is akin to an (ahead of time) dispatch instruction, but this is not expressly stated in the standard, and ambiguity like this would create complexity for implementers to understand and reach a clear understanding as to an instruction.

The UMEI standard [10] is also a specification designed around market constructs. While a "Reservation Market" is defined for ahead-of-time offers of Flexibility Services before dispatch, the standard states *"The actual dispatching can be done outside the market or in a consecutive activation market."* The standard does not define an activation market or dispatch method, beyond setting out an activation market as a *"Flexibility market where the assets are dispatched for the period covered by a trade."* No details is given as to how dispatch should be carried out in UMEI. Further reviewing other adjacent documentation about UMEI, including available literature, LongFlex (i.e. future activation of service) contracts are set out as being activated either through ShortFlex, or *"bilateral(ly) through other means"*, which is presumed to mean outside of the scope of the standard. ShortFlex is defined as being *"activated based on trade confirmation"* (which is presumed to mean considered dispatched upon the order being placed), and InstantFlex is *"activated based on automatic power settings (frequency, voltage or reactive)"*, which is presumed to mean self-dispatch. In Deliverable 2.6 of the EUniversal Project, the API interfaces are defined, and they move from the "Pre-trading phase" to the "trading phase" (i.e. posting and reading orders), without considering the dispatch or activation method. While "Flexibility activation" was claimed to be in the present version of UMEI in that report, no further details could be found of their flexibility activation method in that report. The glossary of UMEI states that, in a Reservation Market, *"The actual dispatching can be done outside the market or in a consecutive activation market."*

It is therefore concluded that neither USEF or UMEI presents a viable GB dispatch solution, as neither includes dispatch functionality as would be required to directly dispatch FSP resources. The complexity of these projects is also likely to introduce challenges for smaller market participants to interact or engage, or even understand the semantics and paradigms used—being rooted from research projects, they appear to be more focused on innovative holistic market design patterns and structures, rather than straightforward and unambiguous industry-ready standards development. Partial adoption would introduce a range of areas of friction, such as a lack of explicit dispatch messages.

As a general observation of both USEF and UMEI, these options appear to be designed to implement a full flexibility marketplace. This is a very different goal to implementing a dispatch protocol. There is significant complexity and breadth in these standards, and the extent to which participants and implementers would need to understand complex market dynamics and functions (such as D-prognoses in the case of USEF) in order to deliver even basic functionality. This is likely to present a significant barrier to entry for new FSPs, as opposed to a simpler and more functional dispatch API focused around clear issuance of unambiguous dispatch instructions. Clear and unambiguous instructions are likely to result in better outcomes for customers, through correct dispatch of flexibility resources where they are required. In addition, by attempting to encompass wider market dynamics and functions, such protocols would likely, were they usable and viable for dispatch (which they do not appear to be) limit future directions of travel of the wider flexibility market by being built around a range of assumptions in how a flexibility market is expected to operate.

The GB approach to flexibility with explicit dispatch differs from many of the assumptions and structures used elsewhere, where implicit approaches to dispatch (i.e. where market participants take actions of their own accord, based on observed network parameters or through price signals) may be favoured. In selecting a tactical dispatch API solution however, it is important to assess standards' fitness for purpose against current Flexibility Products and service definitions, which do not operate in explicit dispatch at present.

IEC Common Information Model (CIM) [11, 12]

CIM is a common information model for the exchange of information about an electrical network, based around a "wires" model. CIM itself is a UML (Universal Modelling Language) model providing a common data vocabulary and an ontology of data, describing data formats, structures, and relationships. CIM itself, as a data model and structure/ontology, does not define a communications protocol itself—other standards or communications protocols are required to then communicate information. For example, IEC 61970 is a set of standards that define APIs (application program interfaces) for energy management systems, and enables communications from the control centre out to external assets. IEC 61970 encompasses CIM. It is also important to note that CIM is not, in itself, a fully defined data model. A series of CIM profiles are then used, which are subsets of the overarching CIM UML model.

A profile is a self-contained data model, which can be used to generate exchangeable artefacts that others can process. CIM profiles are themselves standardised. For

8.2 Delivering Interoperable Dispatch

example, IEC 61970-452 defines an Equipment Profile; IEC 61970-453 describes a Schematics Layout Profile, and IEC 61970-456 defines an Analog Measurements Profile, Discrete Measurements Profile, State Variable Profile, and Topology Profile. One challenge encountered in attempting to review CIM-based dispatch models is that, since CIM represents a data model and ontology, there is not one single representation or protocol to review. CIM does not represent a specific API-based approach or similar, and therefore an additional standard is required, to overlay the concepts of dispatch on top of a CIM data model. OpenADR 2.0 is aligned with the CIM information model, and is interoperable with IEC 61968 (Distribution Management) and IEC 61970 (Energy Management). OpenADR 2.0 is based on SOAP/XML and XMPP-based transports, which are now generally considered to be obsolete in modern IT systems, in favour of RESTful APIs, which OpenADR 3.0 makes use of. Given that stakeholder feedback to date has pointed towards a preference for the use of modern web-standards like HTTP/REST (which have a greater available number of developers familiar with the technologies concerned, as opposed to legacy options like SOAP/XML APIs), implementing a CIM-based API solution facing FSPs is likely to present barriers to adoption, due to FSPs needing to work with both IT and web standards, as well as legacy technologies.

There is also a significant burden likely to be encountered by FSPs, even if other approaches were taken, and there are indicators to suggest that a number of standards for CIM-compatible dispatch have limited support. For example, IEC 62325-301:2018 presents a CIM extension for markets, including dispatch functionality. It constitutes 446 pages of standard document, featuring more than 750 tables. The standard itself is based on 4 packages, comprising common, management, and operations aspects of markets, plus environmental considerations. This would present a significant burden for FSPs to interact with or implement, and gaining familiarity with it would take a significant amount of time. This standard is also XML-based, built upon CIM, and much of the standard is not necessary for flexibility dispatch. Where functionality is designed for flexibility, it is very limited—for example, the DispatchInstReply (i.e. a reply to a dispatch instruction) defined in Section 6.5.8.17 is defined to support only dispatches of active (i.e. real) power, measured in Megawatts. A requirement to be able to communicate both real and reactive power has been identified from the minimum dispatch requirements, alongside the ability to use relative or absolute values for dispatches.

According to EPRI's CIM Primer (8th Edition) [3], most guidance on CIM uses XML examples, as IEC 61968-100's main focus was on SOAP, rather than HTTP APIs. It is possible to convey CIM information over JSON, as well as over REST, but this would entail adopting a divergent version of a significant suite of standards, which would place significant burden on FSPs and implementers. While there have been other attempts to standardised CIM communications over HTTP, such as IEC 62325-504, which defines a framework for web services communication of CIM data, this was withdrawn in 2023 and there does not appear to be a newer revision of this standard. This was a SOAP-based API, which is again in contrast with the approach that FSPs indicated a clear preference towards. Despite this, CIM is likely to be used by network operators and dispatch platform vendors as a model for interchange of underlying information "behind the scenes" of a dispatch system, and could certainly prove useful for ancillary information exchange about network schematics (This work has explored the use of CIM in the FSP-facing interface facing for dispatch of Flexibility Services). In this context, adopting CIM would effectively lead to creation of a new standard by default, and a standard which is likely to be considerably more complex for FSPs to implement than other options. It therefore does not offer a credible technical solution at this time for flexibility dispatch, even if it is used in the wider information exchange of network information internally.

NESO Dispatch-Relevant Specifications

In exploring relevant existing options, it was identified that NESO has existing dispatch systems for their own transmission Flexibility Services. Since these are already in use in the GB market, these were explored these for completeness, as if usable, there could be advantages in alignment of flexibility dispatch interfaces between DSO and NESO markets, in order to reduce barriers to entry in adjacent markets. NESO's relevant dispatch systems for the Balancing Mechanism were evaluated. EDL and ASDP were explored, and had a preliminary discussion about the Wider Access API (WAAPI). Electronic Dispatch Logging (EDL) is a legacy ASCII-based text control protocol, and is not applicable to HTTP APIs, and would not be suitable for adoption in a web-based API. The Ancillary Service Dispatch Platform (ASDP) API was also reviewed, which is an XML SOAP-based API, rather than REST-based API. ASDP was stated to be end of life from NESO's perspective, and not something they would recommend implementing at this point. A preliminary discussion took place about the Wider Access API (WA API), which NESO stated is being transitioned to a new platform as part of a move to a new balancing platform, and built with some platform assumptions that mean it is carrying technical debt that would not be viable to deploy as a new initiative. NESO highlighted that their Open Balancing Platform (OBP) work aims to replace each of these components, and that it is not yet ready to be adopted at-scale, but may be in a few years' time. On this basis, there did not appear to be an existing available API standard/specification from NESO that would be appropriate at this time for interoperable flexibility dispatch functionality.

Existing Flexibility Dispatch Platform

As a part of the review, the pro and cons of adopting an existing flexibility dispatch platform all network operators was explored. Since multiple network operators have already adopted interim dispatch platform API solutions, this option explores selecting one existing platform and adopting it (as a platform) as a common industry solution for dispatch. The following recommendations and observations are made.

Commercial/Risks:

- If an existing platform was to be adopted, there should be a way for the "platform" or server implementation to be hostable by network operators, or a pathway to this being feasible before deployment, as opposed to relying on vendor-hosted (and risk-aggregated) centralised platform servers being provided as a service.

8.2 Delivering Interoperable Dispatch

- For a tactical solution, while it may be possible to run a single platform instance in the short term (i.e. a multi-tenant architecture), there should be a smooth migration pathway planned towards a single-tenant architecture. This will require routes to identify service URLs and their discoverability, in order to allow for network operators to operate their own platform instances in future.
- Careful consideration should be given to commercial implications, and how to ensure value for money, since this approach would effectively select a "winner" in the market. This will reduce competition, and likely have a negative impact on other suppliers, and potentially introduce significant barriers to switching platform vendor in the future.

Technical:

- No one platform currently meets the minimum service parameter requirements. Of the currently available options a direct adoption of an existing commercial solution do not appear to necessarily deliver an interoperable solution across network operators.
- For a tactical dispatch solution, to reduce the burden on FSPs and facilitate market entry for small providers, as well as individual 'pro-sumers', and reduce requirements on FSPs to have their own server infrastructure or substantive cloud footprint, there should be a viable route for an FSP to provide services by polling a network operator-provided API as a client, without having to provide and expose server or web-hook APIs to the internet.
- A tactical dispatch solution should still offer sufficient security to reduce the risk of exploitation of a single internet-connected component and limit the attack surface. This would need to be explored for each platform in more detail, were this option considered further.

Adoption of an existing platform would therefore not present an immediate technical solution to delivering interoperable dispatch, given that none of the options explored currently meets all of the minimum requirements. Furthermore, given the commercial risks of this approach highlighted above selecting a single platform vendor to be adopted by all network operators is likely to deliver poor outcomes for the market, and increase pricing over time, as well as introducing barriers to switching platform vendor.

An option to adopt only the API specification from as a common standard was also explored. The vendor's platform implementation would not necessarily be adopted, but their API semantics would be adopted as the GB dispatch standard. Other dispatch API platform vendors would be able to implement the API specification, in order to allow for competition in service provision, while adopting one vendor's API. However, this does not present a materially distinct option and would result in potentially shifting more towards resembling a platform adoption, with similar commercial risks.

8.2.3 Next Steps

At the time of publication of this book, all GB network operators had agreed that OpenADR 3.0 is likely to be the most appropriate standard available, noting the strong desire from key stakeholders to see an internationally aligned standard used for interoperable flexibility dispatch systems. Recognising the aspiration for true interoperability and interchangeability, rather than notional but incomplete interoperability (such as through a standard for wire communications, but not for the wider ecosystem), ENA was working to deliver a wider, holistic, whole-of-dispatch standard, encompassing not only the API messages and communications, but also to specify assumptions and expectations on the supporting infrastructure, as well as a standard architecture. This should reduce barriers to entry for FSPs and aggregators providing services to multiple parallel flexibility markets, as well as enabling easier switching between technical solutions by all market participants going forward.

The Open Networks together with the Market facilitator and NESO will develop a standard that should encompass a wide range of other critical information to enable meaningful, real-world interoperability and interchangeability of solutions and platforms. As such, it is envisioned that the work will include the following outputs and requirements:

- A standardised solution architecture for the overall flexibility dispatch API solution, to ensure that network operators deploy consistent and interoperable architectures of implementations of the flexibility API, and that the overall solution design is suitably secure and able to be deployed in manners that meet the security expectations of stakeholders, including network operators and NCSC.
- A standard security architecture and threat model against which implementations are developed, to reduce barriers for FSPs to enter into new flexibility markets, and to ensure that the overall solution sets out clear and explicit expectations on actors around security.
- A standard network operator-side architecture, to enable easier switching by network operators between flexibility dispatch platform vendors and deliver better value for bill-payers by avoiding vendor lock-in and proliferation of non-interoperable solutions that do not facilitate migration between vendors.
- Standard messaging and communication protocols (including, where possible, common and extensible data models), which should, as a minimum, be able to express the required information for the standard Flexibility Products, plus NESO-relevant distribution connected dispatch, for how a system operator will notify an FSP or aggregator of an OpenADR programme (i.e. forthcoming dispatch), and then dispatch that service.
- Standard reporting profiles for the messaging from FSPs/aggregators to system operators to report status, including stateful logic and standard means of specifying the intervals or events that trigger these messages.

8.2 Delivering Interoperable Dispatch

- An architecture and specification of the expected behaviours of an FSP/aggregator implementation of OpenADR 3.0, including that of stateful logic, so implementers of FSP/aggregator-side software have a clear understanding of the exact expectations on them.
- Technical security specifications for FSP/aggregator implementations based on different scales of flexibility capacity, and aggregated risk profile across the overall GB portfolio.
- Clear definition of behaviours for an initial unconfigured client (FSP/aggregator) endpoint, including an interoperable approach to bootstrapping.
- A clearly defined and interoperable onboarding and configuration process for each FSP/aggregator implementation, to allow a user to change from an existing to a new network operator or aggregator, or to do an initial setup towards a network operator or aggregator.
- A clearly defined specification for cryptographic algorithm agility and replaceability, in order to allow future revisions of the standard to deprecate current cryptographic implementations, and set the expectation for implementers to be prepared to implement new cryptographic algorithms in future due to, say, compromise or the need to implement post-quantum cryptographic techniques.
- A definition of the stateful information to be retained and stored by FSPs or aggregators at their side of the implementation; means for recovery of this state in the event of a system failure at their side; and documented state machine for the whole lifecycle of a dispatch across the system.
- A set of interoperable and predictable fallback behaviours to be carried out by any FSP/aggregator in the event of a loss of communications at any stage in the stateful system, such that a network operator can tell unambiguously how an FSP/aggregator will behave in the event of a loss of communications, and whether a Flexibility Service delivery will be carried out or not in absence of communications. The requirements around this need to consider both software and solution requirements, as well as electrical and network requirements.
- A clear version and revision lifecycle management standard, including around versioning and progressive enhancement, to allow the standard to evolve, including the use of semantic versioning, and associated protocol messaging– to inform consumers of an API when they are on a deprecated or unsupported API version and how they can remedy this.
- Development, from technical standard drafts, of a "gold standard" reference implementation of the resulting technical standard (in scope of a VEN/VTN implementation), and development of a test suite to test.
- Development should also strive for alignment with PAS 1878 [13] where feasible

A visual summary of the summary of the work, in the context of wider development is illustrated in Figure 8.7

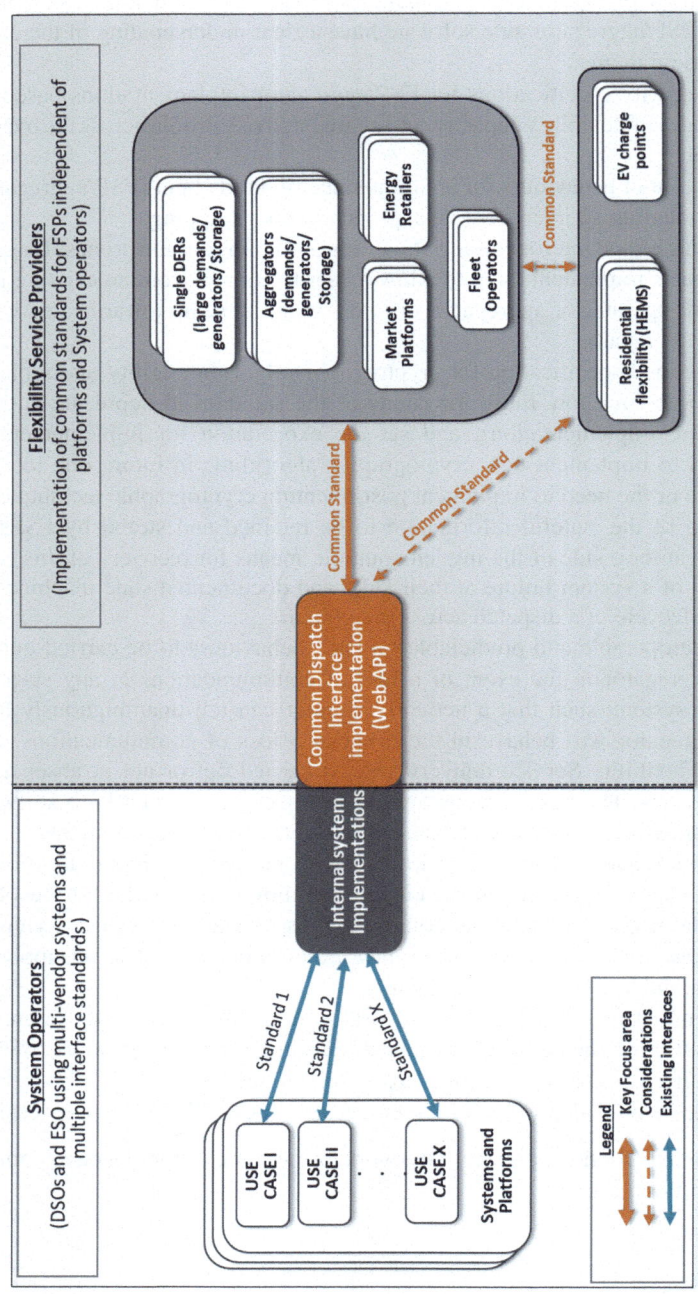

Fig. 8.7 High level description of the scope of the planned work

8.3 Suggested Further Reading

Listed below are the links to Open Networks publications relevant to this chapter in reverse chronological order. Please note, some of the documents/spreadsheets may now by outdated or superseded as the topics have evolved.

1. Flexibility Service System Interoperability Review of options around APIs and Standards for the Dispatch of Flexibility Services Publication (2024).
2. Flexibility Service System Interoperability Comparative Analysis of Solutions for the Dispatch of Flexibility Services (2024).
3. Dispatch interoperability and settlement review of existing practices and gap analysis (2022).
4. Dispatch Interoperability and Settlement Key Service Parameters (2022).

References

1. Ofgem's future insights paper 6–Flexibility platforms in electricity markets, Ofgem, 2019
2. ANSI/IEEE Std 754-2019, IEEE, 2019
3. The Floating-Point Guide, comparison, [Online]. Available: https://floating-point-gui.de/errors/comparison/
4. Secure-by-design, CISA.gov-An official website of the U.S. Department of Homeland Security, 2024. [Online]. Available. https://www.cisa.gov/resources-tools/resources/secure-by-design
5. Design and build a privately hosted public key infrastructure-The PKI principles, National cyber security centre, 2020. [Online]. Available. https://www.ncsc.gov.uk/collection/in-house-public-key-infrastructure/pki-principles
6. The guide to NIS, Information commissioner's office (ICO) (2018) [Online]. Available. https://ico.org.uk/for-organisations/the-guide-to-nis/security-requirements/
7. Special category data, Information commissioner's office (ICO) (2024) [Online]. Available. https://ico.org.uk/for-organisations/uk-gdpr-guidance-and-resources/lawful-basis/a-guide-to-lawful-basis/special-category-data/
8. Commission implementing regulation (EU) 2018/151, legislation.gov.uk (2018) [Online]. Available. https://www.legislation.gov.uk/eur/2018/151/contents
9. USEF foundation-publications, USEF (2024) [Online]. Available. https://www.usef.energy/flexibility/
10. Innovation-universal market enabling interface (UMEI), European commission-EU-funded innovations (2023) [Online]. Available. https://innovation-radar.ec.europa.eu/innovation/44267
11. Common information model (CIM) primer: Eighth edition, EPRI: program 161: Information and communication technology, 2022
12. IEC 61970-457-Energy management system application program interface (EMS-API)–Part 457: dynamics profile, IEC
13. PAS 1878–Energy smart appliances–System functionality and architecture, BSI, 2024. [Online]. Available. https://www.bsigroup.com/en-GB/insights-and-media/insights/brochures/pas-1878-energy-smart-appliances-system-functionality-and-architecture/
14. *OpenADR Alliance launches OpenADR 3.0*, OpenADR Alliance, 2023.

Open Access This chapter is licensed under the terms of the Creative Commons Attribution 4.0 International License (http://creativecommons.org/licenses/by/4.0/), which permits use, sharing, adaptation, distribution and reproduction in any medium or format, as long as you give appropriate credit to the original author(s) and the source, provide a link to the Creative Commons license and indicate if changes were made.

The images or other third party material in this chapter are included in the chapter's Creative Commons license, unless indicated otherwise in a credit line to the material. If material is not included in the chapter's Creative Commons license and your intended use is not permitted by statutory regulation or exceeds the permitted use, you will need to obtain permission directly from the copyright holder.

Chapter 9
Flexible Connections and Flexibility Services

9.1 Principle of Flexible Connections

As part of delivering net zero terrestrial greenhouse gas emissions by 2050, the amount of electricity generated from renewable sources will increase. A significant proportion of this will be connected to the lower voltage distribution networks (at 132 kV and below) rather than to the transmission network at 275–400 kV, as would be the case historically for larger generation assets. These new connections will change the traditional role of distribution networks from being passively managed to requiring more active management of power flows. DSOs are planning for the effects of increasing volumes of distributed, and often intermittent, generation and demand (e.g. Electric vehicles, and heat pumps) on their networks.

Generators wanting to connect to distribution networks under the historic regulatory access rules (also known as "shallow-ish" access rules) may find the connection costs too expensive. When generators connect to an (export) constrained network they can trigger reinforcement costs; currently under the "shallow-ish" access rules generation customers are liable for the network reinforcement costs at their connection voltage plus further into the network (at one voltage level above). Similarly demand connections can find they are subject to high reinforcement costs or long timescales for connection. Many of these new demands come from increasingly variable demands such as electric vehicle charging, heat pumps, and electric energy storage.

Finding solutions that provide faster, more cost-effective access to the distribution networks without reinforcement has been a major focus for DSOs because accommodating the growth and management of distribution connected flexible assets is a prerequisite for the GB transition to net zero carbon. This is the core principle underpinning Active Network Management (ANM) enabled Flexible connections. Its use can speed up connections and materially reduce connection costs, primarily for the benefit of those customers who are able to operate their assets flexibly with a Flexible Connection facilitated by an ANM scheme.

Note: In December 2018, Ofgem launched the Access and Forward-Looking Charges Significant Code Review (Access SCR) [1, 2], *part of a package of reforms to how different parties access and pay charges for the electricity network. The objective of the Access SCR was to ensure that electricity networks are used efficiently and flexibly, reflecting users' needs and allowing consumers to benefit from new technologies and services while avoiding unnecessary costs on energy bills in general. As the Access SCR decision in 2022 were "forward looking", they were not appliable retrospectively. The contents of the chapter are relevant to Flexible connections energised prior to the implementation of Access SCR. While the fundamental principle and the types remain the same, few key elements such as the description of the curtailment information discussed in Section 9.2.3 is only applicable to ANM enabled Flexible Connections prior to the implementation of the outcomes of the SCR.*

9.2 Flexible Connection (ANM)

In constrained areas of the electricity distribution network, the time and cost to reinforce the network can be a significant deterrent to customers wanting to connect their assets. To mitigate this, customers may be offered, or request directly, a more flexible (constrained) connection (sometimes described as a "non-firm" connection) that either limits the times in which the generator can export, or the capacity that can be exported. Similarly for demand customers with a (constrained) connection, the times in which a site can import or the level of import may also be constrained. Where assets on flexible connections have their export or import constrained by the DSO this is commonly termed curtailment.

These (constrained) connections are often facilitated through the network control system known as Active Network Management (ANM) and are termed Flexible Connections (ANM). Whilst there are other types of flexible connections, this Chapter is focused primarily on Flexible Connections (ANM).

Customers with Flexible Connections (ANM) can avoid paying a large proportion of the network reinforcement costs and avoid delays whilst waiting for the reinforcements to be completed. Flexible Connections (ANM) can be implemented as both enduring or temporary solutions.

Operationally, the DSOs use Flexible Connections (ANM) to connect more customers (typically generators/storage) through active management of the network loads, often in close to real time. This automated management of network loads prevents a constrained part of the network from exceeding its firm capacity.[1] ANM schemes requires the use of communications equipment and centralised control systems to limit generation for short periods. The ANM systems continually monitor all the constraints on the network in real time and allocate the maximum amount of capacity available to customers in that area based on the Principle of Access and curtailing other assets as required to avoid exceeding the network limits.

---

[1] A network's firm capacity is the load it can carry following one or more faults (i.e. less than the total capacity). An "N-1" network operating within its firm capacity is able to continue delivering power to customers without loss of service following a single network fault.

9.2 Flexible Connection (ANM)

9.2.1 Types of Flexible Connections

Whilst this chapter is primarily concerned with Flexible Connections enabled through ANM schemes, there are several different types of flexible connections that may be offered by the DSOs. The types of flexible connections available can vary depending on the DSO.

1. **Timed export/import connections**: these offer customers the possibility of connecting to the network but with limited export or import during certain periods of the day, week, month or year. A typical example could be wind generation limited to exporting during non-daylight hours in areas with significant PV generation.
2. **Single Generation Active Network Management (SGANM)**: is similar to a full ANM scheme above, except instead of managing multiple constraints and multiple generators it manages only one generator and up to two constraints.
3. **3rd party ANM connections**: some customers can also consider shared capacity and demand management, both of which are installed and managed by the customer.
4. **Export Limited Connections**: these are used where the installed generation asset has a greater export capability than that which has been agreed to be exported onto the distribution system. For example, a site which utilises the majority of energy generated within onsite processes and therefore limits the export of electricity to the network. These sites incorporate failsafe protection devices to ensure that the level of export is limited to a pre-agreed threshold.
5. **Import Limitation Connections**: these are used where the installed asset has a greater import capability than that which has been agreed to be imported from the distribution system. For example, a site which has multiple electric vehicle charge points may utilise an import limitation scheme to control smart charging of their electric vehicle fleet so as to not exceed their maximum agreed import capacity (MIC). These sites incorporate failsafe protection devices to ensure that the level of import is limited to a pre-agreed threshold.

9.2.2 Applying for a Flexible Connection (ANM)

In the first instance, customers should access the local DSO's website where more information detailing the potential flexible connection options available can be found, together with advice on how to apply.

If a customer decides to apply for an unconstrained connection (sometimes described informally as a "firm" connection), and there is significant reinforcement required to connect their asset, then alternative flexible connection options may be suggested by the DSO. Options will be dependent on the voltage level of connection, and the reinforcement works required.

Times and costs for connection applications for both flexible and unconstrained connections can vary by DSO and also by voltage, connection offer type and the work required. The DSO will support customers in understanding the options and the relative benefits and risks.

Customers can, at any time, request changes to their connection arrangement and a formal process already exists. Customers can request additional security or additional capacity via the normal connection application process. This application (via the DCUSA G99 Form) is subject to Assessment and Design fees and any subsequently identified charges to provide the additional requirements.

There is no planned formal review of customer contracts by DSOs, however, DSOs offer Connection Surgeries and/or stakeholder events where such options can be discussed ahead of a formal request for an unconstrained connection quote. Customers are advised to raise concerns about their contracts by directly contacting their DSO in the first instance.

9.2.3 *Curtailment of Flexible Connections (Pre-SCR)*

Investment in network capacity is driven largely by peak usage but where the outputs from customers' generation assets are variable, the overall network requirements can be less (due to diversity in when the different assets are running). How often an asset is curtailed depends on how often the site's annual output profile coincides with the times constraints are likely to appear on the network.

To maintain safe operation of the network the worst-case scenario is used in the network modelling; i.e. coincident operation is assumed in the DSOs' network capacity assessments. This may result in a conservative curtailment estimation for the customer (in other words the network capacity assessment may forecast more curtailment than is experienced in practice).

Finally, ANM systems have an integral failsafe mode to eliminate any technical or regulatory risk to the DSO of associated unsafe network operation. The DSO sets the technical parameters/triggers for the ANM scheme, but the actions are automated and not generally at the discretion of the DSO in real time. The actions are supervised by control room staff in the event that manual intervention is required to prevent an unsafe condition.

Curtailment of the generation assets will occur at varying levels based on a real-time network load assessment on the distribution system. The level of curtailment will depend on a number of factors including, but not limited to, those listed below and may increase or decrease over time:

1. Changes in operational running of the distribution system
2. Changes in the level of demand on the distribution system
3. Increases in the number of connecting small scale embedded generators
4. Reinforcement of the distribution system triggered by demand
5. Reinforcement of the distribution system triggered by unconstrained generation connections

9.2 Flexible Connection (ANM)

6. Any Active Network Management system or associated communications systems outage
7. Any reduction in the normal ability of the distribution system to absorb generation export and/or supply load
8. Technology of the generation assets and their standard profile
9. In some ANM enabled distribution networks, transmission constraints at the Grid Supply Point (GSP) level also play an important part.

Prior to the implementation of the outcomes of the Access and Forward-Looking Charging SCR there were three approaches exist for determining curtailment of Flexible Connections (ANM) to manage congestion on the distribution system, all three are deterministic and rules based: However, each DSO employs a preferred method and as such customer do not have an option to select an alternative method.

(a) **Last In First Out (LIFO)** where any binding network constraint is resolved by curtailing generators in reverse order of their connection applications. In this way generators are insulated against greater curtailment caused by other generators connecting after them, as illustrated in Fig. 9.1a.
(b) **Pro-rata** where curtailment in an impacted ANM zone is shared equally across all generators exporting during the constraint, as illustrated in the Fig. 9.1b. Pro-rata curtailment resolves constraints based upon each generator's proportional contribution.
(c) **Curtailment Index** where customers with flexible connections are assigned a forecasted index value, and a maximum cap value of curtailment they should expect to see during the course of a year. This is used to rank the future curtailment stack in which flexible connection will be curtailed by the DSO, as illustrated in Fig. 9.2.

If a customer applies for a formal quote for a flexible connection, the DSO will also provide a Curtailment Assessment report that gives customers an estimate of how often their connection may be curtailed over the course of one year. This estimate depends on factors such as historical network power flows, typical load, generation profiles, and the defines how assets that contribute to the same constraint get curtailed when that constraint for example asset's position in the priority stack (if the DSO employs this method).

Last-In First-Out (LIFO) priority stack

Last-In First-Out (LIFO) is a "Principle of Access" that defines how assets that contribute to the same constraint get curtailed when that constraint materialises. Under LIFO, each generation asset is assigned a position within a priority stack based on application date. When new generators apply for a connection in the area, they are given a position at the bottom of the priority queue i.e. these assets will be curtailed first when a constraint is binding. A customers' position in the priority stack is reserved unless the connection offer expires or is withdrawn.

During a constraint event the generator at the bottom of the priority stack will be curtailed first. This means that a generator with a lower priority will always be fully

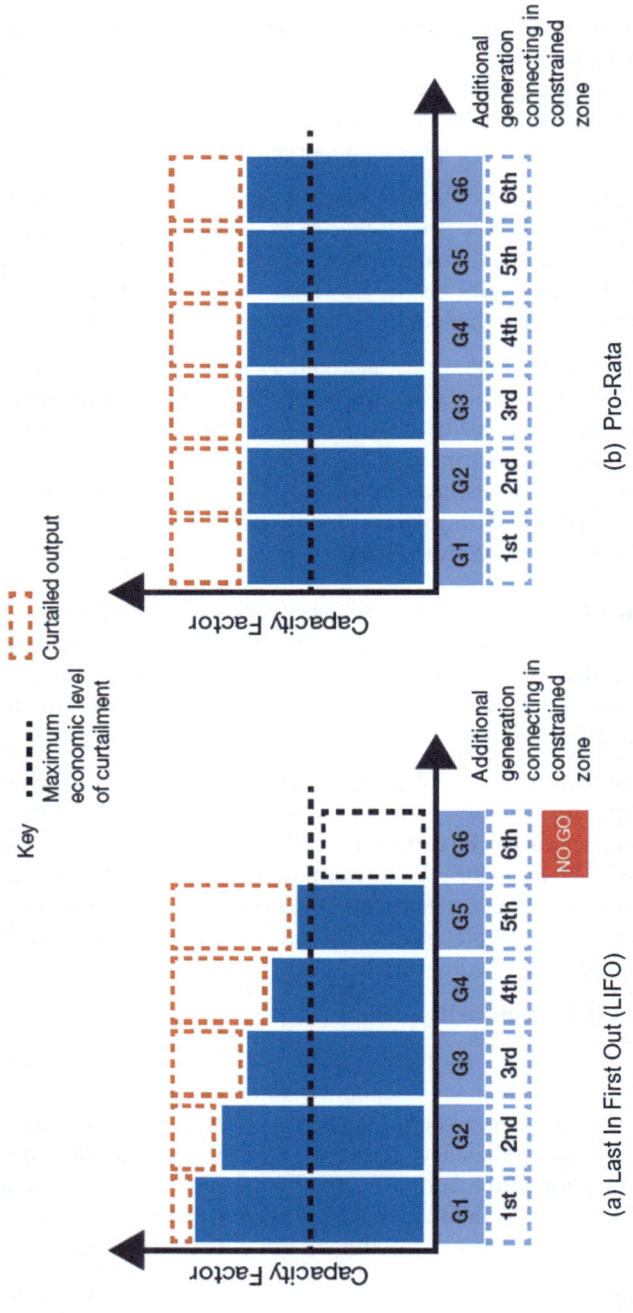

Fig. 9.1 Approach to determine curtailment of flexible connections (ANM)

9.2 Flexible Connection (ANM)

Fig. 9.2 Curtailment index

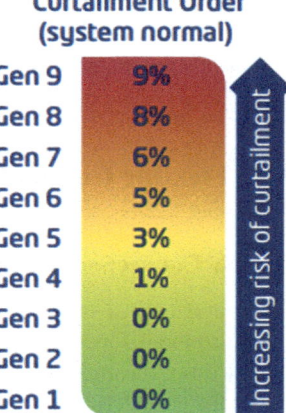

curtailed before the generator one position higher in the priority stack is curtailed. LIFO ensures that the curtailment for a given generator will not be impacted by generators with a lower position in the priority stack.

Curtailment Index priority stack mean

The aim of the Curtailment Index is to provide assurance to network users that the network will be available for use for an average time per year. A Curtailment Index is presented as the percentage of time that the network is unavailable per site. To determine the Curtailment Index, DSOs consider all the scenarios when the system is abnormal and unavailable, for instance during faults, construction, and maintenance outages. The actual curtailment experienced by the customer is monitored, and if this level approaches or exceeds the index value DSOs will investigate and potentially seek to intervene. Network users with a Curtailment Index will receive an annual review letter, providing the curtailment experienced over the previous year. The 'actual' curtailment is typically calculated based on a six-year rolling average. Any new site being connected to the network which has a flexible connection will be assigned a Curtailment Index, to allow it to be incorporated into the ANM system; legacy sites already connected to the network which already have the facilities to operate flexibly will also be assigned a Curtailment Index. It is envisaged that some legacy sites may choose to sign up to flexibility bilateral trading or may wish to alter their connection agreements to make savings on their electricity bills by accepting a flexible connection agreement; these sites are also enrolled into the Curtailment Index system. The Curtailment Index system provides a safety net for customers with flexible connections to protect them from being excessively curtailed by manual and/or automated control processes.

All sites with Curtailment Indexes are entered into a stack which puts those with the lowest Curtailment Index at the top of the stack for future curtailment (higher likelihood). When a customer's import/export is curtailed as a result of a distribution network constraint the index is incremented to reflect this curtailment and the customer's position within the curtailment stack will be reassessed.

9.3 Flexible Connections (ANM) Versus Flexibility Services

Flexible Connections (ANM) are a connection option where:

- In return for a faster cheaper connection, the customer accepts a contractual, requirement for their usual power flows to be changed by the DSO remotely, in real time, through automation. The amount of change, or curtailment, varies as per the connection agreement.
- The flexible connection is curtailed (if and when required) and uncompensated as set out in the connection agreement.
- The up front connection charges are reduced and faster connection facilitated
- The customer chooses to have specific, binding (curtailment) terms added to the connection agreement.
- Usually, the connection choice is driven by the need to achieve a faster, cheaper connection
- The key beneficiary of this scheme is the connecting customer who benefits from reduced connection costs/timescales in return for reduced network access rights.

Flexibility Services is a commercial arrangement which requires participants to deliver a change in their usual power flows where:

- A market-led initiative, through procurement exercises, finds customers' assets located within constrained networks that are able to deliver flexibility to help manage constraints.
- The customers can choose not to respond to the dispatch signal (albeit with commercial consequences in some areas).
- The participation is voluntary and the Flexibility Services provided are rewarded as set out in the commercial contract.
- Contracted customers are required to deliver Flexibility Services as defined and requested by the DSO (Market led).
- The key beneficiaries of this scheme are the DSOs, and connected customers who pay 'Distribution Use of System (DUoS) charges due to reduced reinforcement expenditure (Table 9.1).

There is no inherent reason why customers with Flexible Connections (ANM) cannot provide Flexibility Services. However, assets with Flexible Connections (ANM) cannot provide Flexibility Services if they do not meet the minimum criteria for participation for the service they are bidding into. For example, a solar farm with a timed connection which can only export during daylight hours, may not meet the criteria for an Operational Utilisation (Scheduled or Variable availability) service

Table 9.1 Comparison of uses for typical Flexible Connections (ANM) and Flexibility Services

	Flexible Connections (ANM)	Flexibility Services
Used for thermal constraints	✓	✓
Used for voltage constraints	✓	✓
Used for fault level constraints	✗	✗ [1]
Controls real power	✓	✓
Controls reactive power	✗ [2]	✓
Export turn-down	✓	✗
Export turn-up	✗	✓
Import turn-down	✗ [3]	✓
Import turn-up	✗	✓
Current likely utilisation periods	Times of high renewable output and low demand	Peak demand, planned outages, network faults

(1) Not currently used; being tested in a range of innovation trials
(2) ANM systems are usually capable of providing the signalling necessary to control reactive power
(3) Done for "Timed" and "SGANM" Flexible Connections

seeking Flexibility during the winter peak of 4–8 pm. Presently, Flexible Connections (ANM) and Flexibility Services are managed primarily as distinct services, so there is unlikely to be any conflict observed. Currently, ANM controlled generation is not turned down or curtailed to solve demand constraints and activating Flexibility Services such as demand turn-down does not increase the level of ANM curtailment.

9.4 Suggested Further Reading

Listed below are the links to Open Networks publications relevant to this chapter in reverse chronological order. Please note, some of the documents/spreadsheets may now by outdated or superseded as the topics have evolved.

1. Improvements to flexible connections offer information V0.3 (2023).
2. Flexibility Connections Explainer and Q&A (2021).
3. Flexible connections (ANM) Legacy contracts review, stakeholder feedback, and recommendations (2021).
4. Flexible connections (ANM) Curtailment Info Principles and Key Requirements (2021).
5. Flexible Resources Connection Guide Publication (2018).

References

1. "Electricity Network Access and Forward-Looking Charging Review—Significant Code Review launch and wider decision," Ofgem, 2018. https://www.ofgem.gov.uk/decision/electricity-network-access-and-forward-looking-charging-review-significant-code-review-launch-and-wider-decision
2. "Access and Forward-Looking Charges Significant Code Review: Decision and Direction," Ofgem, 2022. https://www.ofgem.gov.uk/decision/access-and-forward-looking-charges-significant-code-review-decision-and-direction

Open Access This chapter is licensed under the terms of the Creative Commons Attribution 4.0 International License (http://creativecommons.org/licenses/by/4.0/), which permits use, sharing, adaptation, distribution and reproduction in any medium or format, as long as you give appropriate credit to the original author(s) and the source, provide a link to the Creative Commons license and indicate if changes were made.

The images or other third party material in this chapter are included in the chapter's Creative Commons license, unless indicated otherwise in a credit line to the material. If material is not included in the chapter's Creative Commons license and your intended use is not permitted by statutory regulation or exceeds the permitted use, you will need to obtain permission directly from the copyright holder.

Chapter 10
Conflict Management (Primacy Rules)

Until relatively recently, the NESO and DSO performed distinct, and largely independent roles. With its responsibility for balancing the electricity system, the NESO would secure turn-up and turn-down services to ensure that demand and generation were balanced in real-time. This flexibility was delivered primarily, if not exclusively, by large generation and storage assets connected to the transmission network. The DNO (prior to introduction of DSO functions), by contrast, was responsible for maintaining the distribution network, managing outages, reinforcing and reconfiguring the network to accommodate changes in demand and new connections. Importantly, the DSO would ensure that there was enough network capacity to allow the peak demand to be met.

A number of trends have changed this relationship:

- The numbers and capacity of generation and storage connected to the distribution network has been increasing steadily.
- The proportion of dispatchable generation (e.g. coal-fired power stations) on the transmission network has been reducing, replaced by variable, non-dispatchable forms of generation (e.g. wind and solar). As a result, the NESO has been increasingly relying on flexibility from distribution-connected customers.
- DSOs are increasingly incentivised to reduce the volume of reinforcement being undertaken on distribution networks. Flexibility is being used to manage peak demand and peak generation. This can take the form of procured services or flexible connections. The more dynamic, real-time management of the distribution network now falls under the function of a Distribution System Operator (DSO).

As a result, the NESO was procuring flexibility from distribution-connected assets that may also be providing Flexibility Services, or may be sited in a constrained area of the network The NESO and DSOs managed the respective transmission and distribution networks in accordance with applicable standards and licence conditions. Each organisation may require one or more services for this purpose. Conflicts between one or more of these services lead to inefficiencies within the whole electricity system.

This would likely increase given the rising procurement of services and limited coordination to date. Hence, in order to manage this potential service conflict and to enable networks to be optimised efficiently and transparently, there was a need to develop a set of clear principles and "primacy" rules. These would enable procurement, planning, scheduling and dispatch of services to be influenced by whole system value and ensure that the division between market/price-driven actions and the electricity system hierarchy of operational needs is clear and transparent.

These rules would look to balance: the local networks' technical requirements; the risks to the overall operability of the whole system; the value for Service Providers through the facilitation of market/price driven actions; the needs of emerging market-based platform developers; and ultimately the overall cost impact on end consumers.

Therefore, a detailed investigation of such potential conflicts was undertaken to

- Identify the NESO and DSO services or 'actions' that may give rise to a conflict.
- Define 'primacy' rules that can alleviate those conflicts.
- Carry out a whole system Cost Benefit Analysis (CBA) to identify the overall impact of each primacy rule, including the impact on each network actor.
- Define the information and control flows between network actors required to enact each of the primacy rules.

10.1 Defining Primacy Rules

Primacy refers to the notion that, under some circumstances, Flexibility Service Providers (FSPs) may be offering Flexibility Services to more than one third party, and that the services offered are in conflict with each other. For example, FSPs may be offering generation turn-down to a DSO and generation turn-up to the NESO in a common area. In any given delivery window, it is not possible for the FSP to deliver both turn-up and turn-down. Equally, the risk that conflicting services within an area may result in neither SO realising the service they had requested needs to be managed. As such, a rule is required to determine whether—in these situations—the DSO or the NESO has 'primacy'.

Each Primacy Rule must (in priority order).

1. Deliver the least Whole Electricity System cost to consumers.
2. Facilitate Fair, Accessible and Efficient Markets.
3. Be clear, transparent, consistent, inclusive and deliverable.

Underpinning these principles there is a requirement to ensure that NESO and DSOs deliver on where and when applicable to (in priority order without compromising the others):

- Efficiently manage national system balance and overall operability
- Ensure Transmission Network Security, and
- Ensure Distribution Network Security.

This should continue to align with the latest industry standards (as they evolve) whilst also allowing the consistent treatment of both asset and non-asset solutions.

10.1.1 High Level Primacy Rules

Primacy rules establish a hierarchy among different types of network services or flexibility options. When multiple options are available, the service or flexibility with higher primacy takes precedence and the other service may be curtailed. Primacy is about establishing the processes to use assets responsibly and in a way that doesn't cause conflicts. In defining primacy rules that work technically, it is important to consider why the DSO needs flexibility. In most cases this will be because the local network is constrained. If the NESO attempts to take a flexibility action that exacerbates that constraint, the DSO must take an action to avoid the network breaching its thermal limits (i.e. the firm capacity of the substation). Whilst the local network will comprise a range of asset types (demand, generation and storage on flexible and non-flexible connections), from the perspective of the NESO, this network can be thought of as a single 'group asset' with a maximum export level.

This leads to two observations:

1. Primacy rules should apply to all assets in a constrained area simultaneously, rather than individual FSPs.
2. Primacy rules should describe the mitigating or enabling actions that are needed for them to work in practice.

At a high level, there are four possible primacy rules that can apply, as summarised below.

1. No Primacy: This describes the case where neither NESO nor the DSO has primacy. Where conflicts arise, the NESO and DSO will take the necessary actions to ensure its network or system requirements are met. For example, if the NESO dispatched generation turn-up from one asset, the DSO's Active Network Management (ANM) system could curtail the generation of a flexibly-connected asset in the same group, unwinding the action, leaving no net change in the export from the group. This is the default state where no formal primacy rules are put in place.
2. DSO primacy: If a NESO flexibility action would cause an issue for the distribution network, or would be unwound by the DSO (see the 'No primacy' rule above), the DSO informs the NESO of the conflict. The NESO then excludes the flexible asset from its dispatch queue, and secures its required flexibility from another source.
3. ESO primacy: For specific combinations of NESO service and distribution constraint, the DSO can take actions to enable the NESO to secure flexibility from a constrained part of the network. Where such mitigation is technically feasible, the DSO takes such action (e.g. headroom creation) on behalf of the NESO.

Assets obliged to change their output to enable such action may be compensated by the DSO or NESO, depending on how this rule is implemented.
4. Joint Primacy: This can be seen as a combination of DSO primacy and NESO primacy. The DSO and NESO agree which primacy rule should apply at a given constrained location, based on the relative costs and benefits of each rule. It is assumed that appropriate compensation is provided to ensure that the selected rule is of mutual benefit to both parties.

Applying Primacy Rules: DSO Actions

DSO actions include any flexibility the DSO uses for the purpose of managing its network—most commonly to address network constraints. DSO actions can be grouped into two types: procured services and Flexible connection (ANM).

- Procured DSO Services: DSO services enable the enactment of both positive and negative flexibility to manage network constraints and optimize distribution operations. DSOs all have independent positions with regard to whether they are pursuing all of the Standard DSO services—Peak Reduction, Scheduled Utilisation, Operational Utilisation, Operational Utilisation with Scheduled Availability and Operational Utilisation with Variable Availability (details in Sect. 3.3).

DSOs are at varying levels of development in adopting and using these services. Generation turn-up and demand turn-down often remain the most frequently procured actions, commonly delivered under the old version of a service—the 'dynamic' product. The actions which now align closely with the scope for Operational Utilisation, which has similar day ahead procurement with real-time responsiveness.

- Flexible connection (ANM): 'ANM curtailment' is one mechanism by which the DSO can effect negative (or positive) flexibility. Flexibly-connected customers are those that do not have unfettered access to the network, but instead have an obligation to accept curtailment when the network is constrained. A FSP in an ANM stack can have its output curtailed when required to manage a network constraint.

An ANM system does not require real-time control by the DSO. Rather, it monitors the real-time export (or import) across a network asset (e.g. a substation), and instructs one or more flexibility-connected customers to vary their output in order to ensure that the export (or import) does not exceed the firm capacity of the network asset. ANM is also used in order to keep the flows within Technical Limits agreed with the NESO to manage a transmission constraint.

Applying Primacy Rules NESO Actions

NESO 'actions' refer here either to actions taken through the Balancing Mechanism (BM) or through procured services. These include system balancing services (i.e. frequency and reserve services) and transmission constraint management services. The services where the data was available to understand the auction stacks from the

10.1 Defining Primacy Rules   195

NESO portal, was firstly considered in scope these include the below services for which the data is presently available via NESO's Single Markets Platform (SMP):

- Balancing Reserve
- Static Firm Frequency Response
- Dynamic Containment
- Dynamic moderation
- Dynamic Regulation
- MegaWatt Dispatch (Data available on SMP but not Auction stack)

And a few services where the data is not available on the SMP, which include

- Quick reserve
- Sow reserve

The services that are not included in the analysis are:

- Lond Term Stability Y_4
- Stability Path Finders
- Mid Term Stability
- Short Term Stability D-1
- Voltage Pathfinders
- Reactive Power Long Term Markets
- Reactive Power Mid Term Markets
- Reactive Power Short Term Markets
- Constraint Management Intertrip Service
- Local Constraint Market
- Constraint Collab Project
- Distributed Restart
- Electricity System Restoration Events

10.1.2 *Consolidated Use Cases*

In principle, combinations of DSO and NESO actions in turn could be considered to define primacy rules for each, evaluate the costs and benefits, and lay out a preferred solution. However, it is more practical and insightful to group these services together in appropriate categories, and develop primacy rules based on those consolidated use cases. This approach ensures scalability and pace of developing the Primacy rules, especially given the total number of permutations of potential conflicts of services are in thousands.

Throughout this discussion two types of actions are described:

- Positive flexibility: refers either to generation turn-up or demand turn-down, as both increase the generation-demand balance on the system
- Negative flexibility: refers either to generation turn-down or demand turn-up, as both decrease the generation-demand balance on the system

The individual NESO and DSO actions all differ from each other in a number of ways, including their value to the system operator, the types of asset that can deliver the service, the timing of procurement and dispatch, and the commercial structures. However, for the purpose of developing and evaluating primacy, the key factor determining which rules make sense is whether the NESO is securing positive or negative flexibility, and whether the distribution network is import- or export-constrained.

It should be noted that most NESO services are structured to allow providers to provide either positive flexibility, negative flexibility, or both (i.e. a symmetrical response). This allows the positive and negative component of these services to be considered separately, which is important since the nature of the conflict with constrained distribution networks is different depending on the direction of flexibility being dispatched.

Put simply, there are two types of conflict that need to be considered:

- NESO flexing *into a* distribution network constraint;
- NESO flexing *away from* a distribution network constraint.

These two situations are summarised in Fig. 10.1, with Cases 1 and 4 relating to the NESO flexing into a distribution constraint, and Case 2 and 3 relating to the NESO flexing away from a distribution constraint.

NESO Flexing into a Distribution Network Constraint

In the case of NESO flexing into a distribution network constraint, it is clear why a conflict arises. If the NESO were to try to secure generation turn-up from an already export-constrained part of the distribution network, this would exceed the network's thermal limits. In order to avoid this breach, the DSO must counteract the NESO action through a generation turn-down action. This is most likely to be through the actions of the ANM system. Note that the NESO may procure generation turn-up from one asset (e.g. a gas engine with a non-curtailable connection), while the DSO's ANM system may counter this by turning down a different asset (e.g. a flexibly-connected solar farm). This case illustrates a number of issues:

- The NESO will not be receiving the change in output it expected to see;
- The gas engine will (rightly) assert that it has delivered the required flexibility, and expect to be paid by the NESO for its services;
- The solar farm will experience additional curtailment, reducing its revenue.

In effect, the DSO already has 'primacy' in this case (i.e. 'No primacy'), which is to say that its right to maintain network integrity supersedes the NESO's right to procure unfettered flexibility. It is, however, an inefficient form of primacy since the NESO is paying for an outcome it does not receive. One solution is for the DSO to notify the NESO when such cases will—or are expected to—arise, allowing the NESO to go elsewhere for its flexibility. This is defined in this chapter as 'DSO primacy'. This is usually better for the NESO and the solar farm and for the consumer when considering whole system costs, but would effectively exclude the gas engine from providing this NESO service (at least while the constraint was in effect).

10.1 Defining Primacy Rules

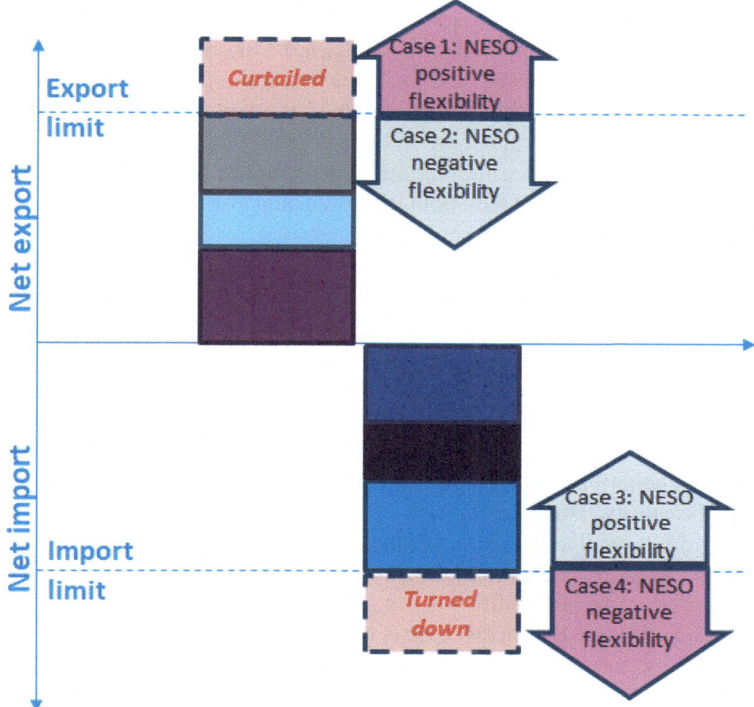

Fig. 10.1 Illustration of flexibility conflict use cases

Depending on the specific service that the NESO is trying to procure, there may be ways of avoiding the conflict. Assuming that the distribution constraint cannot be exceeded, this can only be done by first creating headroom on the distribution network. For example, this could be done by curtailing the solar farm ahead of time, and instructing the ANM not to use up the headroom created. This would allow the gas engine to offer turn-up balancing services to the NESO by operating within that newly-created headroom (again, with the ANM instructed to allow this to happen). This would almost certainly require functionality to be added to ANM systems, as this is not a mode of operation that is used today. This is an example of 'NESO primacy', where the DSO acts on behalf of the NESO to create the conditions to allow it to procure flexibility from an otherwise constrained part of the network.

NESO Flexing Away from a Distribution Network Constraint

In the case of NESO flexing away from a distribution network constraint, it is less clear why there is a conflict. If the NESO wishes to secure generation turn-down from an export-constrained part of the network, doing so would not risk exceeding the thermal limits of the network. On the contrary, it would reduce the thermal loading of the network.

However, the potential conflict arises if the DSO is already securing generation turn-down from that part of the network. This is most likely to occur where there is an ANM scheme operating. If the ANM scheme is curtailing one or more curtailable assets, when the NESO triggers the turn-down of another asset on the same part of the network (e.g. a firm generator), the ANM will 'see' this as additional headroom. Its own logic will then reduce the curtailment of one of the curtailable assets to compensate. This is consistent with both its design, and also the rules by which these assets agree to curtailment when they connect.

As before, this is effectively a case of the DSO having primacy, but with the NESO being unaware that such a situation is in effect. As before, the NESO would not receive the flexibility it expected to see, and the gas engine would (rightly) expect to be compensated for delivering turn down. The solar farm, having previously been facing one level of curtailment, ends up experiencing less curtailment, so sees an increase in its revenue.

As before, a formalised 'DSO primacy' rule would have the DSO notify the NESO that this situation was likely to occur, allowing the NESO to go elsewhere for its required flexibility. This would be better for the NESO, and would mean that the solar farm did not receive its extra revenue. However, it would effectively exclude the gas engine from the NESO market while the constraint was biting, which it may not have realised was a possibility when it first connected.

The 'NESO primacy' rule in this case is more straightforward than in the 'flexing into the constraint' case. At the time of instructing generation turn-down, the NESO could also inform the DSO that this was occurring. The DSO would then instruct the ANM not to fill up the created headroom. Note that whilst this is relatively simple in principle, it would still require ANM functionality beyond that which exists today. There would also be additional complexities in enacting this for aggregated assets due to visibility.

Summary of DSO and NESO Primacy Under Each Use Case

In summary 4 cases were identified, relating to NESO and DSO flexibility conflicts, as shown below in Table 10.1.

Outage Cases

In addition to the above cases, the interaction between NESO flexibility and outage actions taken by the DSO should also be considered. An outage on the network affects the ability of the NESO to access flexibility from a FSP in one of two ways:

1. The FSP may be taken off supply, meaning that it is unable to import or export, and is therefore unavailable to provide flexibility.
2. The network may be reconfigured during the outage, meaning that the Grid Supply Point (GSP) through which FSP flexibility takes effect is changed.

Two types of outage:

10.1 Defining Primacy Rules

Table 10.1 Flexibility conflict use cases

	NESO positive flexibility		NESO negative flexibility	
Export-constrained distribution network: DSO/ANM is, or is close to, effecting negative flexibility	Case 1		Case 2	
	Conflict case: DSO counters NESO action with negative flexibility to manage thermal constraint		Conflict case: DSO counters NESO action with positive flexibility to absorb newly-created headroom, reducing the curtailment of customers in the area	
	DSO primacy: NESO opts out of dispatching positive flexibility from the area	NESO primacy: NESO can only dispatch positive flexibility if headroom has first been created on the constrained distribution network	DSO primacy: NESO opts out of dispatching negative flexibility from the area, so the DSO takes no action and no curtailment reduction occurs	NESO primacy: NESO dispatches negative flexibility, and the DSO ensures that is does not release the newly-created headroom to customers being curtailed
Import-constrained distribution network: DSO is, or is close to, effecting positive flexibility	Case 3		Case 4	
	Conflict case: DSO counters NESO action with negative flexibility to absorb newly-created headroom, reducing demand turn-down in the area		Conflict case: DSO counters NESO action with positive flexibility to manage thermal constraint	
	DSO primacy: NESO opts out of dispatching positive flexibility from the area, so the DSO takes no action and no demand turn-down reduction occurs	NESO primacy: NESO dispatches positive flexibility, and the DSO ensures that it does not reduce demand turn-down in response or release the newly-created headroom to customers being curtailed	DSO primacy: NESO opts out of dispatching negative flexibility from the area	NESO primacy: NESO can only dispatch negative flexibility if headroom has first been created on the constrained distribution network

- Planned outages: Where the DSO takes a network asset offline (e.g. for repairs), with prior knowledge that this is going to occur and, typically, the ability to choose when this occurs.
- Unplanned outages: Where a network fault is unanticipated, and the DSO takes action to restore the network to its intact state.

Whilst both types of outage affect the ability of the NESO to dispatch FSP flexibility, the DSO only has any real choice over the unplanned outage. Therefore no meaningful primacy rules were identified that would apply in this case.

For planned outages, however, the DSOs have some discretion as to the nature and timing of its actions. In principle, therefore, primacy rules could apply to ensure that these planned outages are carried out in a way that maximises the whole system benefit. However, this level of analysis was not undertaken.

10.2 Cost Benefit Analysis

10.2.1 Cost Benefit Analysis Approach

A Whole System Cost Benefit Analysis (CBA) was undertaken to assess the overall benefit of each primacy rule under each case, and how these rule affects individual parties on the system. The scope of the CBA is limited to the costs of the flexibility actions themselves, including:

- Procurement costs for the NESO and DSO.
- Lost revenue/opportunity costs for the connected customers.

Implementing these primacy rules will require new data, processes and systems to be implemented. Establishing, operating and maintaining these will require the NESO and DSO to incur additional costs. These are excluded from this CBA. The CBA is, therefore, intended to indicate where there may be a benefit of implementing a given primacy rule. Additional analysis will then be required to determine whether the investment required to bring it into effect is economically justified.

For the CBA, a notional baseline was established under which the network is not constrained. The NESO procures its flexibility as if the local network were not at its thermal limit. It has access to all distribution-connected flexibility providers, and those FSPs are able to offer their services unencumbered. Note that this baseline cannot exist in reality, since these local networks are constrained. It is, however, a useful baseline to use to illustrate the relative cost impacts of the different primacy rules.

The costs of the different primacy rules were considered under the four cases identified earlier in the Sect. 10.1.2. 'DSO primacy' rule is common to all four cases: the NESO is informed that the network is constrained, and that the flexibility it requires from the distribution-connected assets in the local area is not available. As such, the NESO must secure an equivalent volume of flexibility from elsewhere on the system. The cost of this primacy rule is driven by the relative cost of that additional flexibility compared to the cost of the distribution-connected flexibility.

Because there are parallels between Cases 1 and 4, and between Cases 2 and 3, at a high level there are two situations the CBA needs to consider for 'NESO primacy':

- Cases 1 and 4—NESO taking an action that will exacerbate the thermal constraint: In this case, the action cannot be taken without first creating the required headroom. Ahead of the delivery window, the NESO must secure turn-down to create headroom. It can then use that headroom to dispatch assets for balancing services.

The cost of these cases are driven by the cost of securing that turn-down, and the length of time for which the headroom needs to be held.
- Case 2 and 3—NESO taking an action that reduces the thermal constraint: In this case, the NESO should be able to secure turn-down. The only barrier is the automated action of the ANM system. Provided this unwinding action can be prevented, there should be no additional cost associated with NESO primacy in this case. There will be additional system and process changes required but, as discussed, these are outside the scope of this analysis.

10.2.2 Cost Benefit Analysis Results

The cases largely consider conflicts at distribution thermal limits. Under technical limits or cases where NESO has specific stability needs (e.g. inertia) the primacy rule that makes sense technically and could be altered because of actions at the point of implementation of the rule being different. For the four core cases of conflict, the CBA revealed that NESO primacy is preferable in some cases, and DSO primacy in others. The results from the CBA are discussed below.

- Conflict case 1: DSO counters NESO action with negative flexibility to manage thermal constraint- Optimal approach: DSO primacy
- Conflict case 4: DSO counters NESO action with positive flexibility to manage thermal constraint- Optimal approach: DSO primacy
- Conflict case 2: DSO counters NESO action with positive flexibility to absorb newly-created headroom- Optimal approach: NESO primacy
- Conflict case 3: DSO counters NESO action with negative flexibility to absorb newly-created headroom- Optimal approach: NESO primacy

For Conflict cases 1 and 4 DSO primacy leads to the best net benefit on a whole system basis. The actions required to enable NESO primacy (holding additional curtailment during the service window, and paying for rebalancing from elsewhere on the transmission network) create significant additional costs. Whilst DSO primacy imposes costs on the NESO, these costs are lower than those under DSO primacy under any reasonable price assumptions considered.

For conflict cases 2 and 3, NESO primacy leads to higher net benefits in this case. Compared with DSO primacy, NESO primacy does not require any additional flexibility or curtailment to avoid unwinding the NESO action. However, both DSO primacy and NESO primacy result in more curtailment than would occur under 'no primacy'. Whilst there is no difference between DSO primacy and NESO primacy in terms of the opportunity cost of curtailed assets, they do have different implications in the compensation that curtailed assets might be entitled to:

- Under DSO primacy, the NESO actions that would otherwise create headroom are no longer taken. Even though this results in more curtailment (compared to 'no primacy') it is unlikely that curtailed assets would expect or be entitled to compensation.

- Under NESO primacy, the headroom is being created, just as it would be under 'no primacy'. The DSO is preventing that headroom from being released to curtailed customers. Depending on how DCUSA and customer connection agreements are interpreted, this may require compensation to be paid to those customers.

Even if compensation were paid under NESO primacy, this would not change the whole system CBA outcomes, since it would represent a transfer of revenue to the curtailed customers from the NESO or DSO, and would not change the overall benefit of the approach.

In the cases where a NESO action moves the loading of the local network away from a constraint, there is little cost incurred by the DSO facilitating that action through NESO primacy. On the assumption that ANM systems can be modified to allow this, both technically and at reasonable cost, there would be justification for enabling these Use Cases. Further consideration will be required to any compensation that would be appropriate for curtailed customers in this situation. Whilst they would not face any additional curtailment compared to a DSO primacy rule (where the NESO would go elsewhere for its required flexibility), NESO primacy in this case would result in their curtailment remaining in effect while there was headroom on the local network. Depending on their flexible connection agreement, it may be appropriate for them to be compensated, in which case they would be net beneficiaries from this arrangement, with the DSO or NESO incurring some cost for covering that compensation payment.

In the case where a NESO action moves the loading of the network into a constraint, it is unlikely that NESO primacy will make economic sense. The need to create headroom throughout the delivery window creates the need for a significant increase in the volume of flexibility being dispatched. Whether this is absorbed by the curtailed party, or paid for by the DSO or NESO as a procured service, this represents a significant additional cost. As such, DSO primacy is likely to be the most suitable approach in these cases. This could change in the future, but would require a number of things to transpire:

- The cost of headroom creation would need to fall. Currently, the flexibility that the DSO is procuring is relatively expensive, including demand turn-up. It is possible that these costs fall in future as the number of flexible assets connected to the distribution network increases, and as more automation is introduced to demand-side flexibility (e.g. EV charging).
- The cost to the NESO of securing alternative forms of flexibility would need to increase. This could arise if flexibility on the transmission network reduced, and if embedded flexibility became the main source for providing NESO services. If the NESO had fewer alternative options for meeting its requirements, this lack of liquidity could push up the cost of DSO primacy relative to NESO primacy.

The only exception to the above could be the use of embedded flexibility to increase inertia on the system. The NESO could secure generation turn-up from a high-inertia generator on an export-constrained part of the network (effectively flexing into a constraint). The DSO (or its ANM system) would then trigger negative flexibility in response to keep the network within its thermal limits. Provided that this negative flexibility came from a low-inertia source (e.g. wind, solar, or demand turn-up), the net result would be of benefit to the NESO. This can be seen as a variant of NESO primacy, where the DSO is obliged to curtail particular asset classes to enable the NESO to increase system inertia.

However, this could be implemented in a number of alternative ways:

1. Better data would allow NESO to turn up a high inertia asset with reasonable confidence that the ANM will respond by turning down a low-inertia asset (based on an understanding of the mix of generators in a given LIFO stack).
2. NESO could explicitly initiate high-inertia turn-up and low-inertia turn-down from behind a constraint in order to ensure the required change in inertia is delivered. This approach would effectively remove the need to modify the ANM.

10.3 Implementing Primacy Rules (Use Case)

Prior to the development of the high level Primacy rules discussed in Sect. 10.1, Open Networks undertook an assessment to develop an extensive framework for the identification and prioritisation of the Use Cases through a mixture of negative and positive prioritisation. Through this analysis, it was agreed to develop and test the Primacy Rules an iterative approach. The use cases chosen for the first iteration in 2022 was the DSO Services, coupled with the DER Transmission Constraint Management service (TCM) being developed via the Regional Development Programmes (RDPs) on different assets in the same area. Due to its scale, complexity and importance, it was agreed to proceed with the implementation of one use case so that progress can be made whilst ensuring learning from each iteration informed subsequent increments.

10.3.1 Transmission Constraint Management and DSO Flexibility Services

This Use Case involved a (possible) scenario in which the NESO was trying to reduce the export of a single/multiple generator(s) to manage a Transmission Export Constraint when, at the same time, the DSO is trying to procure a Generation Turn Up (GTU)/Demand Turn Down (DTD) service from different assets in the local area.

Through the ongoing Regional Development Programmes between the NESO, NGED and UKPN, the development of the "MW dispatch" service was underway. This is a Transmission Constraint Management compensated service aimed at DER of 1MW or more whose active power output may be curtailed following an instruction from the NESO. This service allows DER to fulfil their connection terms and conditions as part of the respective Bilateral Connection Agreements between the DER and DSO and then subsequently between the ESO and the DSO. The original RDP study work determined that is likely to be more cost efficient overall for GB consumers in allowing DER to connect under 'Visibility and Control' terms and conditions, in lieu of large transmission reinforcement works. This service was aimed at those parties that don't wish to provide high levels of flexibility within larger market arrangements (such as the BM) however, these alternative options remain open as a means of fulfilling these connection terms and conditions.

As part of the first release, this new service employed existing DSO network infrastructure up to the DER point of connection to facilitate TCM service. Instructions from the NESO control room would flow through to DSO control room and then on to DER. This will ensure efficient use of existing infrastructure between the DSO and DER with communication or dispatch routes as needed. The NESO has since continued to work with DSOs to develop various communication and coordination arrangements, which include the use of third party routes as part of future releases across of their respective RDPs.

In addition to existing BM participants, this new TCM route complemented traditional options for the management of transmission constraints and, should DER choose to provide visibility and commercial control to the NESO via one of these existing routes, this would also satisfy the connection terms and conditions. The deployment of such a service, that is fully coordinated between NESO and DSO processes, enabled the 'connect and manage' principles to be applied on a deeper level, whilst continuing to enable connections in areas where there is high DER activity.

A requirement to participate in the service was included in the connection agreements of generators in the specific RDP areas (South West Peninsula and South Coast). Initial deployment focused on DER with obligations to provide the service but was extend to voluntary participants in future releases to improve the liquidity of the market and the efficiency of coordination between the NESO and DSOs. The service was part of the first roll-out of a fully co-ordinated constraint management service between the NESO and DSOs.

10.3.2 Process and Data Flows for Primacy Rules Deployment

Note The rules considered for deployment in the first increment were bespoke for the selected use case and differed slightly from the rules discussed in Sect. 10.1.1. The learning from the deployment of increment 1 use case ultimately informed the generic high level primacy rules approach presented earlier in this chapter.

The level of conflict mitigation depended on the rules selected (discussed in further detail below) as a result of the level of information sharing between licensees and their forecasting assumptions. Where conservative assumptions were built into DSO Flexibility Service dispatch (building on planning assumptions as set out in ER P2/8 and EREP 130) then impact of conflict was minimised.

10.3 Implementing Primacy Rules (Use Case)

For example, if the DSO assumes minimal contribution to network security from non-contracted generators,[1] it will dispatch more contracted assets. The TCM actions of the NESO will counteract the services but will not impact network security. Forecasting DSO dispatch requirements closer to real time would improve the overall efficiency of DSO services as it can consider a more realistic perspective on non-contracted generation. Feeding in additional data from the NESO, and wider markets on the positions of generators would further enhance dispatches, and whole system outcomes. The additional targeting of services could reduce the periods of conflict. However, this does create more dependencies between parties and more coordinated primacy scenarios. With less leeway (i.e., margin for error between forecasts) in the scenario, the impact of conflict could be higher. It was therefore noted that, as closer to real time forecasting rolls out further, it will be important to adequately consider the uncertainty of data shared (for example BM participant-submitted data at the Day Ahead is indicative and subject to change).

It is broadly envisioned that 'Ahead of time' related more to timelines up to Week Ahead and 'Closer to Real Time' involved exchange of information at Day Ahead or beyond during control room timescales. During the initial deployment/testing of the did not place any judgement of merit on the rules. At that stage the aim was to understand the available options.

Primacy Rules Considered

The deployment was based on two core rules.
Rule 1: DSO Priority.

- Variant (a): Information shared ahead of time
- Variant (b): Closer to real time information sharing

Rule 2: Management of Planned Outages.

- Variant (a): Additional coordination of planned outages
- Variant (b): Closer to real time planned outage cancellation

It was noted that, there was no option for NESO priority in shorter term operational timeframes. The exchange of information and coordination through RDPs would enable the DSO and the NESO to identify the DER that can provide an efficient and cost-reflective TCM service. The TCM service is a location-specific service, and this has been considered when optioneering some of the Rules. If the DSO has committed to the use of Flexibility Services to manage the network, then any option that would limit the ability of the DSO to procure services in the local area would impact the security of the network.

Given the wider geographic bounds of transmission network, the NESO should have alternative options to dispatch in other areas, either adjacent DSO areas or from transmission connected assets, which the DSO does not require, or cannot access.

---

[1] Non-Contracted from the DSO perspective. This means the DSO does not have an active commercial contract with the asset for the period of time considered.

The rules considered are highlighted in the process flow charts below. The flow charts are for illustrative purposes only and do not reflect a detailed delivery model for RDPs, since they present a high-level approach to achieve mitigation of conflicts and coordinated dispatch of NESO and DSO services. It should be noted that there could also be other approaches/variables (not presented as part of this section) that could offer similar or more efficient solutions. As RDPs are further developed, and Primacy Rules begin trial, the NESO and DSOs (together with the Marker Facilitator) will develop further the details of these processes.

1. Rule (1a): DSO Priority—Information shared ahead of time

In this rule, the DSO services hold priority over the NESO TCM service due to local nature of DSO flexibility and limited alternative options. This rule involves the sharing of a commonly agreed "Risk of Conflict" forecast between the DSO and NESO. This would reflect the DSOs approach to forecasting (as noted earlier) and translate it into the identification of risk of conflict. This may initially be quite simplistic but would evolve as DSO processes mature. Enhancing the required data elements from the NESO to the DSO may be necessary to improve this forecasting.

At the time of publication of this book, the RDP work concluded that the "Forecast Risk of Conflict" would be fed into the NESO's planning processes for the TCM service, with the NESO rejecting TCM sites where the DSO has identified a risk of conflict. This allowed for a consistent, simple implementation for the NESO, with the onus for conflict identification residing with the DSO in a regular data sharing process in planning timescales. The data sharing processes in this use case are relatively simple. As they are not near real time, they can rely on the upload and download of data from an online portal, or the sending of CSVs via Email (Fig. 10.2).

2. Rule (1b): DSO Priority—Closer to Real Time Information Sharing

This Rule builds on Rule (1a). It adds:

- Additional required information flows from the NESO to the DSOs (in agreed timescales) to allow for better forecasting of flexibility requirements and potential risk of conflict.
- Enhanced real time systems that allow for the assessment of distribution of network conditions and constraints at Day Ahead. This will inform the understanding of conflicts by the DSO and the sharing of this information with the NESO.

As such, this combines the longer term information from (1a) (which can serve as an indication from and to the NESO), with the closer to real time view from the DSO serving as the final view on whether conflict is likely to materialise. This should be much more accurate as it improves forecasting accuracy and takes account of real-time network dynamics. The DSO holds the ability to enforce the rules, due to the current implementation of this service and the use of DSO equipment to implement dispatch (Fig. 10.3).

3. Rule (2a): Management of Planned Outages—Additional Coordination of Planned Outages

10.3 Implementing Primacy Rules (Use Case)

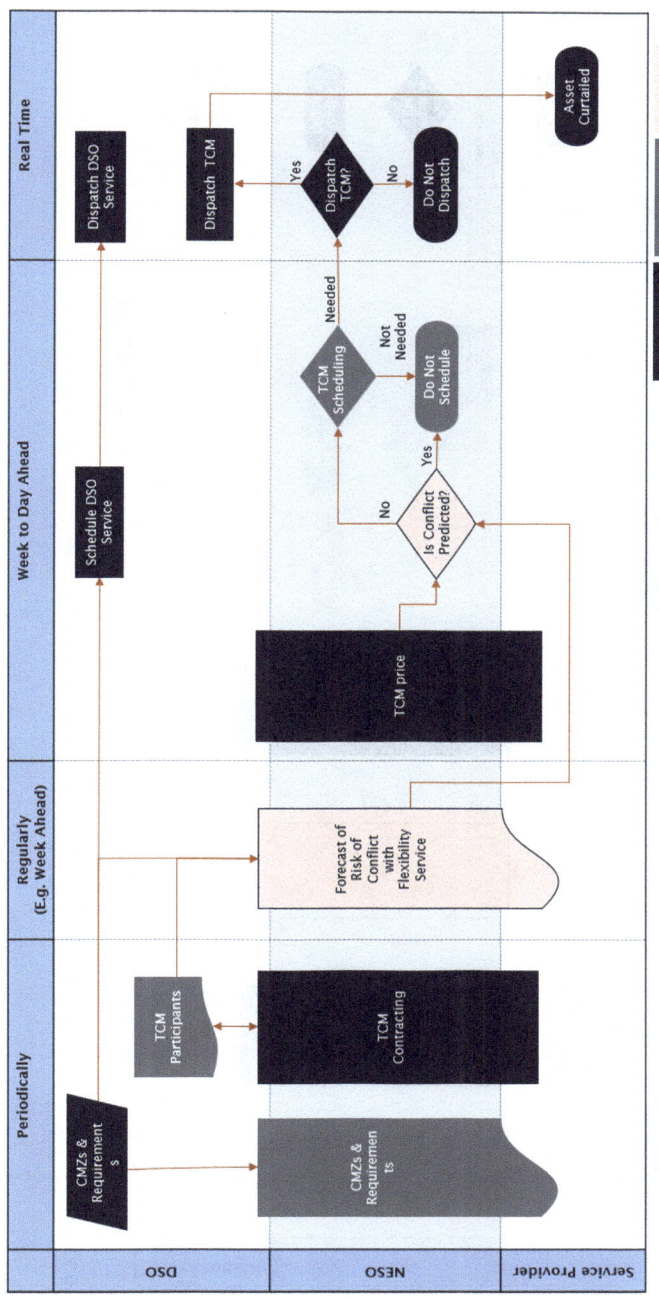

Fig. 10.2 Process flows for Rule (1a): DSO priority-information shared ahead of time

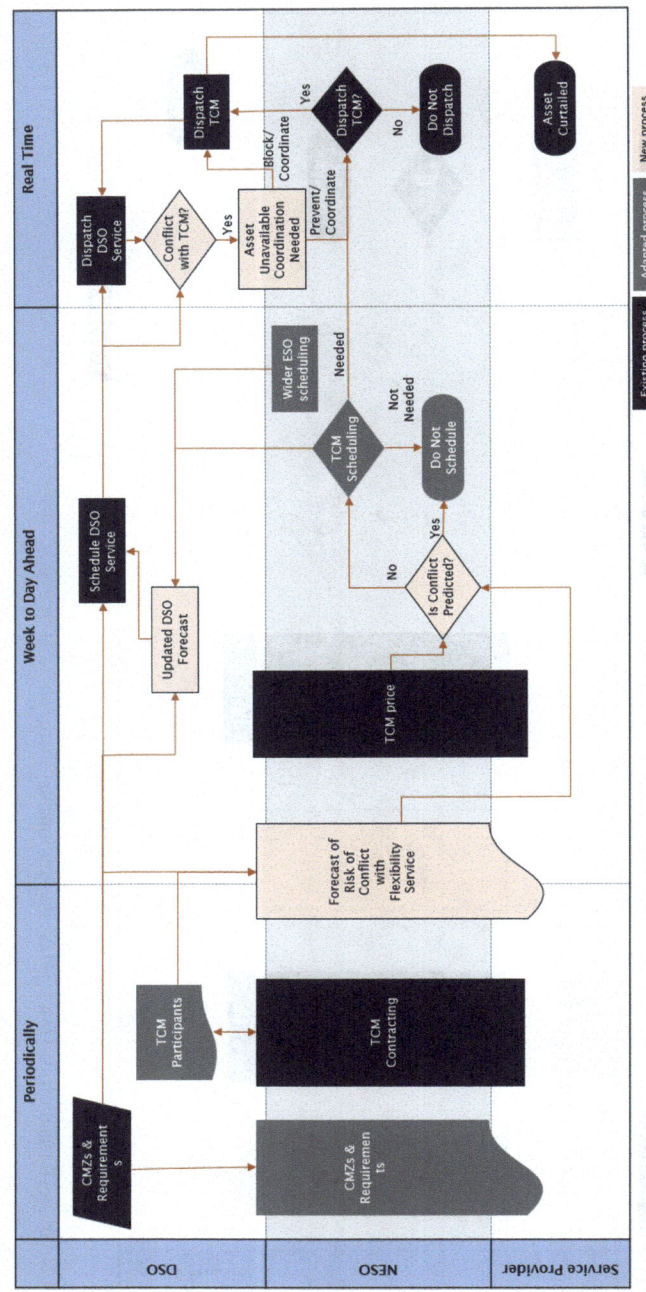

Fig. 10.3 Process flows for Rule (1b): DSO priority—Closer to real time information sharing

10.3 Implementing Primacy Rules (Use Case)

This Rule builds on (1b). It adds an additional process to existing planned outage coordination processes to allow for the assessment of the whole system cost impact of the outages. This builds on the fact that some requirements for both TCM and DSO services are linked to both transmission and distribution outages. Where this is the case, and the release of an outage is likely to increase the overall risk of conflict, the cost impact of allowing the outage to proceed should be weighed up against the cost of rescheduling the outage through a joint CBA. (An equivalent Rule could be created building on the simpler Rule (1a.)) (Fig. 10.4).

4. Rule (2b): Management of Planned Outages—Closer to Real Time Planned Outage Cancellation

This Rule removes the advanced planning of Rule (2a.) and instead focusses on closer to real time coordination. This would require the NESO to liaise with the TO and possibly cancel/move their outage where high dispatch costs are expected. This would allow for operability risks to be managed, was noted to not be done efficiently. Due to the timescales, a robust CBA of all options was not possible and so sub optimal outcomes are to be expected (Fig. 10.5).

Ultimately Rule (1a.) DSO Priority—Information shared ahead of time was taken forwards for testing and GB-wide roll out. Given the additional value unlocked by Rule (1.b), it was noted that the rule should also be investigated, acknowledging the additional processes and timescales necessary for delivery.

The rules associated with the TCM use case trailed and tested as part of RDPs. Rule (1a) was tested within the NGED RDP in the South West. UKPN and the NESO continued to investigate the elements of Rule (1b) that was delivered as part of the RDP in the South East, recognising a longer timescale due to that new processes that needed to be developed in relation to the additional data sharing requirements within this Rule. Due to the limited expected conflict the trial focused on the creation of mutual robust data sharing processes between the NESO and DSOs and was implemented to the extent where these are achievable by the end of ED1 (Regulatory year 2022/2023).

Key Data to Be Exchanged Under Each Rule

Alongside the process flowcharts, a review of the key data to be exchanged under each option was undertaken. This was used to understand the likely data that will need to be collated and exchanged during the Rules implementation phase (See Table 10.2).

210 10 Conflict Management (Primacy Rules)

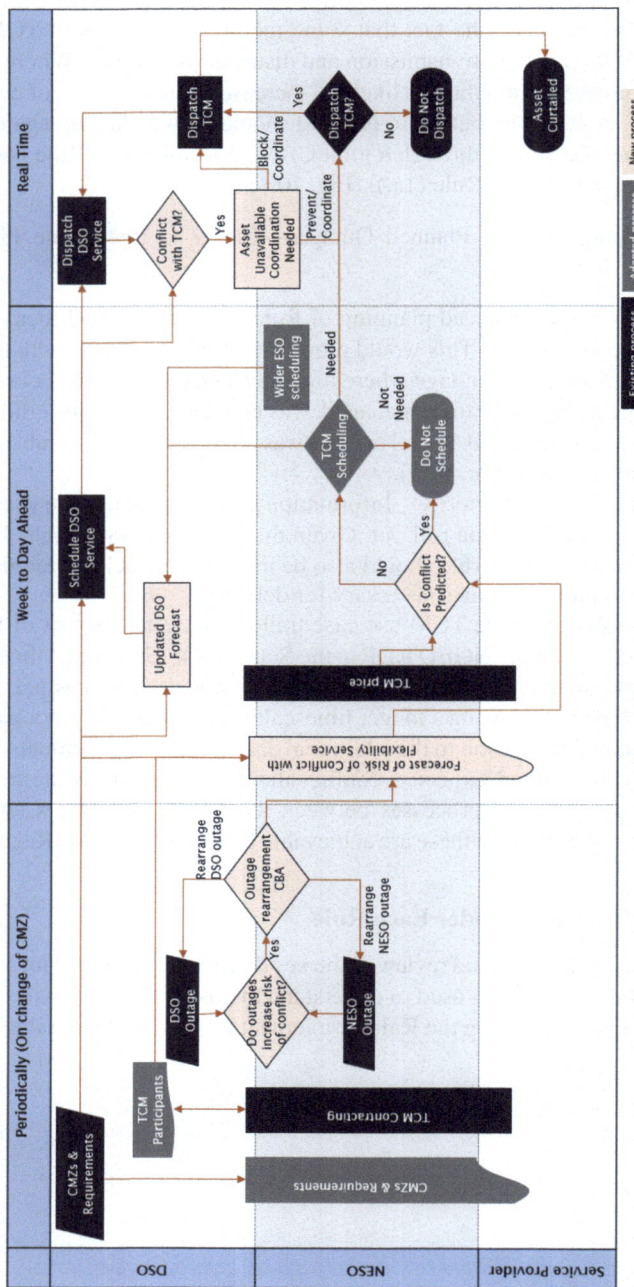

Fig. 10.4 Process flows for Rule (2a): management of planned outages—additional coordination of planned outages

10.3 Implementing Primacy Rules (Use Case)

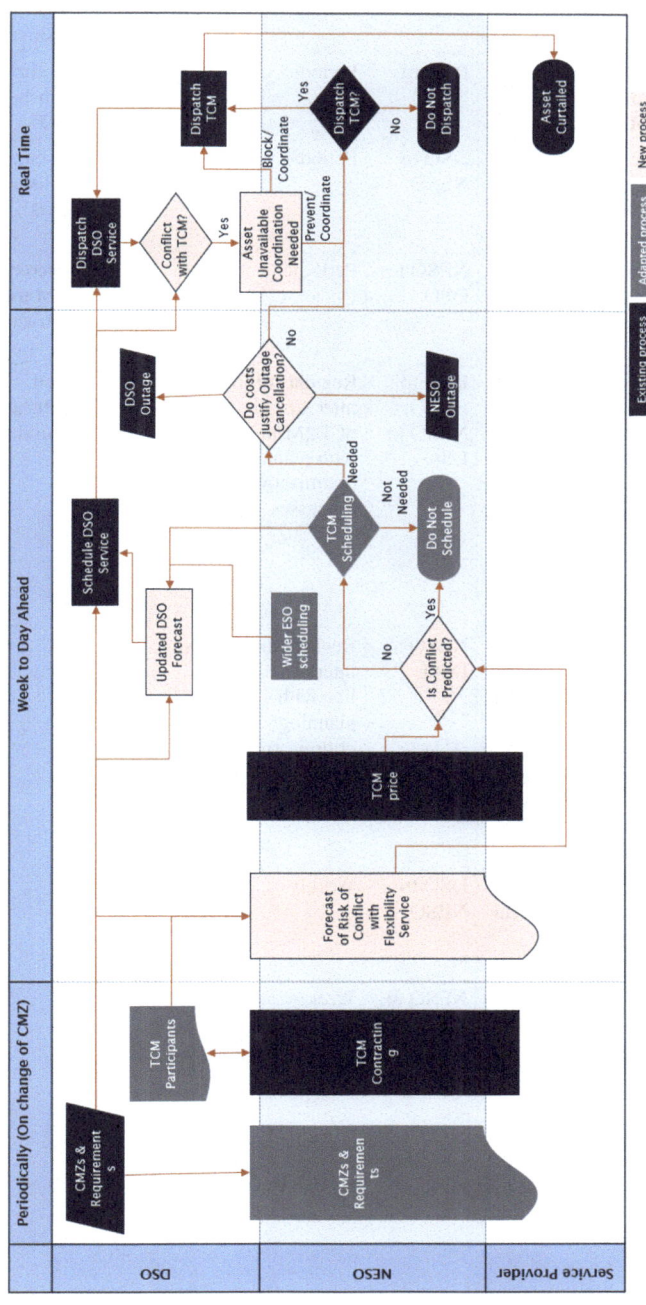

Fig. 10.5 Process flows for Rule (2b): management of planned outages—Closer to real time planned outage cancellation

Table 10.2 Likely data that will need to be exchanged during the Rules implementation

Data being shared	Description	Direction	Timing	Currently available?	Commercially sensitive?
CMZ locations (for 1a, 1b, 2a, 2b)	The geographic boundaries of CMZs	DSO to NESO	Periodic	Yes	No, currently published by DSOs
CMZ requirement (for 1a, 1b, 2a, 2b)	The temporal and magnitude of service requirement	DSO to NESO	Periodic	Yes	No, currently published by DSOs
List of TCM generators (for 1a, 1b, 2a, 2b)	Which generators are subject to TCM	NESO to DSO	Periodic	No	Covered in the TCM tripartite contract
Forecast of conflict (for 1a, 1b, 2a, 2b)	A view as to which areas (linked to CMZs) would see conflict if the ESO were to curtail a TCM generator TCM info from the ESO (including service scheduling) (for 1b onwards)	DSO to NESO, NESO to DSO	Regular rolling intervals, ahead of TCM/in line with planning/ optimisation processes	No	Not if published beyond NESO
ESO planning outputs (for 1b, 2a, 2b)	Day Ahead NESO planning assumptions. Used to improve DSO scheduling and dispatch for both DSO Flexibility Services and TCM dispatch	NESO to DSO	Regular rolling intervals, in line with planning/ optimisation	No	TBC (Subject to legal review)
DSO outages (for 2a, 2b)	List of DSO outages that would trigger Flex procurement	DSO to NESO	Weekly	No	No
Transmission outages (for 2a, 2b)	List of transmission outages that would trigger TCM procurement	NESO to DSO	Weekly	Partly	No

10.3.3 Learnings and Considerations for Implementation of Primacy Rules

Building Additional Capability and Processes

Ideally, primacy rules should be based on specific conflict conditions, requiring DSOs and NESO to collaborate on conflict forecasting by collecting real-time data. Both DSO primacy and NESO primacy require investment in order to enact them. This includes improved data, forecasting and information sharing between parties. In some cases it also involves making changes to systems, such as ANM systems to allow them to adjust their notional headroom based on NESO actions, as well as changing the Principles of Access to prioritise some assets over others. Further work would be required to understand the technical changes required, and the costs involved. Additional data would be required to understand the prevalence of DSO-NESO conflicts, and hence the value that implementing these primacy rules could create.

The CBA analysis gave a strong indication of the preferred primacy rule for the different use cases. The available data on the frequency of conflicts, the costs of mitigation is limited. Furthermore, additional work would be required to understand the detailed implementation requirements, and the costs associated with implementing these primacy rules.

The importance on a robust process to assess the risk of conflict in planning and closer-to-real-time timescales to understand and manage risk was also noted. The exclusion of service providers from the NESO market will reveal the true Whole System cost of the service while catering for DSO constraints. The lack of coordination could have led to nullification of services and the potential for increased costs. This needs to be picked up as part of the planning process such as NOA or DNOA, to identify solutions for limiting the scale of conflict (like network reinforcement) and conduct suitable cost benefit analysis.

Impact on Different Customer Types

The whole CBA showed the overall cost and benefit of the different primacy rules, including how each rule creates costs or benefits for different network users. However, when considering the feasibility of implementing these rules, there is also a need to consider how these rules might conflict with network users' existing connection agreements. Potential conflicts identified include:

- Non-curtailable customers on a constrained network: Such customers will have secured a distribution connection with full access rights. This means that they have the right to export at their maximum capacity (provided the network is intact). They may also reasonably expect such access to extend to the ability to offer Flexibility Services to the NESO, just as they would if they were connected to the transmission network. The emergence of a constraint on the network means that the value of their flexibility to the ESO may be significantly diminished.

Without an explicit primacy rule, their flexibility will be unwound by the ANM system, meaning that the NESO will be expected to reduce the perceived value of that flexibility;

Under DSO primacy, the NESO will opt not to procure from the asset at all, since the DSO will have informed them of the conflict;

Under NESO primacy, the asset will be able to provide flexibility for most NESO services (although not generation turn-up for managing transmission constraints). However, either the DSO or NESO will be responsible for paying for creating the required headroom. It is reasonable to expect that this additional cost will reduce the perceived value of flexibility from this asset.

Regardless of the primacy rule, then, non-curtailable assets sited behind constrained parts of the distribution network are expected to see the value of their flexibility reduce. This is a risk that they may not have been made aware of at the point of connection, and indeed may be something they remain unaware of. This risk needs to be better communicated, and may need to be factored into the decision to opt for ANM-managed zones rather than reinforcing the network.

- Curtailable customers being curtailed for headroom creation: NESO primacy requires that the DSO instruct the ANM to create headroom to allow the Flexibility Service provider to deliver its flexibility. Under such circumstances, there will be curtailable customers under curtailment whist there is technically headroom on the constrained network. Their level of curtailment will not be higher than it would otherwise be (i.e. if the NESO were not procuring flexibility, or if DSO primacy were implemented). However, DSOs will need to verify that these customers' connection agreements allow curtailment when the network is not actually fully constrained. Even if this is contractually permitted, it may be seen not to be within the spirit of the curtailable connection agreement

Regulatory and Commercial Changes

As well as the technical requirements for implementing Primacy rules, a number of regulatory and commercial changes are expected to be needed. The following potential areas of change were identified:

- Licence conditions, Grid and Distribution Codes, Connection and Use of System Code (CUSC) and Distribution Connection and Use of System Agreement (DCUSA): The implementation of both DSO Primacy and NESO primacy will impose additional responsibilities on DSOs and the NESO to each other and to connected customers. Operator licences and codes may need to be updated to codify those new obligations. Required changes may need to include:

    Obligations on the DSO to produce accurate curtailment forecasts, and to communicate those to the NESO (or make the live data available through an OpenData portal).

    Modifications to NESO's procurement principles to allow connection of customers to be consistent with delivering an optimal whole-system outcome.

10.3 Implementing Primacy Rules (Use Case)

Obligations on the DSO towards connecting customers, including curtailable connections, but also towards non-curtailable customers who may ultimately be sited in a constrained area.

- Commercial and contractual changes: DSOs and NESO may need to review their connection and service agreements with customers to account for Primacy rules. This could include:

    Changes to future or existing curtailable connection agreements to allow the DSO to change the Principles of Access in line with NESO Primacy rules (e.g. holding headroom or deviating from the curtailment stack), with appropriate compensation as required.
    Changes to NESO service agreements to allow the NESO to block bids and offers under constrained conditions, to opt not to dispatch these assets, or to disarm their dispatch.
    To account for any compensation that may be required under the implementation of one or more of the Primacy rules, such as for those customers who may face additional curtailment, or who may be curtailed under circumstances that go beyond their original connection agreement.

- Regulatory incentives: Although beyond the scope of this chapter, consideration may need to be given to the ways in which the DSO and NESO are incentivised to implement Primacy rules. For example, in order for the DSO to provide the NESO with live curtailment data and curtailment forecasts, investment in their capabilities will be required. Furthermore, some interpretations of NESO primacy could see the DSO or NESO facing additional costs (e.g. in compensating curtailed customers). Whilst doing so may be in the interest of the whole system, it may be necessary to create regulatory measures to ensure that incentives on each operator are aligned.

Downstream Investability

Downstream investability for parties in the energy market as a result of primacy rules will depend on how they are implemented. If NESO primacy facilitated actions that maintain or even improve DER revenues (e.g., through compensation for curtailment or changes to DCUSA/SCR which impact curtailment limits), this could enhance the financial attractiveness of DER projects. However, uncertainty in how primacy rules will be applied—particularly if DSO primacy leads to increased perceived curtailment—could deter investment in DERs in constrained areas due to perceived risks of revenue instability.

In the case where DSO service flexibility could be used to create headroom, this could drive demand for distribution Flexibility Services, particularly in areas where constraints are common. This presents opportunities for service providers to secure contracts. The relatively high cost of DSO-procured flexibility currently could limit market growth unless costs decline with scale and automation.

The uncertainty around primacy rules—particularly the distinction between NESO and DSO primacy—could complicate financial modelling and reduce the predictability of returns on DER investments. The potential for "negative flexibility" obligations (e.g., curtailment triggered by the DSO in NESO primacy) adds an additional layer of operational complexity for DER operators.

The value of providing flexibility is closely tied to the timing and location of the response in relation to NESO/DSO needs, as well as the level of market saturation. This will continue to be a key factor influencing the investability of DER projects, especially where flexibility revenues are critical to their commercial success.

If the cost of flexibility decreases due to automation and scaling of DERs, and if regulatory frameworks ensure fair compensation for curtailed assets, both DERs and flexibility providers could see increased investment. Coordination between NESO and DSOs to clearly define obligations and compensation mechanisms could reduce uncertainty and make investments more predictable.

Good Governance and Efficacy

An evolving governance strategy for primacy rules is essential due to the rapid transformation of the sector Whilst the future governance of the Primary will be determined by the Market Facilitator, a few points of learnings were noted.

- Need to evolve, adapting to technological advances. The energy industry is integrating new technologies, such as advanced ANM technologies, and digital grid management tools, at an unprecedented pace. A flexible governance approach allows for the rules to keep up with these innovations, ensuring efficient use and management of new resources.
- Addressing changing market conditions: As market dynamics shift—driven by changes in energy demand, decarbonisation goals, and regulatory updates—primacy rules must adapt to reflect these conditions. This prevents outdated rules from hindering the market's ability to respond to new challenges and opportunities.
- Ensuring whole-system optimisation: With increasing interconnections and dependencies across the energy system, governance must evolve to support whole-system benefits, such as enhanced resilience, cost-effectiveness, and environmental sustainability. Static governance and rules may create inefficiencies or conflicts between different parts of the system, whereas dynamic governance can better align with overall system goals.
- Responding to Uncertainties and Risks: The energy transition involves uncertainties, such as fluctuating fuel prices, regulatory shifts, and climate impacts. An adaptable governance framework allows for timely adjustments to primacy rules, enabling the industry to respond more effectively to unforeseen risks and maintain system reliability and security.
- Need to establish a structured review process/framework to regularly assess the effectiveness of primacy rules to ensure they align with whole-system goals. This framework should include periodic evaluations, stakeholder feedback mechanisms, and transparent reporting on outcomes.
- Need to define outcome-based metrics to evaluate whether primacy rules are delivering expected benefits across the system, such as cost efficiency, reliability, and

Flexibility Service provider/connected customer feedback. Adjust these metrics as needed to reflect evolving energy system priorities.
- Need to implement adaptive governance to maintain flexibility in governance processes to allow for adjustments to primacy rule processes based on observed impacts, emerging technologies, and market changes. This could involve piloting new variations e.g. moving from day ahead to real time primacy decisions, to assess their effects before broad implementation.

10.4 Suggested Further Reading

Listed below are the links to Open Networks publications relevant to this chapter in reverse chronological order. Please note, some of the documents/spreadsheets may now by outdated or superseded as the topics have evolved.

1. Primacy Rules High level Framework (Increment 3) (2025).
2. Primacy Rules High level Framework CBA (Increment 3) (2025).
3. Primacy Rules for NESO/DSO Coordination (Increment 2) (2024).
4. Primacy Rules for NESO/DSO Coordination (Increment 1) (2023).
5. Primacy Rules (Increment 1) Testing Report (2023).
6. Primacy Draft Rules development framework (Increment 1) (2022).
7. Independent cost–benefit analysis on primacy rules for (STOR versus ANM) Flexibility Service (2023).
8. Primacy Rules (STOR versus ANM) Cost–Benefit Analysis Overview (2022).
9. Primacy Rules (STOR versus ANM) Cost–Benefit Analysis (2022).
10. Primacy Rules for Service Conflicts—Use Case Prioritisation Framework (2021).
11. Primacy Rules for Service Conflicts—Use Case Prioritisation Supporting Slides (2021).

Open Access This chapter is licensed under the terms of the Creative Commons Attribution 4.0 International License (http://creativecommons.org/licenses/by/4.0/), which permits use, sharing, adaptation, distribution and reproduction in any medium or format, as long as you give appropriate credit to the original author(s) and the source, provide a link to the Creative Commons license and indicate if changes were made.

The images or other third party material in this chapter are included in the chapter's Creative Commons license, unless indicated otherwise in a credit line to the material. If material is not included in the chapter's Creative Commons license and your intended use is not permitted by statutory regulation or exceeds the permitted use, you will need to obtain permission directly from the copyright holder.

Chapter 11
Stackability of Flexibility Services

11.1 Primacy Versus Stackability

Operators of flexible assets seeking to maximise value from their assets have numerous options for their trading strategy. Sources of revenue include trading energy on the wholesale market; providing balancing services to the NESO; capitalising on opportunities created by imbalance pricing; and more recently providing Flexibility Services to the DSOs. Each revenue stream comes with associated commercial and regulatory complexity, and almost all of them are subject to ongoing development.

One point to clarify is the distinction between 'primacy' and 'stackability'. Both these concepts relate to the ability of FSPs to offer multiple services, so there is a risk of conflation in their usage. For the avoidance of confusion, the following distinction were made (See Table 11.1).

In principle, the application of stringent stackability rules could minimise the numbers of occasions on which primacy rules are required (by avoiding situations where a FSP might be expected to offer two conflicting services). Conversely, lax stackability rules may increase the frequency of such conflicts, requiring primacy rules to be enforced more often.

It should be noted, however, that even the most strict stackability rules will not eliminate the need for primacy rules. The main reason for this is that stackability rules apply to individual FSPs, whereas primacy rules need to consider cases where the dispatch of one FSP affects the flexibility of other FSPs.

11.2 Revenue Stacking of Services

A key issue for providers of flexibility when choosing which revenue streams to use is the extent to which those revenues can be "stacked" with revenues from providing other services. Revenue stacking broadly refers to the ability for FSPs to contract

Table 11.1 Distinction between primacy and stackability

	Stackability	Primacy
Summary	Stackability rules tell a FSP whether they are allowed to offer two services for the next, say, 1-month period, or whether they need to choose which service to offer Concerns whether two services can be offered by an FSP in a single delivery window	Primacy rules apply in real- or near-real time, and define how the system operators are allowed to dispatch a service Concerns which system operator can dispatch a service in the event that two services conflict
Timing	Applies at the **procurement or pre-procurement** (i.e. eligibility) stage	Applies at the **operational dispatch** stage (i.e. within-day or day-ahead)
Who the rule applies to	Applies to the **FSP** looking to offer multiple services. Not affected by the actions of other FSPs	Applies to the **system operator (or other 3rd party)** looking to dispatch the Flexibility Service. May not only consider a single FSP, but also the interaction between different FSPs

across multiple revenue streams to both maximise their own revenues as well as their benefit to the electricity system. However, there are multiple types of stacking which can be subdivided into revenue 'splitting', 'jumping' and 'co-delivering'. There are nuances to how and when assets can earn revenues from each of these services co-optimally. For the purpose of this chapter, the stacking of revenues under two of the common stacking definitions are assessed:

- **'Splitting'**—earning revenue and being able to deliver multiple services from the same asset in the same time period, but not from the same MW. The asset can provide different MWs at the same time, providing the ability of the asset to deliver in all contracted service(s) is not impeded.
- **'Jumping'**—earning revenue from the same asset and the same MW, but during adjacent or different time periods.
- **'Co-delivery'**—refers to being able to deliver multiple services and earn revenue from the same MW in the same time period. Our research found that there were many challenges with co-delivery, and that there needed to be a wider industry discussion about whether co-delivery should be achievable for FSPs on an intentional basis. As a result, a detailed assessment of co-delivery was not included. However, the current issues and challenges associated with co-delivery is discussed briefly later in this chapter.

A summary of the key distinction between the categories of Stacking is shown in the Table 11.2.

To assess the 'stackability' of services, combinations of the DSO services and key wider flexibility revenue streams available to FSPs are discussed to determine if and how they could be stacked with another individual service. However, it is noted that, in most instances, there is no single definition on the explicit interaction

11.2 Revenue Stacking of Services

Table 11.2 Distinction between different categories of stackability

	Splitting	Jumping	Co-delivery
Asset	Same	Same	Same
Capacity	Different	Same/different	Same
Time	Same	Different	Same
Direction	Same/different	Same/different	Same

between services and how these would be concurrently provided by FSPs. Therefore, in order to distinguish between the different levels of clarity in stacking, the below classification between services as employed:

- **Explicitly stackable**—service terms, rules, guidance, wider industry barriers, or clear market/technological reasons render services stackable.
- **Implicitly stackable**—based on our understanding of market rules, regulations and processes, the services are likely to be stackable without significant issue or barriers, and there are no service terms such as exclusivity that explicitly prevent stacking.
- **Implicitly unstackable/technical issues arise**—based on our understanding of market rules, regulations and processes, the services are likely to be unstackable, inter-operational challenges mean FSPs are unlikely to be able to or want to stack the services.
- **Explicitly unstackable**—rules or guidance explicitly state that revenues cannot be stacked across services e.g. the service requires exclusivity from the provision of all other services for the duration of the agreed contract. Or otherwise, clear reason why the services cannot be stacked.
- **N/A**—this option is included where service splitting or jumping is not applicable. This is most notably for the Capacity Market which has some unique considerations, and a long-term product centred on a capacity (i.e. form of availability) payment. Revenues are not typically split or jumped between in a similar manner to other services.

Within the five DSO services, some contain product variants. These usually vary by having different timing of utilisation instructions; this can impact the ability to stack the services. Therefore the headline product variants for the services as detailed in Sect. 3.3 are investigated. Whilst individual DSOs may take different approach to implementation of the services, but commonality of the services results in an assessment that is broadly applicable across the DSOs.

The views regarding stacking are our best view at the time of drafting (May 2024) based on the understanding of service terms, potential operability challenges, industry rules, and industry practice where other information is less available. Where new NESO services are included in the tables, information has been taken from latest service designs or direction taken from similar recently developed services. While the tables and information presented are an informed view of stacking arrangements, there may be instances where industry views or experience deviate from our own including due to specific asset circumstances.

11.2.1 Flexibility Service Provider Perspective

There are a wide range of different types of FSPs with a breadth of strategies for maximising their customer portfolios and optimising revenue streams. FSPs also have varying risk appetites which will impact how they engage in and prioritise different services.

For new DSO services being able to stack (i.e. either splitting or jumping in this chapter) between revenue streams as a result of service designs and rules (as considered in this chapter) is one of the key considerations; however, it does not mean that the FSP will decide to stack the services in their trading and optimisation strategies.

Some key additional considerations from an FSP's perspective include:

1. **Location**—DSOs only procure for flexibility if there is a requirement in a specific location. Due to the highly locational nature of DSO services, FSPs will only be able to stack with DSO services should they be situated in a location where there is a need. The requirements also vary over time, meaning FSPs need to ensure they track DSO tenders to ensure they compete when opportunities arise.
2. **Technology types**—the technology or the technologies an FSP is operating in its portfolio must be capable of entering the services. As discussed previously, certain DSO services may be implemented to attract a particular technology type (e.g. slow responding demand reduction), whereas an NESO service such as frequency response is more likely to see battery storage. However, the importance of services being technology agnostic is important as the types of FSPs change in the future.
3. **Risk appetite**—delivering multiple services must not create risk which doesn't yield the level of reward required. There is an element of interpretation when it comes to stacking DSO services with NESO services with limited concrete guidance on what can or can't be done. FSPs may perceive a risk when stacking services in the absence of guidance and will want to avoid any potential penalties for non-delivery. Furthermore, honouring Capacity Market obligations will also be key as DSO services are not a listed Relevant Balancing Service. FSPs will need to assess any risk of providing DSO services against any CM agreements in place.
4. **Price, liquidity, and transparency**—FSPs are more likely to focus on the markets with the greatest revenue potential, as well as markets which are liquid and transparent. As NESO services typically have a large market size with good levels of liquidity, these are likely to remain a key focus for FSPs. The typically smaller size requirements for DSO services, combined with them being both locational and potentially temporarily procured in any given location, means that how FSPs who are operating large portfolios (or aggregated units) view the ability to 'split' will be important, as the DSO service may only utilise a small part of their overall unit capacity.
5. **Competition with the Balancing Mechanism**—there is a growing number of FSPs gaining access to the Balancing Mechanism, with the NESO improving

its dispatch processes (e.g. Open Balancing Platform) and market participants such as Virtual Lead Parties and aggregation/optimisation parties looking to this revenue stream. The BM has strict requirements for submitting Final Physical Notifications at Gate Closure and any real time response to DSO services will likely be in breach of these requirements.

6. **Knowing the value of flexibility**—the value of flexibility is often not known far in advance of the event but based on on-the-day system conditions. Many of the DSO services see their prices set 'at trade' for both utilisation and availability. The time of procurement will likely impact an FSP's approach and pricing. Some FSPs may see value on fixing the value of their flexibility further in advance, however, others may see this as a risk and instead focus on short-term markets. It is noted that some of the new DSO services allow for availability to be refined closer to the event, while utilisation instructions can be known in advance (e.g. week-ahead or day-ahead) which will act to support FSPs in stacking with NESO services that procure closer to real time (i.e. mostly day-ahead).

7. **Administrative burden**—stacking can require a detailed knowledge of the rules and regulations (including baselining and performance monitoring requirements), while participation in DSO services requires additional sign-up requirements, market tracking and platform registration. Depending on how DSOs choose to implement the services, there may continue to be (more minor) differences between services in different DSO regions which FSPs will need to understand. There is also an administrative burden on FSPs entering NESO services with strict requirements, pre-qualification and metering data requirements. The more services in which FSPs engage, the greater the burden.

11.3 Stacking Assessment for DSO Services

The new suite of standard Flexibility Products were rolled out by the six GB DSOs in 2024. Each of the DSOs have varying flexibility needs and requirements due to the physical characteristics of their networks; DSO services are inherently locational. Each service procured by each DSO is only tendered in the specific locations for which they are required. Therefore, FSPs will only be able to participate in the service(s) should they be situated in a location within the relevant DSO region being tendered for.

(The service uptake for each DSO, as observed May 2024, is shown in the Sect. 3.4).

Stacking combinations of each of the new DSO services (including their product variants) were assessed against key wider revenue streams available to FSPs. This includes the wholesale market alongside NESO services covering energy balancing, system security, thermal constraint management, reserve, and frequency response.

Table 11.3 details the key for the stacking tables including what each of the colour assessments mean for service stacking.

Table 11.3 Key for flexibility service stacking tables (applicable to splitting and jumping)

Key		Short explanation
E-US	Explicitly unstackable	Service terms, rules, guidance, wider industry barriers, or clear market/technological reasons render services unstackable.
I-US	Implicitly unstackable	While not explicit in the service terms or guidance, something (e.g. operational or contractual conflicts) implicitly means FSPs either can't or would unlikely attempt to stack the services.
I-S	Implicitly stackable	While not explicit in the service terms or guidance, it is likely that FSPs would be able to, or choose to, stack these services.
E-S	Explicitly stackable	Service terms, rules, guidance, or clear market/technological reasons means they are stackable.
N/A	Not applicable	This is specifically for the CM which has some unique considerations which are explore separately when considering co-delivery of services.

Table 11.4 outlines whether services can be delivered by jumping from one service to another in adjacent or nearby settlement periods. This typically has a greater stacking ability. Limitations typically arise when a service has long or enduring delivery windows, are written into connection agreements, or registration and/or the ability to participate in one market excludes an asset from another market (e.g. Demand Flexibility Service).

Table 11.5 summarises the ability of different services to be delivered by the same asset but different MW in the same settlement period. The column headers link to each of the respective sections containing details.

By cross examining the headline designs of the new DSO services with the range of relevant NESO services and or wider revenue streams, there are a number of key findings and trends to highlight from our research concerning stacking.

11.3.1 Jumping of Services

Revenue jumping is a more readily available option for stacking services than revenue splitting. There are fewer potential delivery considerations for FSPs when looking at options to jump between services, as the delivery or availability for one DSO or NESO service is unlikely to impede the ability to dispatch into another DSO or NESO service in adjacent time periods. Our findings regarding revenue jumping include:

- The new DSO services have typically moved to delivery (availability or utilisation) periods that align with settlement period or Electricity Forward Agreement (EFA) blocks, which supports alignment with the approach taken for many NESO services. This supports the ability to jump between services with minimum waiting periods. However, although utilisation or availability periods may have a settlement period of EFA block granularity, it is unclear how DSOs will implement these services and if they will seek or favour FSPs able to provide across multiple periods.

11.3 Stacking Assessment for DSO Services 225

Table 11.4 Ability to jump different services

	Service jumping	Scheduled Utilisation	Peak Reduction	Operational Utilisation**		Scheduled Availability + Operational Utilisation		Variable Availability + Operational Utilisation	
				2 min & 15 min	Week ahead	2 min	Day ahead	2 min & 15 min	Day ahead	Week ahead
Energy balancing & system security	Wholesale market	E-S	E-S	I-S	E-S	E-S	E-S	E-S	E-S	E-S
	Balancing Mechanism	E-S	E-S	I-US	E-S	E-S	E-S	E-S	E-S	E-S
	NIV Chasing/imbalance	E-S	E-S	I-S	E-S	E-S	E-S	E-S	E-S	E-S
	Capacity Market	N/A	N/A	N/A	N/A	N/A	N/A	N/A	N/A	N/A
	Demand Flexibility Service*	E-US	E-US	E-US	E-US	E-US	E-US	E-US	E-US	E-US
Thermal	Local Constraint Market	E-S	E-S	I-US	E-S	E-S	E-S	E-S	E-S	E-S
	MW Dispatch Service	E-S	E-S	I-US	I-US	E-S	E-S	E-S	E-S	E-S
Reserve services	Short Term Operating Reserve	E-S	E-S	I-US	E-S	E-S	E-S	E-S	E-S	E-S
	Slow Reserve	E-S	E-S	I-US	E-S	E-S	E-S	E-S	E-S	E-S
	Quick Reserve	E-S	E-S	I-US	E-S	E-S	E-S	E-S	E-S	E-S
	Balancing Reserve	E-S	E-S	I-US	E-S	E-S	E-S	E-S	E-S	E-S
Frequency response	Dynamic Containment	E-S	E-S	I-US	E-S	E-S	E-S	E-S	E-S	E-S
	Dynamic Moderation	E-S	E-S	I-US	E-S	E-S	E-S	E-S	E-S	E-S
	Dynamic Regulation	E-S	E-S	I-US	E-S	E-S	E-S	E-S	E-S	E-S
	Static Firm Frequency Response	E-S	E-S	I-US	E-S	E-S	E-S	E-S	E-S	E-S

*The assessment of DFS is based on the service design for winter 2023–2024 and guidance published in December 2023. However, it is noted that in a DFS update from the NESO, it identified enabling stacking as a critical area for improvement. In a webinar in June 2024 the NESO said it is looking to allow service stacking with the CM and DSO services for winter 2024–2025
**Operational Utilisation parameters (notably real time variants) mean FSPs may not get foresight of when they will be required, inhibiting jumping. However, it is noted OU may be implemented by DSOs to replace the historic 'Restore' product, restoring or supporting the network following an unplanned fault. NESO service terms typically include provisions for unavailability due to unforeseen technical circumstances, which may mean that FSPs can provide OU provided the NESO service allows for unavailability. However, it is understood that Restore has historically not been viewed stackable in this way, and ability to stack OU may depend on what the DSO is using the product for

Table 11.5 Ability to split different services

Service splitting*	Scheduled Utilisation	Peak Reduction	Operational Utilisation		Scheduled Availability + Operational Utilisation		Variable Availability + Operational Utilisation		
			2 min & 15 min	Week-ahead	2 min	Day-ahead	2 min & 15 min	Day-ahead	Week-ahead
Energy balancing & system security									
Wholesale market	I-S	I-S	I-US	I-S	I-US	I-S	I-US	I-S	I-S
Balancing Mechanism	I-S	I-S	E-US	I-S	E-US	I-S	E-US	I-S	I-S
NIV Chasing/imbalance	I-S	I-S	I-US	I-S	I-US	I-S	I-US	I-S	I-S
Capacity Market	N/A	N/A	N/A	N/A	N/A	N/A	N/A	N/A	N/A
Demand Flexibility Service**	E-US	E-US	E-US	E-US	E-US	E-US	E-US	E-US	E-US
Thermal									
Local Constraint Market	E-US	E-US	E-US	E-US	E-US	E-US	E-US	E-US	E-US
MW Dispatch Service	E-US	E-US	E-US	E-US	E-US	E-US	E-US	E-US	E-US
Reserve services									
Short Term Operating Reserve	I-US	I-S	E-US	I-US	E-US	I-US	E-US	I-US	I-US
Slow Reserve	I-S	I-S	E-US	I-S	E-US	I-S	E-US	I-S	I-S
Quick Reserve	I-S	I-S	E-US	I-S	E-US	I-S	E-US	I-S	I-S
Balancing Reserve	I-S	I-S	E-US	I-S	E-US	I-S	E-US	I-S	I-S
Frequency response									
Dynamic Containment	I-S	I-S	E-US	I-S	E-US	I-S	E-US	I-S	I-S
Dynamic Moderation	I-S	I-S	E-US	I-S	E-US	I-S	E-US	I-S	I-S
Dynamic Regulation	I-S	I-S	E-US	I-S	E-US	I-S	E-US	I-S	I-S
Static Firm Frequency Response	I-S	I-S	E-US	I-S	E-US	I-S	E-US	I-S	I-S

*While many services are assessed as being implicitly splitable, this is currently only applicable when providing services in the same direction. None of the applicable services were assessed as being splitable when providing services in opposing directions

Note (Refer to Table 11.3 for Key)

**The assessment of DFS in this chapter is based on the service design for winter 2023–2024 and guidance published in December 2023 [1]. However, it is noted the NESO in its latest update have enabled service stacking with the Capacity Market (CM) and DSO services for winter 2024–2025

11.3 Stacking Assessment for DSO Services

- NESO services which contain strict exclusivity clauses [e.g. Demand Flexibility Service (DFS)] remain an issue regarding revenue jumping and splitting. However, these clauses now appear to be limited to a few services. (Note that for DFS, the NESO in its latest update have enabled service stacking with the Capacity Market (CM) and DSO services for winter 2024–2025).
- DSO services which have pre-scheduled utilisation or availability periods (i.e. known ahead of real time, preferably at least one hour before the start of a Settlement Period or even day-ahead) are also more readily jumpable. This is because, when jumping, it is important for FSPs to know when they will (or could) be utilised in order to reliably move between services. In the absence of any pre-defined utilisation periods or availability windows (i.e. the real time Operational Utilisation variant), an FSP risks being called on by the DSO when choosing to provide another service.
- ESO services which are 'evergreen'—meaning they are a permanent requirement for an FSP (e.g. MW Dispatch is written into an FSP's connection agreement)—raise additional considerations for FSPs. However, Primacy Rules that were developed in 2023 have helped to be able to stack the MW Dispatch service with DSO services. For MW Dispatch, Open Networks Primacy rules state the DSO flexible services hold priority over the NESO Transmission Constraint Management service (with MW Dispatch used as an example), and, therefore, FSPs are able capture revenue from both DSO and NESO services as a result via service jumping (but not splitting).
- Although jumping remains more readily available, there is still some interpretation regarding service rules and requirements in order for FSPs to confidently stack services. The NESO's Local Constraint Market service (LCM) is an example where the service is procured over a longer term (six-month agreements), but the service's ability for FSPs to declare availability/unavailability a day in advance or price themselves out the market aids jumping, while NESO documentation has clarified that providers can participate with other services as long as it's not in the same Settlement Period.
- Being able to jump requires clarity over availability or utilisation periods. NESO or DSO services where instruction is not known in advance (at least day-ahead) can create challenges in planning service delivery via jumping. The DSO Operational Utilisation Product, specifically variants with real time instructions, is an example where there is no clear indication of whether an FSP will know what periods they may be called upon. However, the impact will depend on how DSOs plan to use the service, with the service seen as similar to the previous Restore product which looks to restore or support the network following an unplanned fault.

11.3.2 Splitting of Services

Revenue splitting appears to be widely available for the new DSO services. However, this remains limited to FSPs providing services in the same direction, while stackability also varies between the DSO services as highlighted in the stacking assessment in this chapter. Our key findings regarding splitting include:

- DSO services which have pre-scheduled utilisation or availability are more readily able to split with NESO revenues. This is because by knowing availability, FSPs are able to participate in NESO auctions, many of which are day-ahead, with greater certainty over any spare volumes available. By knowing utilisation in advance, it further allows FSPs to comply with the rules of many NESO services which either require Physical Notifications to be submitted by Gate Closure (for BM Units) or an equivalent for non-BM units via another platform (e.g. the Ancillary Services Dispatch platform).
- Key to splitting DSO services with NESO services is whether or not each of the DSOs allows for 'over-delivery'. Many NESO services are based on real time utilisation, such as the BM, dynamic frequency response services, and the new reserve services. Therefore, if called upon by the NESO, an FSP is likely to overdeliver against any DSO service (assuming services are being provided in the same direction). This is because with the DSO services, FSPs often pre-agree to deliver a certain volume in a specific time period, and an NESO instruction will send them beyond the DSOs requirements. The DSOs clarified that over-delivery of DSO services is currently acceptable across all regions. However, this could change in the future should it result in adverse consequences due to much greater volumes of service delivery than planned.
- Real time instructions from the DSO are prohibitive to service splitting. DSO services with 2- or 15-min response times (i.e. real time instruction) are not typically compatible with splitting with NESO services. This is because NESO services (notably including the BM, dynamic frequency services, and new reserve services) often require physical notifications (PNs) at Gate Closure for BM Units or equivalent for non-BM Units for baselining purposes. Therefore, real time DSO instruction will impact the position of the FSP against the NESO's expectation. Furthermore, there are likely to be operational challenges in responding to both real-time instructions from both the NESO and DSO in the same time periods, with challenges against measuring delivery against either service.
- Opposite direction services are typically not splitable between NESO and DSO services. Although our stacking tables do not strictly distinguish between the direction of the services being provided (i.e. upward or downward generation or demand), this will be an important consideration for FSPs when determining which services to trade (in order to optimise revenue opportunities for their assets). In most circumstances, it is considered that an FSP would not be able to contract for a DSO service which required it to move in a certain direction, and at the same time contract for an NESO service potentially requiring it to move in the opposite. This would likely result in under-delivery against a DSO service if instructed by

the NESO. At present, none of the applicable services in this chapter were assessed as being splitable when providing services in opposing directions.
- Some NESO services continue to have strict exclusivity clauses about revenue splitting, examples include the DFS and LCM services. As discussed, change for DFS from winter 2024–2025 to open up the service to stacking with the CM and DSO services, has been confirmed.
- Service delivery requirements can also prevent service splitting. For example, if an asset is switched off for one service it cannot split volume to provide another. MW dispatch is an example of this, where participating assets are required turn-down generation to zero.

11.3.3 Co-delivery of Service

In the context of Flexibility Service provision, co-delivery can be thought of as the ability of an asset to earn revenue from the same unit of capacity in multiple revenue streams in both the same time period and direction. For example, participation in a DSO service and the wholesale market, or participation in a DSO service and the Capacity Market.

The "same time period and direction" in the description above elude to simplicity when discussing co-delivery. However, as noted earlier and in our previous work, the area rapidly becomes complex. Setting aside technical asset considerations, this is because:

- There is no acknowledgement that, as a high-level principle, co-delivery is an acceptable practice where it is technically possible. There remains good reason in many instances why co-delivering should not be acceptable, however, there are instances where co-delivery could be beneficial but service requirements or rules will need to adapt to accommodate this.
- Co-delivery is more nuanced than the "same time period and direction" caveat implies, because variations include the potential to earn revenues from different streams but in opposite directions, or be co-available for services but not necessarily utilised in either or both. These complex interactions can mean that, in practice, co-delivering is challenging to implement.
    1. Service and scheme guidance documents lack clarity on co-delivery (in part driven by the points above) and this leads FSPs to reach implicit conclusions about combinations of revenue streams.
    2. Where co-delivery is possible, the lack of explicit guidance leaves the question of an asset's starting point for revenue calculation open to interpretation.
- Special cases exist, such as the ability to earn revenue from the Capacity Market, while retaining the ability to participate in, and earn revenue from a DSO service (in the same time period and direction). However, DSO services are not Relevant Balancing Services for the purpose of the CM.

The combination of these factors means that in general, there are currently few situations where the ability to co-deliver is easily identifiable and can be undertaken with confidence by the FSP. This chapter focusses on the ability to jump between, or split capacity across revenue streams. The position between the Capacity Market and DSO services are discussed below. There is scope for a wider industry debate to reach a consensus on the principle of co-delivery and when it is acceptable. This would pave the way for individual scheme structures to explicitly accommodate the practice where principles allow. As a result, the better clarity should enable better decision making on asset deployment, leading to more efficient system operation and better value for money for consumers.

11.3.4 Other Key Findings for Service Stacking

There are several wider key findings that were identified including:

- Visibility of the ability to stack services still lacks clarity in many instances. There remains an onus on FSPs to review and interpret legal text or operational conflicts/misalignment between services. In the absence of clear guidance, this can make it difficult for FSPs to determine which services can be stacked when optimisation their assets. An example of where this is improving is stacking guidance for NESO services procured on the Enduring Auction Capability platform, and recent proposals from the NESO to allow stacking of DFS with the CM and DSO services with guidance on how they intend it to work.
- Stacking with DSO services is inherently locational. As discussed, understand DSOs are each procuring a slightly different range of the services, and are using different product variants depending on the needs of the network. This means that the stacking options available for FSPs are highly locational, depending on the needs of the network where assets are located and the product variants used by the DSO.
- Although certain services were assessed to be stackable (based on services designs and rules), in practice it may be the case that some services are rarely stacked. This can be because the specific DSO or NESO services are trying to achieve different outcomes or will benefit different technologies. An example of this is the DSO's Peak Reduction which appears to benefit demand reduction, whereas the NESO's MW dispatch is for generation turn down. Furthermore, some of the new DSO service descriptions indicate they will benefit FSPs which cannot respond quickly to market signals (notably Scheduled Utilisation and potentially Peak Reduction), whereas many NESO services require assets which can respond quickly, such as battery storage in frequency response services.

Furthermore, there remain several wider industry rules and service requirements which mean that explicit stacking capabilities are not clear or impact FSPs views on revenue stacking. These include:

11.3 Stacking Assessment for DSO Services

- DSO service volumes are not considered in Applicable Balancing Service Volume Data. This means that delivering a DSO service will result in the FSP (or balancing responsible party) being in imbalance if it is unable to trade its wholesale position to reflect activities undertaken.
- Grid Code and BSC requirements for BM Units mean that accurate FPNs must be submitted at gate closure (i.e. an hour before the start of each SP), and any deviations from this not as a result of an NESO instruction means an FSP is in breach of their BM obligations. This remains prohibitive for FSPs operating BMUs to participate in any DSO services with real time instructions, due to the fact that a real time DSO instruction would force the BMU to deviate from its FPN which would not be allowed.
- DSO services are not Relevant Balance Services (RBS) for the purpose of the Capacity Market. The RBS's allow delivery during a CM-relevant System Stress Event to be discounted from provider's obligation under the CM, ensuring the provider is not penalised under the CM for having not fully delivered (because it was providing another service). As a wide range of FSPs look seek to secure CM agreements, this may prove prohibitive to FSPs taking up the new DSO services should they believe there is a risk of CM non-delivery. It is noted this risk will usually be relatively low, while many DSO services that operate in the same direction as the CM may limit or mitigate the risk.
- Interaction with Enduring Auction Capability rules for the stacking of bids will likely be of interest to FSPs. FSPs will need to look at EAC rules, because it is a key marketplace for providers, and interpret them as best as possible to determine whether or not they can stack with DSO services. In the absence of direct guidance, FSPs may act conservatively and apply EAC stacking rules more broadly to DSO services (i.e. interpret as not being able to stack, or only move in certain directions).

In reality, many services are implicitly able to be stacked together in some form but require FSPs or aggregators to make commercial decisions about the services they wish to pursue to maximise profitability. Therefore, this places the onus on FSPs to identify any operational challenges in splitting or jumping of different services.

Note that stacking DSO services with other DSO services is not part of this assessment. In most instances, FSPs are not expected to commonly seek to stack DSO services together due to the highly locational aspect of DSO services, while DSOs are unlikely to procure two services in the same area. However, this could be a consideration in the future.

Capacity Market Interactions

The Capacity Market is a key revenue stream for many FSPs, giving participants the opportunity to secure a degree of revenue certainty as CM agreements are typically for one or 15 years. CM payments are made on a £/kW basis, paid monthly across the year. FSPs with agreement are required to deliver their 'derated' capacity (i.e. capacity under agreement) during a System Stress Event. CM units are also required to meet three Satisfactory Performance Days tests over the winter period, demonstrating the FSPs ability to meet its obligation over a single settlement period; FSPs choose these three SPDs.

With the CM typically being an important revenue stream for many FSPs, its interaction with the revised suite of DSO services is important. The design of the CM means it is generally possible to stack it with the majority of other revenue streams. However, this is typically limited to NESO services and the BM, due to CM participants' obligations being reduced in line with any requirements to deliver flexibility under a defined list of Relevant Balancing Services (RBS). This list is regularly reviewed to reflect the launch of new services.

However, these defined RBS are presently limited to NESO services, with DSO services currently excluded. While an FSP might choose to stack the CM with DSO service revenues, the provider could be exposed to CM penalties payments for underdelivery if called upon for two services at once. There is no obligation not to provide other services under the CM but providing a service that is not covered by Relevant Balancing Services could expose a CM provider to penalties should a CM Stress Event occur.

The risk of CM (or DSO) non-delivery will vary depending on the DSO service being provided. A DSO service that results in the FSP delivering on their CM obligation due to it being in the same direction (by increasing generation or reducing demand) will mean risk is minimised or even null; however, providing a DSO service to reduce generation or increase demand would more likely pose a risk. The risk would be greatest when System Stress Events are more likely (i.e. winter peaks). Interactions with CM baselining should also be considered by FSPs, notably for demand side response units. This is because delivering a DSO service (notably turning down demand) could impact the baseline that the NESO uses to assess an FSPs delivery against its CM obligation.

However, CM Stress Events are likely to be limited and asset owners might consider the risk of participating in certain other services to not impede its ability to deliver a CM obligation (e.g. if the service is in the same direction). However, this would not constitute service splitting or jumping, and fall more in line with co-delivery (or co-availability).

11.4 Individual DSO Service Assessments

This section explores stacking assessments in further detail, including individual service assessments for the new DSO services and reasons for their stackability with each of the wider revenue streams available to FSPs. Individual DSO service assessments. Each of the new Flexibility Services (or products) are assessed in detail, outlining what they are, key service parameters, and their stackability against wider system services and revenue streams available to FSPs.

11.4 Individual DSO Service Assessments

11.4.1 Peak Reduction

The Peak Reduction (PR) service is where an FSP contracts to reduce its electricity usage during a set period or periods (usually times of overall peak demand) regularly the term of the contract. This response can soften the high peaks in daily demand and prevent networks going beyond firm capacity limits.

- Derived from historical product "Sustain" with similar essential function and parameters.
- FSPs likely to benefit: End-users or storage assets with capacity to reduce usage during typical peak demand periods on a regular basis.
- Use case example: "This product could also be used to reduce a sites overall electricity consumption across the year, specifically during high peak periods".

Options for stacking

Peak Reduction, has been designed for "long-term energy efficiency activities", which may be carried out over a year to reduce electricity consumption. Based on the service definition, Peak Reduction is likely to be highly stackable with most NESO services. However, this will depend on how each DSO implements the service, when it is procured for, and the allowance for DSO service over-delivery.

As the service is designed to reduce long-term electricity consumption over peak periods, it is likely that expected delivery volumes will be known far in advance. However, this will depend on when the service volumes are agreed which is current set as 'at trade'. Further participation via splitting and revenue jumping in NESO services (which often procure at the day-ahead stage) should broadly be achievable by knowing the delivery requirements in advance. However, when splitting revenues any further instruction from the NESO will likely result in over-delivery of the Peak Reduction service. It is understood that over-delivery is currently acceptable with the DSOs, which should enable service splitting for many NESO services. However, rules of DSO service over-delivery should continue to be monitored by FSPs to ensure there is no breach of any service rules. It is noted that for many this has been assessed as 'implicitly stackable' because it requires interpretation from FSPs of the rules and requirements.

Provided utilisation is known in advance, FSPs should be able to trade in the wholesale market, and even capture additional revenue while avoiding imbalance, as well as participate in the BM—as FSPs should be able to submit accurate FPNs to the NESO.

Furthermore, Peak Reduction utilisation periods align to settlement periods, enabling short-term revenue jumping but this may depend on the number of settlement periods within a day for which the DSO requires the service (i.e. the shorter the procurement time, the more quickly FSPs can revenue jump).

The ability to split revenues when providing Peak Reduction is also likely to be limited to providing NESO services in the same direction (i.e. likely to be demand reduction)—as utilisation from the NESO in the opposite direction may counteract the service being offered to the DSO. Therefore, reserve services in the 'negative' direction, and high frequency response services, are likely to not be splitable with Peak Reduction.

Peak Reduction is also not a Relevant Balancing Service for the purpose of the Capacity Market, meaning FSPs will need to determine any risk of non-delivery in the CM when contracting for Peak Reduction. It is noted Peak Reduction is typically aimed at reducing demand, which would be the same direction as is required under the Capacity Market (i.e. demand turn down). While this could mean that, in theory, Peak Reduction and CM could be stacked well together, it is noted delivering a Peak Reduction contract may impact a demand response FSP's baseline with regards to the Capacity Market (and therefore ability to deliver). As Peak Reduction is a service required over peak periods on a regular basis over a long period of time, it is very possible that Peak Reduction would lower an FSPs ability to provide capacity into the CM and deliver. However, an FSP could limit this risk by only offering partial capacity into Peak Reduction and/or the CM—so that it can ensure delivery on its CM obligations at all times Table 11.6.

11.4.2 Scheduled Utilisation

Scheduled Utilisation (SU) delivers pre-agreed flexibility for specific period or periods on an ad hoc basis. This product will benefit FSPs that cannot respond in real-time or near to real-time (i.e. the day of delivery), alongside more flexible providers. The service sees service parameters agreed the day- or week-ahead of delivery. Scheduled Utilisation has two distinct versions, "Settlement Periods" where the delivery period begins and ends aligned with a market standard settlement period, or "Specific Periods" where delivery period can begin and end at any time.

- DSOs have indicated procurement of this service to manage a variety of needs, with network companies saying the product will be used to replace a variety of previously procured services (including historical dynamic, sustain and secure products).
- FSPs likely to benefit: FSPs that cannot respond in real-time or near to real-time, alongside more flexible providers.
- Use case example: "This service can be used by the Network Companies to manage seasonal peak demands and defer network reinforcement".

Options for stacking

Scheduled Utilisation, has been designed to manage seasonal peaks in demand or defer network reinforcement. However, each DSOs may use the product to manage different network needs.

11.4 Individual DSO Service Assessments

Table 11.6 Service stacking summary for Peak Reduction

Peak Reduction Product	Splitting Summary	Jumping summary	Notes
Wholesale market	I-S	E-S	Provided volumes are known in advance this should enable wholesale trading without the FSP or balancing responsible party being exposed to imbalance charges, and accurate FPNs to be submitted for BM. For splitting, it only appears viable for same direction services. DSO over-delivery is likely.
Balancing Mechanism	I-S	E-S	
NIV Chasing/imbalance	I-S	E-S	
Capacity Market	N/A	N/A	Peak Reduction is not a Relevant Balancing Service. However, the long-term nature of the product and fact it's for demand reduction means there is likely to be some compatibility with CM delivery, but with several key considerations
Demand Flexibility Service*	E-US	E-US	Services contain relatively strict exclusivity clauses or are not compatible with Peak Reduction (notably for revenue splitting). MW dispatch is for generation turn-down to zero, limiting splitting ability. LCM does not permit stacking with any other service in the same settlement period, while DFS rules state providers cannot be providing any other services including the CM and DSO services.
MW Dispatch Service	E-US	E-S	
Local Constraint Market	E-US	E-S	Service jumping more readily available, except for DFS due to the prohibition on providing other services. LCM markets allows FSPs to declare unavailability for any SPs it is not available a day in advance, while MW dispatch is subject to Primacy Rules giving priority to the DSO.
Short Term Operating Reserve	I-S	E-S	Committed STOR has strict service terms to be available for service windows, and PNs for purpose of baselining must be equal to or less than zero. Although there could be limited circumstance for splitting, instances for splitting may be rare and limited to demand reduction. STOR allows for service jumping.
Slow Reserve	I-S	E-S	Provided Peak Reduction volumes are known in advance this should enable participation in NESO response and reserve services on the Enduring Auction Platform. For splitting, it only appears viable for same direction services (i.e. positive versus negative reserve, and high and low reserve products). DSO service over-delivery is currently acceptable. There should not be many or any limitations on service jumping.
Quick Reserve	I-S	E-S	
Balancing Reserve	I-S	E-S	
Dynamic Containment	I-S	E-S	
Dynamic Moderation	I-S	E-S	
Dynamic Regulation	I-S	E-S	
Firm Frequency Response - Static only	I-S	E-S	Static FFR is procured day-ahead and Peak Reduction should be known in advance of this in order to plan for FFR participation. For splitting, FFR activities would need to be in same direction and would likely cause DSO over-delivery.

*The assessment of DFS in this chapter is based on the service design for winter 2023–2024 and guidance published in December 2023 [1]. However, it is noted the NESO in its latest update have enabled service stacking with the Capacity Market (CM) and DSO services for winter 2024–2025
Note (Refer to Table 11.3 for Key)

Based on the service definition, Scheduled Utilisation is likely to be highly stackable with most NESO services; however, this will depend on how each DSO implements the service, when it is procured for, and the allowance for DSO service over-delivery.

As the service is scheduled, it is likely that expected delivery volumes will be known in advance. The point at which volumes are known will depend on when the service volumes are agreed which is current set as 'at trade', where DSOs have indicated a range of approaches from day-ahead to month-ahead procurement. By knowing the delivery requirements in advance, further service participation via splitting and revenue jumping in NESO services (which often procure at the day-ahead stage) should broadly be achievable.

However, when splitting revenues any further instruction from the NESO will likely result in over-delivery of the Scheduled Utilisation service. It is understood that over-delivery is currently widely accepted across the DSOs, which should enable splitting with many NESO services. However, rules of DSO service over-delivery should continue to be monitored by FSPs to ensure there is no breach of any service rules. Therefore, for many service combinations an assessment of 'implicitly stackable' were given.

Provided utilisation is known in advance, FSPs should be able to trade in the wholesale market and even capture additional revenue while avoiding imbalance, and participate in the BM—as FSPs should be able to submit accurate FPNs to the NESO and comply with Grid Code or BSC requirements.

Furthermore, Scheduled Utilisation periods should align to settlement periods or EFA blocks (according to ENA's designs), enabling short-term revenue jumping, but this may depend on the number of settlement periods within a day for which the DSO requires the service.

The ability to split revenues when providing Scheduled Utilisation also appears to be limited to providing NESO services in the same direction—as utilisation from the NESO in the opposite direction may counteract the service being offered to the DSO. For example, if looking at demand reduction, reserve services in the 'negative' direction, and 'high' frequency response services, would not be splitable.

Some NESO services contain explicit rules or exclusivity clauses making stacking with any DSO service unviable. This is notably for DFS (where stacking rules are subject to change for winter 2024–2025), and LCM (which only restricts service splitting).

Scheduled Utilisation is also not a Relevant Balancing Service for the purpose of the Capacity Market, therefore FSPs will need to determine any risk of non-delivery in the CM when contracting for Scheduled Utilisation. The direction of the Scheduled Utilisation product will have an impact; if demand turn up or generation turn down is required, this is likely to risk CM delivery in a System Stress Event. Delivering a Scheduled Utilisation contract may also impact a demand response FSP's baseline with regards to the Capacity Market (and therefore ability to deliver). Therefore, it is possible that Scheduled Utilisation would lower an FSPs ability to provide capacity into the CM and deliver. However, an FSP could limit this risk by only offering partial capacity into Scheduled utilisation and/or the CM—so that it can ensure delivery on its CM obligations at all times Table 11.7.

11.4.3 Operational Utilisation

Operational Utilisation (OU) offers flexibility in real time (same day) or week-ahead. This product allows for a DSO to agree on trade parameters (price and period of provision) ahead of time through monthly, weekly, or day-ahead tenders but with instructions issued either in real time, with a 2- or 15-min response notice, or week-ahead response time. The assets can be dispatched, as per instruction, for the required

11.4 Individual DSO Service Assessments

Table 11.7 Stacking summary for Scheduled Utilisation

Scheduled Utilisation Product	Splitting Summary	Jumping summary	Notes
Wholesale market	I-S	E-S	Provided volumes are known in advance this should enable wholesale trading without the FSP or balancing responsible party being exposed to imbalance charges, and accurate FPNs to be submitted for BM. For splitting, it only appears viable for same direction services. DSO over-delivery is likely.
Balancing Mechanism	I-S	E-S	
NIV Chasing/imbalance	I-S	E-S	
Capacity Market	N/A	N/A	Scheduled Utilisation is not a Relevant Balancing Service. In instances where Scheduled Utilisation is in same direction as CM obligation, then it's unlikely to impede CM delivery. However, there are several considerations FSPs including impact on CM baseline.
Demand Flexibility Service*	E-US	E-US	Services contain relatively strict exclusivity clauses or are not compatible with Scheduled Utilisation (notably for revenue splitting). MW dispatch is for generation turn-down to zero, limiting splitting ability. LCM does not permit stacking with any other service in the same settlement period, while DFS rules state providers cannot be providing any other services including the CM and DSO services.
MW Dispatch Service	E-US	E-S	
Local Constraint Market	E-US	E-S	Service jumping more readily available, except for DFS due to the prohibition on providing other services. LCM markets allows FSPs to declare unavailability for any SPs it is not available a day in advance, while MW dispatch is subject to Primacy Rules giving priority to the DSO.
Short Term Operating Reserve	I-US	E-S	Committed STOR has strict service terms to be available for service windows, and PNs for purpose of baselining must be equal to or less than zero. Although there could be limited circumstance for splitting, instances for splitting may be rare and limited to demand reduction. STOR allows for service jumping.
Slow Reserve	I-S	E-S	Provided Scheduled Utilisation volumes are known in advance this should enable participation in NESO response and reserve services on the Enduring Auction Platform. For splitting, it only appears viable for same direction services (i.e. positive versus negative reserve, and high and low reserve products). DSO service over-delivery is currently acceptable. There should not be many or any limitations on service jumping.
Quick Reserve	I-S	E-S	
Balancing Reserve	I-S	E-S	
Dynamic Containment	I-S	E-S	
Dynamic Moderation	I-S	E-S	
Dynamic Regulation	I-S	E-S	
Firm Frequency Response - Static only	I-S	E-S	Static FFR is procured day-ahead and Scheduled Utilisation should be known in advance of this in order to plan for FFR participation. For splitting, FFR activities would need to be in same direction and would likely cause DSO over-delivery.

*The assessment of DFS in this chapter is based on the service design for winter 2023–2024 and guidance published in December 2023 [1]. However, it is noted the NESO in its latest update have enabled service stacking with the Capacity Market (CM) and DSO services for winter 2024–2025
Note (Refer to Table 11.3 for Key)

level of service based upon real-time network measurement data. Therefore, a DSO does not pay for flexibility it does not need (or an FSP does not provide an unnecessary reduction), compared to Scheduled Utilisation where reduction volume is set at the point of trade.

- Derived from historical product "Restore" with same essential function and parameters.
- FSPs likely to benefit: FSPs able to respond quickly (for 2 and 15 min product variants). Week-ahead service may benefit FSPs less able to respond quickly.

- Use case example: "A DSO may utilise this product in order to restore network supplies following an unplanned outage/fault where the regulatory funding does not allow for availability payments e.g. customer interruptions".

Options for stacking

Operational Utilisation has been designed for use cases where flexibility delivered is agreed close to real time, with utilisation instructions either 2, 15 min, or a week ahead of delivery. Based on ENA's defined parameters (i.e. no availability windows), it appears that FSPs providing the service will not have much foresight of when they might be called upon (the exception being week ahead response).

Service variants which require real time utilisation instructions (2- or 15-min response) appear more challenging to stack for several reasons. Notably, any FSP which is BM participating will not be able to provide Operation Utilisation with 2- or 15-min response due to the requirement to submit accurate FNPs at Gate Closure. Deviating from this FPN for the purpose of providing a DSO service will result in the FSP breaching its BM obligations. This barrier is similar to many other NESO services for both BM and non-BM units, with many using a baseline of the FPN (or equivalent) as part of the service. Real time response will also create technical delivery challenges; responding to real time signals from both the NESO and DSO would likely provide challenging and not be possible and make measuring performance against each of the services difficult.

Real-time utilisation also causes challenges for FSPs to trade on the wholesale market to match their delivered volumes (wholesale trades can be made up to the start of a settlement period), meaning the FSP, or balancing responsible party, may be exposed to imbalance charges for any instructed volumes. This is because DSO services are not included in Applicable Balancing Services Volume Data.

However, for the week-ahead variant, the Operational Utilisation is slightly more readily stackable. This is because FSPs would be able to submit accurate FPNs for the BM or baseline positions to NESO services. However, there are reasons why week-ahead Operational Utilisation may still be challenging to stack with NESO services, such as the 'minutes' utilisation period potentially not aligning to NESO delivery periods (i.e. settlement periods or EFA blocks) and making jumping less efficient, and that the nature of the service being to manage network faults or restore the network. And similar to previously discussed services, delivery of NESO service would need to be in the same direction as the DSO service and could result in DSO over-delivery, which is widely acceptable across DSOs at present. The ability to stack will therefore depend on what product variant each of the DSOs implements.

Regarding revenue jumping, the week-ahead service should be readily jumpable between most services. However, for real-time variants it is unclear how jumpable the service is. This is because under the service descriptions FSPs may not know when they may be called on to deliver Operational Utilisation, in the absence of advanced availability windows or pre-determined utilisation periods. This may create risk when entering into NESO services, particularly if the DSO agreements are over any extended period.

It is also noted that Operational Utilisation may be used to restore network supplies following an unplanned outage or fault (i.e. akin to the previous Restore product), in which case the FSP may not be able to provide service to the NESO due to the conditions on the distribution network. Many NESO service terms have provisions for unavailability; for example, Dynamic Response service terms have provisions for 'unplanned outage or other unforeseen technical circumstances'. In this instance, an FSP may be able to provide Operational Utilisation to restore the network, but this will come down to individual service terms and whether the NESO allows for service unavailability for an unplanned network outage while supporting the DSO in restoring supplies. Delivering OU under these circumstances would constitute revenue jumping rather than splitting as the NESO service would not be delivered or paid for.

Furthermore, regarding MW dispatch, established Primacy rules for the service exclude the historic Restore product. Therefore, should the Operational Utilisation product be used to replace Restore as some DSOs have specified, stacking with MW Dispatch is likely to be limited, particularly for real time product variants.

Operational Utilisation is also not a Relevant Balancing Service for the purpose of the Capacity Market, therefore FSPs will need to determine any risk of non-delivery in the CM when contracting for Operational Utilisation Table 11.8.

11.4.4 Scheduled Availability + Operational Utilisation

This service is uses operational utilisation but with the added feature of procuring, ahead of time, the assured availability of an FSP to deliver operational utilisation for pre-defined periods if required. This availability will be defined at the point of procurement and cannot be modified once the contract has been agreed. The FSP will then be dispatched for the required level of service at either the day-ahead stage or with 2-min notice. This service helps ensure that a DSO only pays utilisation rates for flexibility procured for the need of the network, presenting overall cost savings even when paying for availability. It equally allows for FSPs to receive availability payment and then engage with another service if that capacity isn't needed by the DSO in the contracted period(s).

- Derived from historical product "Dynamic" with same essential function and parameters.
- FSPs likely to benefit: A wide range of FSPs may benefit; faster response product variants will benefit more flexible FSPs, while the day-ahead variant will also benefit slower responding assets.
- Use case example: "An example use case for this product is when a DSO is planning for sufficiency of flexible services contracts based upon short-medium range forecasting of network constraints."

Table 11.8 Stacking summary for Operational Utilisation

Operational Utilisation Product	Splitting Summary		Jumping summary		Notes
	2 min & 15 min	Week ahead	2 min & 15 min	Week ahead	
Wholesale market	I-US	I-S	I-S	E-S	Real-time instructions mean accurate wholesale market optimisation is challenging, but week-ahead service more readily stackable. Jumping is more achievable, but real-time utilisation without pre-agreed availability creates challenges in trading wholesale markets.
Balancing Mechanism	E-US	I-S	I-US	E-S	Close to real time instructions would result in deviating from FPN, prohibiting splitting. Week-ahead variant more likely to comply with BM requirements, but actions need to be in same direction and result in DSO over-delivery.
NIV Chasing/imbalance	I-US	I-S	I-S	E-S	Active NIV chasing is challenging for real-time utilisation, although providing service will put FSP (or balancing responsible party) into imbalance anyway as service is not in ABSVD.
Capacity Market	N/A	N/A	N/A	N/A	Operational Utilisation is not a Relevant Balancing Service. In instances where Operational Utilisation is in same direction as CM obligation, then co-delivery may be possible. but there will be carefully considered for FSPs.
Demand Flexibility Service*	E-US	E-US	E-US	E-US	MW dispatch is for generation turn-down to zero, limiting splitting ability. LCM does not permit stacking with any other service in the same settlement period, while DFS rules state providers cannot be providing any other services including the CM and DSO services. Real time Operational Utilisation also creates additional challenges as it may impede NESOs service delivery. Service jumping is not currently available for DFS due to the prohibition on providing other services. Jumping may not be viable for real time MW dispatch due to Primacy rules excluding the 'Restore' product. LCM markets allows FSPs to declare unavailability for any SPs it is not available a day in advance.
MW Dispatch Service	E-US	E-US	I-US	I-US	
Local Constraint Market	E-US	E-US	I-US	E-S	
Short Term Operating Reserve	E-US	I-US	I-US	E-S	Committed STOR has strict service availability requirements, and PN's for purpose of baselining must be equal to or less than zero. There could be limited circumstance for splitting week-ahead DSO product, but real-time instructions risk STOR delivery. STOR allows for service jumping, but unknown DSO real-time delivery requirement could create challenges.
Slow Reserve	E-US	I-S	I-US	E-S	Real time utilisation of DSO service would impact 60-min/FPN baselining requirement for NESO services, prohibiting splitting. For 'week-ahead' DSO service, this should theoretically enable splitting, however, this may depend on the precise service implementation by each DSO, such as aligning service periods. For splitting, it only appears to be viable for same direction services. DSO service over-delivery is currently acceptable. There should not be many limitations on service jumping for week-ahead product, but real time variant could create risks if FSPs don't know then they will be used.
Quick Reserve	E-US	I-S	I-US	E-S	
Balancing Reserve	E-US	I-S	I-US	E-S	
Dynamic Containment	E-US	I-S	I-US	E-S	
Dynamic Moderation	E-US	I-S	I-US	E-S	
Dynamic Regulation	E-US	I-S	I-US	E-S	
Firm Frequency Response - Static only	E-US	I-S	I-US	E-S	Real time DSO instruction inhibits FFR participation in same time period with numerous risks and challenges. DSO real-time delivery requirement could also create challenges for jumping should the FSP not have visibility of when the DSO might instruct them.

*The assessment of DFS in this chapter is based on the service design for winter 2023–2024 and guidance published in December 2023 [1]. However, it is noted the NESO in its latest update have enabled service stacking with the Capacity Market (CM) and DSO services for winter 2024–2025
Note (Refer to Table 11.3 for Key)

11.4 Individual DSO Service Assessments

Options for stacking

Our stacking assessment for Scheduled Availability + Operational Utilisation (SA + OU) shares many of the same stacking characteristics as both Schedules Utilisation and Operational Utilisation. SA + OU has been designed so that DSOs can procure flexibility ahead of time but only call on FSPs based on actual needs of the networks.

The variant which has a real time (2 min) instruction, shares many stacking features of operational utilisation. Any FSP which is BM participating will not be able to provide SA + OU with 2-min response due to the requirement to submit accurate FNPs at Gate Closure. Deviating from this FPN for the purpose of providing a DSO service will result in the FSP breaching its BM obligations. This barrier is similar to many other NESO services for both BM and non-BM units, with many using a baseline of the FPN (or equivalent) as part of the service. Real time response will also create technical delivery challenges; responding to real time signals from both the NESO and DSO would likely provide challenging and not be possible and make measuring performance against each of the services difficult.

Real-time utilisation also causes challenges for FSPs to trade on the wholesale market to match their delivered volumes (wholesale trades can be made up to the start of a settlement period), meaning the FSP, or balancing responsible party, may incur imbalance charges for any instructed volumes. This is because DSO services are not included in Applicable Balancing Services Volume Data.

However, for the day-ahead variant, the SA + OU is more readily stackable. This is because FSPs would be able to submit accurate FPNs for the BM or baseline positions to NESO services. However, there are reasons why day-ahead SA + OU may still be challenging to stack with NESO services, such as the 'minutes' utilisation period potentially not aligning to NESO delivery periods (i.e. settlement periods or EFA blocks) and making jumping less efficient, and day-ahead utilisation instructions will need to time well with the NESO's day-ahead auctions so FSPs can plan their activities. Similar to previously discussed services, delivery of NESO service would need to be in the same direction as the DSO service and could result in DSO over-delivery. It is understood that over-delivery is currently widely accepted across the DSOs, which should enable service splitting for many NESO services. However, rules of DSO service over-delivery should continue to be monitored by FSPs to ensure there is no breach of any service rules. The ability to stack may therefore depend on how each DSO implements the service.

Furthermore, for the day-ahead variant, it equally enables FSPs to receive availability payments and then engage with another service if that capacity isn't needed by the DSO in the contracted period(s). This is a positive aspect of the service that may be attractive to FSPs.

Regarding revenue jumping, the day-ahead service should be readily jumpable between most services. Furthermore, the real-time variant should also be readily jumpable across most NESO services. This is because the inclusion of advanced availability windows (which are across settlement periods) means that FSPs can plan their delivery in NESO markets with foresight of DSO requirements. This does, however, depend on when the utilisation is instructed, where knowing utilisation before NESO day-ahead auctions will be beneficial.

Scheduled Availability + Operational Utilisation is also not a Relevant Balancing Service for the purpose of the Capacity Market, therefore FSPs will need to determine any risk of non-delivery in the CM when contracting for SA + OU. The direction of the SA + OU product will have an impact; if demand turn up or generation turn down is required, this is likely to risk CM delivery in a System Stress Event. Delivering a SA + OU contract may also impact a demand response FSP's baseline with regards to the Capacity Market (and therefore ability to deliver). Therefore, it is possible that SA + OU would lower an FSPs ability to provide capacity into the CM and deliver. However, an FSP could limit this risk by only offering partial capacity into SA + OU and/or the CM—so that it can always ensure delivery on its CM obligations. Table 11.9

11.4.5 Variable Availability + Operational Utilisation

This service builds further on Operational Utilisation by allowing DSO to procure a greater level of contracted available capacity and then refine the requirement closer to the event. Availability requirement can be refined by a DSO month or a week ahead of delivery.

This allows DSOs to further maintain flexible redundancy in case of unexpected faults or planned but disruptive activity further in advance, without needing to pay higher availability or utilisation payment rates, and for FSPs to engage with flexible services by opting in to a long-term availability service and receive the associate payments.

- Derived from historical product "Secure" with the same essential function and parameters.
- FSPs likely to benefit: A wide range of FSPs may benefit; faster response product variants will benefit more flexible FSPs, while the day-ahead and week-ahead variants will also benefit slower responding assets.
- Use case example: "An example use case for this product is when a DSO is planning for sufficiency of flexible services contracts based upon long range forecasting of network constraints."

Options for stacking

Our stacking assessment for VA + OU is very similar to Scheduled Availability + Operational Utilisation. The key difference is that VA + OU has greater service flexibility in that availability can be refined close to the time of delivery (although only refined by the DSO and not by the FSP).

The product variants which have a real time (2 or 15 min) instruction are more challenging to stack in terms of revenue splitting. Any FSP which is BM participating will not be able to provide VA + OU with real time response due to the requirement

11.4 Individual DSO Service Assessments

Table 11.9 Stacking summary for Scheduled Availability + Operational Utilisation

Scheduled Availability + Operational Utilisation Product	Splitting summary		Jumping summary		Notes
	2 min	DA	2 min	DA	
Wholesale market	I-US	I-S	E-S	E-S	Real-time instructions (2min) mean accurate wholesale market optimisation is challenging, but day-ahead more readily splitable. Jumping is more stackable particularly with availability known in advance.
Balancing Mechanism	E-US	I-S	E-S	E-S	Close to real time instructions would result in deviating from FPN, prohibiting splitting. Day-ahead variant more likely to comply with BM requirements, but actions need to be in same direction and result in DSO over-delivery.
NIV Chasing/imbalance	I-US	I-S	E-S	E-S	Active NIV chasing is challenging in same SPs that real-time DSO utilisation occurs, although DSO utilisation will put FSP (or balancing responsible party) into imbalance anyway as service is not in ABSVD. Scope to split with day-ahead service
Capacity Market	N/A	N/A	N/A	N/A	SA+OU is not a Relevant Balancing Service. In instances where SA+OU is in same direction as CM obligation, then it's unlikely to impede CM delivery. However, this will need to be carefully considered by FSPs including impact on CM baseline.
Demand Flexibility Service*	E-US	E-US	E-US	E-US	MW dispatch is for generation turn-down to zero, limiting splitting ability. LCM does not permit stacking with any other service in the same settlement period, while DFS rules state providers cannot be providing any other services including the CM and DSO services. Real time SA+OU creates additional challenges as it may impede NESO requirements.
MW Dispatch Service	E-US	E-US	E-S	E-S	
Local Constraint Market	E-US	E-US	E-S	E-S	Service jumping more readily available, except for DFS due to the prohibition on other services. LCM markets allows FSPs to declare unavailability for any SPs it is not available a day in advance, while MW dispatch is subject to Primacy Rules giving priority to the DSO.
STOR	E-US	I-US	E-S	E-S	Committed STOR has strict service availability requirements, and PN's for purpose of baselining must be equal to or less than zero. There could be limited circumstance for splitting day-ahead DSO product, but real-time variant risks STOR delivery. STOR allows for service jumping.
Slow Reserve	E-US	I-S	E-S	E-S	Real time utilisation of DSO service would impact 60-min/FPN baselining requirement for NESO services, prohibiting splitting. For 'day-ahead' DSO service, this should theoretically enable splitting, however, this may depend on the precise service implementation, such as aligning delivery periods and time of day-ahead instructions. For splitting, it only appears to be viable for same direction services. DSO service over-delivery is likely which is currently acceptable with the DSOs. There should not be many limitations on service jumping.
Quick Reserve	E-US	I-S	E-S	E-S	
Balancing Reserve	E-US	I-S	E-S	E-S	
Dynamic Containment	E-US	I-S	E-S	E-S	
Dynamic Moderation	E-US	I-S	E-S	E-S	
Dynamic Regulation	E-US	I-S	E-S	E-S	
FFR – Static only	E-US	I-S	E-S	E-S	Real time DSO instruction inhibits FFR participation in same time period with numerous risks and challenges. Where availability and utilisation is known further in advance, splitting is more achievable but may depend on precise DSO requirements (e.g. utilisation/availability length, and time of day-ahead instructions).

*The assessment of DFS in this chapter is based on the service design for winter 2023–2024 and guidance published in December 2023 [1]. However, it is noted the NESO in its latest update have enabled service stacking with the Capacity Market (CM) and DSO services for winter 2024–2025
Note (Refer to Table 11.3 for Key)

to submit accurate FNPs at Gate Closure. This barrier is similar to many other NESO services for both BM and non-BM units, with many using a baseline of the FPN (or equivalent) as part of the service. Real time response will also create technical delivery challenges; responding to real time signals from both the NESO and DSO would likely provide challenging and not be possible and make measuring performance against each of the services difficult.

Real-time utilisation also causes challenges for FSPs to trade on the wholesale market to match their delivered volumes (wholesale trades can be made up to the start of a settlement period), meaning the FSP (or balancing responsible party) will be exposed to imbalance charges for any instructed volumes. This is because DSO services are not included in Applicable Balancing Services Volume Data.

However, for the day-ahead and week-ahead variants, the VA + OU is more readily stackable. This is because FSPs would be able to submit accurate FPNs for the BM or baseline positions to NESO services. However, there are reasons why day-ahead VA + OU may still be challenging to stack with NESO services, such as the 'minutes' utilisation period potentially not aligning to NESO delivery periods (i.e. settlement periods or EFA blocks) and making jumping less efficient, and day-ahead utilisation instructions will need to time well with the NESO's day-ahead auctions so FSPs can plan their activities. Similar to previously discussed services, delivery of NESO service would need to be in the same direction as the DSO service and could result in DSO over-delivery. Over-delivery is currently widely accepted across the DSOs, which should enable service splitting for many NESO services. However, rules of DSO service over-delivery should continue to be monitored by FSPs to ensure there is no breach of any service rules. The ability to stack may therefore depend on how each DSO implements the service.

Furthermore, for the day-ahead and week-ahead variants, it equally enables FSPs to receive availability payments and then engage with another service if that capacity isn't needed by the DSO in the contracted period(s). This is a positive aspect of the service that may be attractive to FSPs.

Regarding revenue jumping, the day-ahead and week-ahead service should be readily jumpable between most services. Furthermore, real-time variants should also be readily jumpable across most NESO services. This is because the inclusion of advanced availability windows (which are across settlement periods) means that FSPs can plan their delivery in NESO markets with foresight of DSO requirements. This does, however, depend on when utilisation is instructed, where knowing utilisation before NESO day-ahead auctions will be beneficial.

It is noted that the variable availability aspect of the product should support FSPs to maximise their activities in other markets, as when they are likely not needed for DSO services, they can use the spare capacity to provide other services.

Variable Availability + Operational Utilisation is also not a Relevant Balancing Service for the purpose of the Capacity Market, therefore FSPs will need to determine any risk of non-delivery in the CM when contracting for Scheduled Utilisation. As with SA + OU, the direction of the VA + OU product will have an impact; if demand turn up or generation turn down is required, this is likely to risk CM delivery in a System Stress Event. Delivering a VA + OU contract may also impact a demand

11.5 Stackability Consideration for New Services (Stackable by Design)

response FSP's baseline with regards to the Capacity Market (and therefore ability to deliver). Therefore, it is possible that VA + OU would lower an FSPs ability to provide capacity into the CM and deliver. However, an FSP could limit this risk by only offering partial capacity into VA + OU and/or the CM—so that it can always ensure delivery on its CM obligations Table 11.10.

11.5 Stackability Consideration for New Services (Stackable by Design)

The NESO and DSOs believe there are significant whole-system benefits to revenue stacking and that any barriers must be challenged during the Product design phase. Changing existing services, sometimes with existing contracts, can be significantly harder than embedding the right principles from the beginning.

By default, all active power services should be readily 'jumpable' and 'splittable' (in the same direction) across NESO, DSO and wholesale electricity markets. Any deviation from this principle should be reviewed the Market Facilitator,[1] to ensure there is a legitimate purpose to the restriction (including minimising system costs) and that purpose could not be achieved in a less restrictive way.

In particular—the following features, identified as common barriers to stacking, should be avoided wherever practical. This checklist should be updated as required by the Market Facilitator and embedded within the design and approval processes for new NESO and DSO services.

1. Services launched without explicit guidance on stacking. All services should propose and make updates to the central stackability register and service technical requirements register, providing clear examples of what is and is not permissible.
2. Use of Exclusivity Clauses. Services should not by default restrict the activities of the Flexibility Provider in delivering other services, unless these can be shown to materially compromise their ability to deliver.
3. Carve outs for particular services which cannot be easily updated. Where possible services should allow stackability with all services, or a list of services that can be readily updated by the Market Facilitator. In particular, reference to 'Relevant Balancing Services', beyond its current use for Capacity Market, should be avoided.
4. Use of historic baselines without a practical way of adjusting for a wide range of service events. Where baselines are calculated based on a provider's recent behaviour, there should be mechanisms in place to discount or adjust observed behaviour during utilisation events for other services.
5. DSO dispatch after Balancing Mechanism gate-closure. Late dispatch services prevent providers making accurate submissions to the Balancing Mechanism. The principle should be for DSOs to handover responsibility for balancing to NESO

---

[1] Conditional on the detailed design for the Market Facilitator role.

Table 11.10 Stacking summary for variable availability + operational utilisation

Variable Availability + Operational Utilisation Product	Splitting Summary		Jumping summary		Notes
	2 min & 15 min	DA & WA	2 min & 15 min	DA & WA	
Wholesale market	I-US	I-S	E-S	E-S	Real-time instructions (2-min) mean accurate wholesale market optimisation is challenging, but day-ahead more readily splitable. Jumping is more stackable particularly with availability known in advance.
Balancing Mechanism	E-US	I-S	E-S	E-S	Close to real time instructions would result in deviating from FPN, prohibiting splitting. Day-ahead and week-ahead variants more likely to comply with BM requirements, but actions need to be in same direction and could result in DSO over-delivery.
NIV Chasing/imbalance	I-US	I-S	E-S	E-S	Active NIV chasing challenging in same SPs that real-time DSO utilisation occurs, although this will put FSP into imbalance anyway as service is not in ABSVD. Scope to split with day-ahead and week-ahead service.
Capacity Market	N/A	N/A	N/A	N/A	VA+OU is not a Relevant Balancing Service. In instances where VA+OU is in same direction as CM obligation, then it's unlikely to impede CM delivery. However, this will need to be carefully considered by FSPs including baseline impacts.
Demand Flexibility Service*	E-US	E-US	E-US	E-US	MW dispatch is for generation turn-down to zero, limiting splitting ability. LCM does not permit stacking with any other service in the same settlement period, while DFS rules state providers cannot be providing any other services including the CM and DSO services. Real time VA+OU creates additional challenges as it may impede NESO requirements. Service jumping more readily available, except for DFS due to the prohibition on other services. LCM markets allows FSPs to declare unavailability for any SPs it is not available a day in advance, while MW dispatch is subject to Primacy Rules giving priority to the DSO.
MW Dispatch Service	E-US	E-US	E-S	E-S	
Local Constraint Market	E-US	E-US	E-S	E-S	
STOR	E-US	I-US	E-S	E-S	Committed STOR has strict service availability requirements, and PN's for purpose of baselining must be equal to or less than zero. There could be limited circumstance for splitting day-ahead/week-ahead DSO product, but real-time variant risks STOR delivery. STOR allows for service jumping.
Slow Reserve	E-US	I-S	E-S	E-S	Real time utilisation of DSO service would impact 60-min/FPN baselining requirement for NESO services, prohibiting splitting. For day-ahead/week-ahead DSO service, this should theoretically enable splitting. However, this may depend on the precise service implementation, such as aligning delivery periods and time of day-ahead instructions. For splitting, only viable for same direction services. DSO service over-delivery is likely which is currently acceptable with the DSOs. There should not be many limitations on service jumping.
Quick Reserve	E-US	I-S	E-S	E-S	
Balancing Reserve	E-US	I-S	E-S	E-S	
Dynamic Containment	E-US	I-S	E-S	E-S	
Dynamic Moderation	E-US	I-S	E-S	E-S	
Dynamic Regulation	E-US	I-S	E-S	E-S	
FFR – Static only	E-US	I-S	E-S	E-S	Real time DSO instruction inhibits FFR participation in same time period with numerous risks and challenges. Where availability/utilisation is known further in advance, splitting is more achievable but may depend on precise DSO requirements (e.g. availability/utilisation length, and time of day-ahead instructions).

*The assessment of DFS in this chapter is based on the service design for winter 2023–2024 and guidance published in December 2023 [1]. However, it is noted the NESO in its latest update have enabled service stacking with the Capacity Market (CM) and DSO services for winter 2024–2025
Note (Refer to Table 11.3 for Key)

(or the designated operator in islanded operating conditions) at BM gate closure, making minimal changes to physical positions after this point. Where requirements could reasonably be foreseen, Distribution System Operators should provide advance notice of dispatch.
6. Long availability windows with no incentive or mechanism to release capacity. Long availability windows can sterilise flexible capacity. System Operators should take reasonable care to optimise the length of availability windows and incentivise themselves to release assets for other services if they are no longer required.
7. Disproportionate penalties for over-delivery. Penalising 'over-delivery' of services can limit opportunities for providers to split asset capacity between multiple services in the same direction. System Operators should take proportionate steps to reflect any cost of over-delivery. Such costs could include, for example, requiring additional compensating actions to restore energy balance.

11.6 Baselining Considerations When Service Stacking

For stacking, understanding the counterfactual that is being used by a DSO or the NESO to measure flexibility is crucial to an FSP effectively managing its obligations when providing flexible services. The need for different methodologies is due to both the needs of the system operator procuring the service and the different characteristics of the FSPs providing the flexibility. Determining an appropriate baseline is an important but sometimes complex task; it can impact the ability for assets to stack services.

The combination of baselining methodologies being used across two services being stacked is an important consideration, as it determines that amount which the FSP is being paid for each service. This in turn impacts whether or not an asset is deemed to be splitting or co-delivering services, as well as impact whether or not an asset is deemed to be over-delivering on a service. However, it is noted that while baselining is an important consideration, wider service requirements (as per service terms) remain paramount as to whether or not two services can be stacked.

The next section lists the key baselining approaches used across DSO and NESO services, and some of the key considerations that need to be thought about when service stacking.

11.6.1 Baselining Methodologies and Considerations for Service Stacking

Baselining method: Historical baselines.

Description and current uses: Based on historical averages. Used for DFS, Capacity Market (DSR only), DSO services.

Key considerations for service stacking

- Historic baselines essentially lock the FSP into the level from which it will be measured for payments based on previous behaviour, unless the historic baseline contains an adjustment mechanism to reflect the provision of other services.
- Key to stacking is whether the system operator (i.e. DSO or NESO) is happy that if the FSP deviates from this historic baseline on the day to provide another service, provided it doesn't impede the asset's ability to deliver the service(s).
- A key consideration for the DSO (or NESO) in this method is whether it is looking for a change in output/demand compared the current level (i.e. manage real time network conditions), or if it is looking from a change in output/demand compared to 'normal' or 'historic' levels (i.e. potentially to manage medium/longer-term network needs).
- The historic baseline does not typically impact the ability for an FSP to provide another service by itself, but instead will impact what the FSP is paid for. The ability to stack a DSO service with NESO services is more impacted here by whether or not the utilisation of the asset is pre-scheduled (allowing visibility to enter into NESO Flexibility Services) or required to dispatch in real-time (less compatible with NESO services). If an FSP has committed to a DSO service and chooses to stack with another NESO service, it could result in the FSP over delivering when using a historic baseline (assuming it is not adjusted for provision of another service). Alternatively, it could result in co-delivery (double payment, which has several wider challenges and considerations), depending on the baseline methodology used by the NESO.
- A further consideration is the impact that providing other services will have on the FSP's baseline over the period for which the historic baseline is set. If an FSP is regularly providing other services, it can impact its historic baseline and ability to meet the service requirements. Historical baselines may use 'adjustments' which can reflect activities outside an asset's normal operating behaviour (e.g. providing other Flexibility Services). This is used in some instances across industry (e.g. changes to the DFS service suggests excluding 'event days' from its historic baseline, while DSO services have a concept of 'non-active' days.
- Other benefits of historic baseline include that it's accessible for a wide range of FSPs, such as demand response units and small-scale FSPs without sophisticated (e.g. real-time) metering requirements.

Baselining method: Zero baselines.

Description and current uses: Requires a starting position of zero. STOR (for generation), DSO services, Capacity Market (for generation).

Key considerations for service stacking:

- Zero baselines have several important considerations for service stacking, but similar to historic baselines this does not impact an asset's ability to stack services alone but instead depends on wider service requirements. The baseline, however,

11.6 Baselining Considerations When Service Stacking

will impact what the FSP is paid for and whether or not it might be 'splitting' or 'co-delivering' services (noting co-delivery has wider challenges).
- A zero baseline can inhibit service stacking, particularly if it is enforced. An example of this is the NESO's STOR service, which requires generators to have a zero baseline (demonstrated via Physical Notifications). This can prevent FSPs from being able to deliver another service as it is only allowed to deviate from zero at the instruction of the NESO.
- However, zero baselines could support stacking if the system operator (DSO or NESO) is OK with the FSP deviating from the baseline for provision of another service, or if the service(s) has pre-scheduled utilisation. In this instance, a zero baseline may result in co-delivery (being paid for two services) and/or over-deliver of one service. However, this will depend on the baselining method of both service(s) being provided and whether the FSP is attempting to co-deliver (noting co-delivery has some wider considerations and challenges).
- Other considerations include that zero baselines are typically easy to implement/monitor and can benefit non-dispatchable assets. However, it is often not suitable for some technology types depending on the use case (e.g. DSR will typically be consuming and not at zero), while it may not be suitable for services with multiple directions (i.e. only suitable for generation turn-up or demand turn up).

Baselining method: Final Physical Notification or equivalent.

Description and current uses: Operational baseline declared 1-h before start of delivery period. BM, dynamic response services, Balancing Reserve, Slow Reserve, Quick Reserve.

Key considerations for service stacking:

- Many NESO services use Final Physical Notifications (FPNs) for baselining purposes. This requires assets to submit their Operational Baseline at Gate Closure (i.e. 1-h before the start of each Settlement Period). For registered BMUs, this is done via the normal BM data submissions, whereas for non-BMUs an equivalent is submitted via another platform [e.g. the Ancillary Services Dispatch platform (ASDP)].
- The common use of FPNs for baselining purposes has a significant impact on the stacking of services. While there are lots of reasons and benefits for using FPNs, they effectively lock an asset into position just ahead of real time. This means the asset can only deviate from this position due to an NESO instruction, but deviating from this position due to a DSO instruction is prohibited. FPNs therefore hinder the ability to stack with real time DSO services (i.e. where DSOs need to issue real-time instructions).
- FPNs are, in our assessment, compatible with stacking for many pre-scheduled DSO services. Provided an FSP knows what they are delivering to the DSO ahead of FPN submission (and preferably prior to the NESOs flexibility auctions too), then they should be able to deliver two services in the same time period (i.e. usually resulting in service splitting). However, this requires the NESO to allow

a non-zero baseline, and delivery of the NESO service can result in over-delivery of the DSO service.
- Other benefits of using the FPN approach are that it (i) provides good visibility for the system operator of FSPs positions close to (but still prior to) real time, and (ii) can be used for a wide range of technologies and the practice is embedded in industry arrangements. Any asset that is BM participating must submit FPNs every half-hour regardless of whether they are participating in other services; this can impact the ability of many FSPs to stack with real-time DSO services at any time.

Baselining method: Nominated baseline.

Description and current uses: Baseline based on expected/forecast operation profile of asset ahead of time (similar/same as the FPN requirement described above). DSO services, Local Constraint Market.

Key considerations for service stacking:

- Nominated baselines are very similar to FPNs as described above. They are effectively a forecast of the generation or demand profile of an FSP in the absence of any Flexibility Service activity. DSOs can use this profile to calculate the deviation of the metered data from the planned profile.
- Many of the considerations for nominated baselines are similar to those of FPNs. However, DSOs may use nominated baselines differently to FPNs, such as requiring nominations further in advance of delivery (e.g. at the point of trade).
- When stacking DSO with NESO services, a nomination baseline for DSO services can be compatible with service stacking. However, the ability to stack is still dependent on wider service requirements while the baselining method will impact the level of flexibility measured for the service and so payment received.
- In the case of pre-scheduled DSO services, a nomination baseline should be compatible with most NESO services and would likely result in service splitting, but potentially also over-delivery of the DSO service. It is assumed that an FSP would nominate its baseline based on its typical profile. If accepted for a pre-scheduled DSO service, it enables the FSP to provide accurate baselines if providing further services for the NESO (i.e. enable accurate FPNs to be submitted). However, this will depend on when the service is scheduled and when the nomination baseline is provided.
- Due to the nomination being based on a 'forecast' of the FSPs typical profile, delivery of a second service on top of the DSO service would result in over-delivery.
- Other benefits of nomination baselines include (i) the ability for the FSP to position its assets as desired for the delivery window, (ii) good visibility for a system operator of an asset's expected operating profile, and (iii) applicability to wide range of asset types. However, they require accurate forecasts, and the timing of nomination is key for FSPs engaging in other markets.

Baselining method: Real time/dynamic baseline.

Description and current uses: Baseline based on real-time position of the asset. Static Firm Frequency Response, MW dispatch.

Key considerations for service stacking:

- Real time baselines have the potential to deliver highly accurate baselines when delivering services but require sophisticated metering requirements not viable for all FSPs.
- A real-time baseline can benefit service stacking as it inherently takes into consideration the position of the asset immediately prior to the event. This means that, should an FSP participate in a DSO service, delivery of that service should not inhibit measuring delivery of another service that uses real-time baselining. However, as discussed, this could result in over-delivery of the DSO service.
- The interaction between baselining methodologies between two services being stacked is important to consider. Interactions can become complex if two services both use real time baselining. If two services are activated at the same time, then the FSP may get paid for both services for the same MWs (i.e. co-delivery). If the services are activated at different but overlapping time periods, then while real-time baselining enables accurate measurement of the second service activated, there could be an over-delivery of the first service. Baselines could be adjusted for the provision of other services; however, this would likely be complex to implement.
- Other benefits include that real-time baselining is reflective of an FSPs actual position, negating the need for forecasting by FSPs. But it is noted that real-time baselines have onerous metering/operational data requirements, and depending on what a particular system operator is procuring for, may not be reflective of the FSP's typical behaviour if providing multiple services.

11.7 Suggested Further Reading

Listed below are the links to Open Networks publications relevant to this chapter in reverse chronological order. Please note, some of the documents/spreadsheets may now by outdated or superseded as the topics have evolved.

1. Revenue Stacking Explainer and FAQ's (2025).
2. Stackable by default statement (2025).
3. Revenue Stacking Assessment Tool (2025).
4. Revenue Stacking Assessment (Aug 2024).
5. DSO Revenue Stacking Publication (2020).

Reference

1. "Demand Flexibility Service- Participation Guidance Document V.11," NAtonal Energy System Operator (NESO), December 2023

Open Access This chapter is licensed under the terms of the Creative Commons Attribution 4.0 International License (http://creativecommons.org/licenses/by/4.0/), which permits use, sharing, adaptation, distribution and reproduction in any medium or format, as long as you give appropriate credit to the original author(s) and the source, provide a link to the Creative Commons license and indicate if changes were made.

The images or other third party material in this chapter are included in the chapter's Creative Commons license, unless indicated otherwise in a credit line to the material. If material is not included in the chapter's Creative Commons license and your intended use is not permitted by statutory regulation or exceeds the permitted use, you will need to obtain permission directly from the copyright holder.

Part V
Additional Resources

Part V
Regulatory Responses

Chapter 12
Carbon Impact Assessment of Flexibility Service

Action 3.6 in the Smart Systems and Flexibility Plan and part of Licence Condition 31E reporting mandated DSOs to develop common methodologies for carbon reporting and monitoring of flexibility markets. As the first step towards the development of a consistent carbon reporting methodology, different carbon reporting methodologies that are prominent in the UK (which may also be applicable in other jurisdictions), were reviewed and summarised. The analysis was undertaken by broadly categorised the approaches by their reporting purpose—GHG inventories, life-cycle assessments, and grid intensity approaches. GHG inventories as adopted by national and corporate inventories tend to have standardised and narrower reporting boundaries (focus on direct impacts) than life-cycle assessments such as PAS 2080.

The following seven methodologies were reviewed

- NESO carbon intensity for GB electricity system
- EU ETS monitoring and reporting [1]
- The GHG protocol for corporate GHG inventories [2]
- The green book and supplementary guidance [3]
- IPCC guidelines for national GHG inventories [4]
- PAS 2080 for carbon management in infrastructure [5]
- Pro low carbon innovation project from NGED.

Publicly and freely available materials were used to undertake this review with an exception of PAS 2080 which relied primarily on internal expertise. The study conducted under the Pro Low Carbon project by Western Power Distribution (WPD, which is now NGED) to help complete the summary, as the project performed a similar review of methodologies as included in their published reports. Additional methodologies, although omitted in this chapter, were considered by the team as part of review of the Pro Low Carbon project report.

12.1 Comparison Between Carbon Reporting Methodologies

See Tables 12.1, 12.2, 12.3, 12.4, 12.5, 12.6, and 12.7.

Table 12.1 NESO carbon intensity of GB electricity system

Category	Grid carbon intensity
Purpose	The carbon intensity dashboard, app, and API forecast is designed to enable consumers or smart devices to optimise their behaviour to minimise their carbon footprint. It provides GB and 14 regional grid carbon intensity forecasts up to 48 h ahead. The dashboard also provides real-time intensity for every 30-min period, historic weekly intensity since 2009, and the change in intensity following NESO actions in the Balancing Mechanism updated every hour
Produced by	NESO, WWF, environmental defence fund Europe and the University of Oxford department of computer science
Boundary	• Electricity generation from all large, metered power stations, interconnector imports, transmission, and distribution losses (to convert to consumption point) • Excludes upstream emissions and indirect land use change impacts • Excludes emissions from unmetered and embedded generators not visible to NESO (but includes estimate for embedded wind and solar)
GHGs	CO_2 only
Calculation	• Generation dispatch is forecasted using historic data and forecasted demand, wind, and solar generation • Carbon intensity is the sum for all generation—carbon intensity (by fuel type) * output, divided by national demand. Then converted to consumption by adjusting for transmission losses • Interconnector intensity uses imported generation mix * carbon intensity (by fuel type) • Effect of balancing actions is the difference between the carbon intensity of all the generators in the BM before and after balancing actions have been applied
Reporting	Data and **forecasts** published on dashboard, app, and API, NESO data portal. This methodology is currently used to publish the carbon intensity of NESO actions as required under RIIO-2 framework
Emission factors	• Carbon intensity, by fuel type, based on output-weighted average efficiency of generation in GB and DUKES CO_2 emission factors (2017, which was derived from total CO_2 emissions per electricity supplied by fuel type) • Interconnector carbon intensity factors (GridCarbon 2017 [6])
Activity data	Generation output uses ELEXON BMRS data, interconnector generation mix from ENTSOE transparency platform
Technologies	Large generation (biomass, coal, gas, hydro, nuclear, oil, other, solar, wind) and interconnector imports. Not considered energy efficiency (impact will be embedded in demand and demand forecasts)
Links	• NESO carbon intensity methodologies • NESO carbon intensity dashboard

Table 12.2 EU ETS monitoring and reporting

Category	Carbon market compliance
Purpose	The EU emission trading scheme (ETS) is a policy instrument to cut GHG emissions within the EU, through a cap-and-trade market mechanism (cap on total GHG emission by sector, reducing over time, each year parties must surrender allowances either allocated or purchased to cover their reportable emissions). The UK ETS replaced the UK's participation in the EU ETS in January 2021. The Monitoring and Reporting Regulation (MMR) and guidance sets out the rules relating to the compliance cycle
Produced by	European Commission
Boundary	Applies to energy intensive industries, power generation, and aviation. Direct emissions only. Operators need to define the monitoring boundaries for each installation
GHGs	CO_2, N_2O, PFCs (also expressed as CO_2 equivalent using GWP)
Calculation	• Monitoring requirements increase based on size of emissions through tiers (like IPCC guidelines). Can be measured (GHG concentration and volumetric flow) or calculated. Calculation method can use emission factors or carbon mass balance • Calculation for combustion emission = activity data (=fuel quantity * net calorific value) * emission factor * oxidation factor (account for incomplete reaction). calculation for process emission = activity data * emission factor * conversion factor
Reporting	Annual monitoring, reporting and verification as part of compliance cycle. Annual emissions report (AER) must be verified by accredited verifier before compliance checks by the competent authority. The competent authority also must approve monitoring plan before implementation. Templates available
Emission factors	Lowest tiers usually apply an internationally applicable default value (such as IPCC standard factors). Second tier uses national factors (used for national GHG inventory). The highest tier usually requires the factor to be determined from laboratory analysis
Activity data	Depending on tier as discussed above
Technologies	Applies to energy intensive industries, power generation, and aviation. Eligible hospitals and small emitters can opt out
Links	Monitoring and Reporting Regulations and guidance documents UK ETS guidance

12.2 Key Observations from Comparison

12.2.1 Purpose of Each Methodology

Each carbon reporting programme has developed guidelines and methodologies to meet their specific purpose. These can be broadly categorised into three main types as noted below (summarised in Table 12.8).

Table 12.3 The GHG protocol

Category	Corporate GHG inventory
Purpose	The GHG protocol—a corporate accounting and reporting standard is a guidance for companies to quantify and report on GHG emissions using standardised approaches and principles. Other methodologies exist such as ISO 14064 but is similar to GHG Protocol. Large UK companies (using > 40 kWh energy, Large defined in Companies Act 2006) need to report their emissions publicly via the Streamlined Energy and Carbon Reporting (SECR) regulations
Produced by	World Resource Institute (WRI) and World Business Council for Sustainable Development (WBCSD). Specific guidance for completing the SECR is available from BEIS which draws on the GHG Protocol methodology
Boundary	• Emissions are categorised into scopes. Scope 1—direct emissions from sources owned or controlled, Scope 2—indirect emissions from electricity purchased for consumption, Scope 3—other indirect emissions consequential of company activities but from sources not owned or controlled. Scope 3 reporting is voluntary, guidance under GHG Protocol Corporate Value Chain Standard • Define operational and control boundaries—financial control boundary (ability to direct financial and operational policies of operations to gain economic benefits from its activities), operational control boundary (full authority to introduce and implement operating policies), and equity share boundary
GHGs	Seven GHG covered by kyoto protocol—CO_2, CH_4, N_2O, HFC, PFC, SF_6, NF_3. Other environmental impacts not considered. Report in CO_2 equivalent based on IPCC's GWP
Calculation	Identify emission sources (focus on most important sources), categorise their scope, select measurement approach (direct measurement, calculated via chemical mass balance or emissions factors), collect activity data, select emission factors, calculate using tools (GHG emissions = activity data x emission conversion factor), collate from all facilities
Reporting	Under SECR, reported via directors report (large companies), annual report (quoted companies), energy and carbon report (LLP). Use of intensity ratio (environmental impact per activity or financial metric) to aid comparison. The GHG protocol methodology is also used by science based targets
Emission factors	Standard does not dictate which factors are used but provides database to aid selection. SECR guidance requires use of UK emission factors updated annually but some organisations can use site specific emission factors if more accurate
Activity data	Scope 1—purchased quantity of commercial fuels, mass balance, or direct measurement through monitoring, Scope 2—metered electricity consumption, Scope 3—calculated from activity data such as fuel use or passenger miles
Technologies	Main emission sources in boundary. Under SECR, companies need to provide narrative on energy efficiency actions taken and if possible provide energy savings but no calculation guidance given. Use of intensity ratios can show efficiencies over time
Links	GHG Protocol SECR Conversion factors 2021

12.2 Key Observations from Comparison

Table 12.4 The green book and supplementary guidance

Category	Policy/programme assessments
Purpose	The green book is used to appraise policies, programmes, and projects. Additional guidance is available to support quantification and valuation of energy use and GHG emissions for options appraisals to build business cases and impact assessments
Produced by	The green book is issued by HM treasury. The supplementary energy and emissions guidance is prepared by BEIS
Boundary	Guidance given on assessing direct impact and indirect impact through planning, land use change, construction, or new products. Guidance only applicable for marginal changes and not wider changes to the market (as it will change values in data tables)
GHGs	Seven GHGs—CO_2, CH_4, N_2O, HFC, PFC, SF_6, NF_3—reported as CO_2 equivalent using GWP. GHG emission factors in data tables incorporate CO_2, CH_4 and N_2O. Can also consider additional wider impacts such as air quality and energy security
Calculation	• Guidance is accompanied by excel-based calculation toolkit and data tables for latest assumptions • Analytical process: identify key drivers of emissions impacted by policy/project; identify energy and emission counterfactual; identify policy interactions; quantify changes in fuel use, emissions (energy change × fuel-specific emission factor, use marginal emissions factor for changes in direct fuel use and average emission factors for footprinting), and other impacts (e.g. air quality); value these changes (split into emission traded/non-traded sectors); calculate cost-effectiveness • Energy efficiency should be net of direct and indirect rebound effects (where available funds are spent on more energy) • Embedded emissions to be considered if appropriate, proportionate, and practical e.g., results in large change in imported emissions
Reporting	Report using template, providing further information e.g., on counterfactual and time profile of emissions, impact by group (households, suppliers). Include impact on UK territorial GHG emissions by sector to understand impact on UK Carbon Budgets. If policy includes material non-CO_2 impacts, should provide impact on each gas separately
Emission factors	• Provided in data tables—marginal and average grid intensity (given in emissions per unit of electricity consumed) and fuel emission factors. Factors mostly from Defra, and DfT for petrol, diesel gas and oil • Bespoke emissions factors should be used where available instead of supplementary guidance emission factors
Activity data	Estimated change in energy use or supply
Technologies	Range of fuel emission factors provided. Guidance includes discussion on energy efficiency calculation incorporating rebound effects
Links	• Green book and supplementary guidance • Data tables

Table 12.5 IPCC guidelines for national GHG inventory

Category	National GHG inventory
Purpose	Guidelines used by countries that report to the UN Framework Convention on Climate Change (UNFCCC) to ensure reported GHG emissions is complete and comparable. The UK's annual National Inventory Report (NIR) meets its commitments as a Party under the UNFCCC, tracks progress against commitments under Kyoto Protocol (including UK's contribution to EU target), and against UK Carbon Budgets
Produced by	The Intergovernmental Panel on Climate Change (IPCC) maintains the guidelines. The NIR is compiled by a consortium led by Ricardo on behalf of BEIS
Boundary	All emission source and removal by sinks since 1990 (excludes historic emissions), within defined geographical territorial boundary, at point of release (excludes emissions from imported goods and international air travel), includes adjustment for trades through the EU ETS
GHGs	Seven direct GHGs under Kyoto protocol—(CO_2, CH_4, N_2O, HFC, PFC, SF_6, NF_3) and four indirect GHGs (NO, CO, NMVOC, SO_2). Each gas is given a Global Warming Potential (GWP) expressed relative to CO_2 equivalent
Calculation	Approach varies by significance of source and availability of data, the more significant or uncertain the higher the "tier". Tier 1 is simplest using activity statistics and default emission factors. Tier 2 uses more country specific data. Tier 3 uses plant specific data. The UK NIR mostly uses Tier2/3 methods. Calculation of direct GHG for sources including power stations = Emissions Factor (EF) × Activity Data (AD)
Reporting	UK's annual NIR includes Common Reporting Format (CRF) tables. Each year the inventory is extended to cover another year and allows updates for new emission sources, revised estimates, and data revisions
Emission factors	Emission factors in 2021 NIR is based on UK specific data. Predominately derived from EU ETS data (2005 onwards), refiner sector reporting (UK Petroleum Industry Association 2020) and from the 2004 Carbon Factors Review (Baggott et al. 2004), with some solid fuel factors derived from UK research (Fynes and Sage 1994); non-CO_2 emission factors are predominately IPCC defaults (IPCC 2006)
Activity data	DUKES for statistics on liquid, solid and gaseous fuels. EU ETS for emissions from installations. UK petroleum industry association for refinery emissions, UK iron and steel industry annual statistics for energy production/consumption in Iron and Steel industry
Technologies	Covers all main categories of source/sinks of emissions including energy industries, industrial processes, agriculture and forestry, and waste. Energy efficiency not considered explicitly
Links	• NIR 2021 • Emissions factors, UK NAEI ("Energy background data") • IPCC 2006 • IPCC 2019 Refinement

12.2 Key Observations from Comparison

Table 12.6 PAS 2080

Category	Standard for infrastructure projects
Purpose	Common framework to help companies reduce volume of carbon used throughout a project, encouraging collaborative working through the supply chain for more effective **whole life cycle carbon** use. Was developed as a result of the 2013 treasury infrastructure carbon review to create a new specification for carbon reduction. Voluntary standard
Produced by	Mott MacDonald and Arup with British Standards Institute
Boundary	Whole-lifecycle carbon management
GHGs	All, normalised to CO_2e
Calculation	• Project level carbon accounting • Should include Whole Life Carbon Emissions of the infrastructure development project • Calculation methodology is not prescribed, although compliant tools exist • Carbon calculations should be done iteratively throughout a project lifecycle with the goal of taking project level action to reduce carbon emissions • Estimation at the design stages, moving to 'as built' emissions as the construction stage
Reporting	Reporting should be undertaken by all value chain members in a project lifecycle at key gateways where carbon emissions can be influenced
Emission factors	None—emissions factors are not specified within framework
Activity data	At discretion of person performing the calculation
Technologies	Emissions sources are reported in line with BS EN 15978:2011 (cradle to grave emissions)
Links	PAS 2080 (BSI)

1. Carbon inventories—standardised methodologies to calculate the carbon impact attributable to a reporting entity such as a country, company, or installation within specified boundaries to comply with legal, regulatory, or voluntary requirements such as the EU ETS or corporate GHG inventories.
2. Life cycle assessments (LCA)—a framework methodology to calculate the life cycle environmental impact of an activity or project, compared to a baseline, affording a greater degree of flexibility to define the scope of the report such as under PAS 2080 for infrastructure projects or evaluating policy interventions using supplementary guidelines for The Green Book.
3. Grid intensity—methodologies to calculate the carbon intensity of the electricity grid, such as the NESO's carbon intensity report, can be used by electricity consumers such as individuals and organisations to calculate their carbon footprint.

A standardised carbon inventory with narrowly set boundaries is simpler to implement and compare between DSO reports, whilst an LCA is inherently more complex with wider boundaries and baselines which increases the scope for divergence on

Table 12.7 Pro low carbon

Category	DSO flexibility services innovation project
Purpose	Pro low carbon (PLC) was part of an innovation project, future flex, which investigated how to increase uptake of domestic flexibility. PLC reviewed methodologies and developed a DSO specific methodology for calculating the carbon impact of procuring flexibility services for different flexible technologies. To date these recommendations have not been adopted
Produced by	PLC was a one-off study conducted by WPD (NGED)
Boundary	Categorises carbon impact into non-operational and operational. Non-operational impact attributed proportion of embodied and end-of-life emissions. Operational impact includes source impact and offset in marginal grid impacts
GHGs	Includes all GHGs in CO_2 equivalent (using GWP) for both operation and non-operational impacts
Calculation	• Operational impact = carbon impact of energy generated/imported minus carbon impact of energy offset (grid intensity) • Non-operational impact = non-operational emissions * % attributed (if investment in asset can be attributed to DSO services, and split based on percentage of DSO services to 10% internal rate of return) • Methodology retains some flexibility based on the technology and how emissions can be attributed to flexibility services
Reporting	Does not specify format or frequency of reporting
Emission factors	• Carbon intensity of fuel emissions from UK Government GHG Conversion Factors for Company Reporting, 2020 • Marginal carbon intensity of grid offset uses independent dataset • Non-operational impact—Carbon Trust life cycle data using data from Ecoinvent through licence
Activity data	Source emissions forecasted via modelling scenarios, whilst non-operational attribution assumes DSO service revenue of £6 k/MW/yr against different technology capital and operational cost
Technologies	Project calculated intensities for batteries (domestic and large scale), DSR (commercial temp-controlled storage, domestic heat pump, EV charger), Diesel, Gas (natural, landfill). Energy efficiency was not considered
Links	Pro low carbon: carbon impact of DSO flexibility services Pro low carbon: carbon assessment methodologies

assumptions and calculations. Even where the framework of an LCA is standardised such as under PAS 2080, the lack of standard datasets increases calculation complexity and divergence.

The LCA is however more suitable for inclusion in options appraisals and could therefore be more useful as an input to the Common Evaluation Methodology (CEM) and whole-systems CBA which are both evaluation tools and methodologies developed through Open Networks.

Under the RIIO-2 framework, the NESO is required to report on the carbon intensity of their actions which is currently using the grid intensity forecasting methodology. Actions taken through the Balancing Mechanism are presented as the

12.2 Key Observations from Comparison

Table 12.8 Comparison of purpose of methodologies

Methodology/programme	Carbon inventory	Life cycle assessments (LCA)	Grid intensity
NESO carbon intensity for GB electricity system			✔
EU ETS monitoring and reporting	✔		
The GHG protocol for corporate GHG inventories	✔		
The green book and supplementary guidance		✔	
IPCC guidelines for national GHG inventories	✔		
PAS 2080 for carbon management in infrastructure		✔	
Pro low carbon innovation project		✔	

change in average grid intensity, and hence reflect the impact they have on electricity consumer's carbon footprint.

Reporting Boundary

Carbon inventories require clearly specified boundaries of what is included and excluded from the accounts. This can be broadly defined at the programme-level for consistency, but the reporting entity needs to specify exact boundaries, such as what is under the DSO's operational control for corporate GHG accounting. Alternatively, the guidelines can accept user defined boundaries based on what is considered by the user as a significant source of carbon emissions. The boundary can be defined based on the type of carbon impact it includes or excludes:

- Direct impact—the carbon impact that can be directly attributed to the activity such as the combustion of fuel or reduction in process emissions. This can be considered Scope 1 emissions under the GHG Protocol, and Scope 2 for electricity consumption, or operational impacts under Pro Low Carbon.
- Consequential impact—the carbon impact that occurs because of the activity such as the change in grid intensity due to the consequential change in the supply–demand balance by dispatching Flexibility Services. This is included in the operational impacts under Pro Low Carbon or can be Scope 3 under the GHG Protocol.
- Upstream indirect impact—the carbon impact of previous activities such as materials extraction and transportation. Known as Scope 3 under the GHG Protocol or non-operational impacts under Pro Low Carbon.
- Downstream indirect impact—the carbon impact of subsequent activities such as end-of-life emissions. Known as Scope 3 under the GHG Protocol or non-operational impacts under Pro Low Carbon.

- Baseline/counterfactual impacts—the carbon impact of the counterfactual(s) which is required under options appraisals and LCA to understand the relative carbon impacts.

The Pro Low Carbon project, which is a WPD Flexibility Services innovation project, recommended inclusion of direct, consequential and indirect carbon impacts for a more complete representation of carbon impact. Whereas the NESO carbon intensity and EU ETS reporting boundaries only captures direct impacts.

Corporate GHG inventories require reporting on direct and electricity consumption impacts, with optional disclosure on upstream and downstream impacts. LCA approaches would include all impacts whilst an options appraisal will require the counterfactual impacts from which to compare.

A suitable reporting boundary for Flexibility Services should be specified in accordance with the purpose of the report as discussed in the earlier sections.

12.2.2 Calculation Approach

The calculation of carbon impact is broadly similar across the methodologies considered, which involve multiplying activity data by an emissions factor and is given in terms of CO_2 equivalent using the IPCC's Global Warming Potential for different GHGs involved.

There were two noteworthy observations on the approach to calculations which may warrant further exploration.

- Methodology hierarchy—under the IPCC and EU ETS guidelines, three tiers are defined, each tier corresponding to the level of complexity of the calculation increasing with the significance or uncertainty of the emissions source. The first and second tiers uses default and national emission factors respectively whilst the third tier uses installation specific calculations such as direct measurements of GHG. Even in LCA approaches, one of the initial steps of the calculation is to identify the most significant emission sources on which to focus the assessment.
- The party responsible for the calculation—the EU ETS monitoring and reporting requirements stipulate that the calculation of emissions is carried out by each installation which is then verified by a third party, whilst the monitoring plan itself needs to be approved by an authority. This differs to a centrally administered calculation such as the NESO grid intensity methodology and the Pro Low Carbon approach.

Emission Factors

Emission factors convert an activity such as consumption of energy or combustion of fuel into a carbon emission value. These are normally given in terms of kg or tonne of CO_2 equivalent per unit of fuel or energy. The following table summarises the

12.2 Key Observations from Comparison

three main types of emission factors that are more relevant to Flexibility Services and the main source of data as used in the methodologies considered.

Some of the methodologies have accuracy as a key principle, and therefore advocate use of bespoke, site-specific factors over default factors where they can provide more accurate results.

- Fuel emission factors (for combustion of fuels)

    BEIS' UK Conversion factors for GHG reporting—used by EU ETS, Pro Low Carbon, and in the Green Book supplementary guidance (via data tables).
    IPCC international default values are used for lowest tier calculations under the EU ETS.
    National GHG reporting under the NIR uses a range of sources including from the EU ETS and the 2004 Carbon Factors Review.

- Life cycle emission factors (for the embodied and end of life emissions)

    Ecoinvent data was used by the Pro Low Carbon project via the Carbon Trust (requires data licence).
    Circular Ecology's Inventory of Carbon & Energy (ICE) V3.0 (embodied carbon only)*.
    Certified product Environmental Product Declarations (EPDs)*.
    (* Not specified within methodology but was identified from DSO internal expertise as standard sources)

- Grid intensity factors (for grid electricity)

    Pro Low Carbon uses the marginal grid intensity based on independent factors not available in public domain.
    DUKES CO_2 emissions from electricity supplied for different fuel types used for NESO Carbon Intensity calculations.
    Data tables in The Green Book supplementary guidance provides marginal and average electricity emissions factors.
    BEIS's UK Conversion factors for GHG reporting

Emission Sources Included

The methodologies that necessarily capture significant volumes of emissions/sinks such as the EU ETS, national GHG inventories, and NESO carbon intensity focuses on the major emission sources such as large generators, interconnector imports, and other energy intensive industries. Arguably some of these large-scale emission sources are less likely to be providing DSO Flexibility Services due the voltage level concerned.

Corporate GHG accounting and LCA allows the reporting entity to select what emissions sources to include, whilst Pro Low Carbon models the most common Flexibility Services technologies—batteries, demand-side response, diesel, and gas generation. Note that the type of technologies that are providing Flexibility Services and the configuration of those technologies, such as a combination of technologies behind the customer meter, are likely to be diverse and may present challenges in specifying standard emission factors.

The treatment of energy efficiency under the methodologies were also considered. Only the Green Book supplementary guidance provided explanations on how to calculate the energy savings. Whilst the SECR guidance for corporate GHG inventories does recommend a narrative explanation on energy efficiency actions taken. Note that measures of carbon intensity as compared between previous years can also offer some indication of energy efficiency.

Whilst not considered in this chapter, the International Performance Measurement and Verification Protocol (IPMVP) is a widely adopted methodology for measurement and verification of energy efficiency projects albeit not extending into standardised quantification of carbon impact.

12.3 Approach Adopted by DSOs for Carbon Reporting

The carbon impact calculation presented in this section incorporates learning from the methodology review and follows the standard methodology that was followed by all DSOs. It is noted that the calculation varies depending on whether the flexibility asset is generation, storage (export), or demand /storage (import). The impacts include direct impacts (such as burning fuel) and consequential impacts (such as demand payback) but not indirect impacts (such as embodied carbon). The conversion factors used are generally industry standard which include grid-intensity, plant efficiencies, fuel emission factors, and payback assumptions. Asset specific factors are not used to maintain consistency between DSO reports which means that the methodology reports an approximation of carbon impacts. The detailed methodology is discussed below.

The purpose of the DSO reports is to make carbon impacts of Flexibility Services actions transparent and consistent across DSOs. It sets the foundation and will inform future developments and interventions that may be required to ensure flexibility markets are consistent with net zero. The methodology in this section covers outturn carbon impacts from Flexibility Services actions. It does not include counterfactual impacts (such as network reinforcement or displacing mobile generators), or even relative impacts (the difference between outturn and counterfactual impacts) which would be required to evaluate between different interventions. Whist the counterfactual calculations are out of scope, the calculations specified in this methodology could form part of that evaluation, such as under the Common Evaluation Methodology discussed in Sect. 4.4.

12.3.1 Carbon Impact Reporting Methodology

This section contains the methodology that all DSOs adopted for consistent reporting in 2022 and 2023. Additional notes, the rationale, limitations of the methodology and potential areas for further work are also discussed.

The calculations apply to Flexibility Services requested for an increase in exports or reduction in imports. This is the most prevalent application of Flexibility Services currently. The calculation is performed by technology category without input from providers, except to confirm the technology category where required. The calculation includes direct (such as fuel combustion) and consequential carbon impacts (such as battery charging) but excludes indirect impacts (such as embedded emissions in the materials). The general formula varies by generation, storage (export), and demand /storage (imports). In the formulae.

In the formulae:

- kWh is the energy delivered (as opposed to requested) measured at the site of the resource;
- η is the energy conversion efficiency of the generator g or storage s;
- EF is the fuel emission factor;
- GI is the grid intensity factor at import i, export e, or at turn-down td; and
- Payback% is the consequential increase in load or generation following a turn-down or turn-up event respectively.
- Generation

    For export increase/import decrease, the carbon impact is:

- Combustion of fuel (direct) $= + \text{KWh}/\eta_g \times \text{EF}$
- displace grid generation (consequential) $= - \text{KWh} \times \text{GI}$

    For export decrease/import increase, the carbon impact is:

- Reduced combustion of fuel (direct) $= - \text{KWh}/\eta_g \times \text{EF}$
- Replaced by more grid generation (consequential) $= + \text{KWh} \times \text{GI}$

When a generator is instructed to export it burns more fuel and displaces the marginal grid generation. When a generator is instructed to reduce exports, it burns less fuel with the reduction replaced by an increase from the marginal grid generation. For a renewable generator assume a zero EF. If the generator is displacing imports, the carbon impact is the same as the equivalent amount exported directly to the grid. For bioenergy, report on both inclusive and exclusive of biogenic CO_2 released during burning of biomass and biofuels by using the relevant emission factors.

2. Storage (export)

    For export increase/import decrease, the carbon impact is:

- Input energy (consequential) $= + \text{KWh}/\eta_s \times \text{GI}_i$ (if from grid) or $+ (\text{KWh}/\eta_s)/\eta_g \times \text{EF}$ (if from generator)
- displace grid generation (consequential) $= - \text{KWh} \times \text{GI}_e$

For export decrease/import increase, the carbon impact is:

- Replaced by more grid generation (consequential) $= +\,\text{KWh} \times \text{GI}$
- Displace grid generation (consequential) $= -\,\text{KWh} \times \text{GI}$

When storage is instructed to increase exports, it displaces the marginal grid generation. The source of that energy and the efficiency of energy conversion is also counted. When storage is instructed to reduce exports, the calculation assumes a temporal shifting of that export. If storage input energy is physically supplied from a renewable generator assume zero carbon, this does not apply to non-physical supplies of low carbon electricity, which should assume grid intensity. If storage discharge is displacing imports, the carbon impact is the same as the equivalent amount exported directly to the grid. Where DSOs are unsure whether storage is providing export increase or import reduction, use the storage calculation. This ensures carbon impacts are not underestimated and incentivises additional information to be provided.

3. Demand or Storage (import)

For export increase/import decrease, the carbon impact is:

- Reduced grid imports (direct) $= -\,\text{KWh} \times \text{GI}_{td}$
- Increased grid imports (consequential) $= +\,\text{KWh} \times \text{GI}_l \times \text{payback}\%$

For export decrease/import increase, the carbon impact is:

- Increased grid imports (direct) $= +\,\text{KWh} \times \text{GI}$
- Reduced grid imports (consequential) $= +\,\text{KWh} \times \text{GI} \times \text{payback}\%$

When demand or storage is instructed to reduce imports the reduction in energy is replaced by the marginal grid generation. There is a consequential rebound in load known as payback. When demand or storage is instructed to increase imports the increase in energy is supplied by the marginal grid generation. There is a consequential reduction in load, an equivalent to "payback". If demand is shifted, such as deferred EV charging, then payback% is 100%. Otherwise, assume an associated payback as a percentage of the turn-down energy of 21%. Where DSOs are unsure, assume load shifting. This ensures carbon impacts are not underestimated and incentivises additional information to be provided. For import increase the payback% assumption is 100%. Where DSOs are unsure whether storage is providing export increase or import reduction, use the storage calculation. This ensures carbon impacts are not underestimated and incentivises additional information to be provided.

Conversion Factors

Conversion factors used in this chapter is a general term to describe factors that are used in the calculations which can include fuel emission factors, grid intensity factors, and plant efficiency assumptions. The carbon reporting methodology is based on generalised assumptions and conversion factors for a type of technology rather than using asset specific assumptions. This maintains consistency in DSO reports but means that the results can only be treated as an approximation.

12.3 Approach Adopted by DSOs for Carbon Reporting

DSOs use the conversion factor sources presented in Table 12.9. The data source used for grid intensity is a static value, all grid intensity factors is therefore equal (except for short-run and long-run factors).

In the case of storage, $GI_i = GI_e$ and in the case of demand, $GI_{td} = GI_i$. The different notations in the formula allows for inclusion of time series grid intensity factors in future. See Supporting Notes for further discussion on use of grid intensity factors.

Grid Intensity refers to the carbon intensity of electricity that is imported from the grid in kg CO_2e/kWh taking into account the carbon emissions from grid generation, interconnector imports, and potentially network losses.

Grid Intensity factor data is also available from subscription-based providers such as *WattTime* and *Electricitymaps*. However, these are not openly available and is therefore were deemed not ideal for transparency and replicability.

A marginal grid intensity factor was considered more appropriate when evaluating the impact of a change whilst an average is generally used for carbon foot printing. A reduction in demand results in an increase in the marginal supply and not the average supply. This means that the Green Book factor was considered most appropriate. However, the Green Book factor is a long-run factor which represents long-term

Table 12.9 Conversion factor sources

Factor type	Source	Notes
Fuel emission factors	BEIS/defra	Conversion factors for company reporting
Efficiency	BEIS electricity generation costs 2020 Coal—DUKES BEIS Storage Costs and Assumptions 2018	Electricity generation costs 2020 [a] Energy statistics-DUKES [b] Storage cost and assumptions 2018 [c] (The DUKES report is updated annually, however the others are one-off reports.)
Grid intensity	Short-run: fuel emission factors and efficiency as per above sources Long-run: green book data tables	Short-run: EF of natural gas divided by efficiency of an OCGT e.g. based on 2022 EF, 183 g CO_2e/kWh / 35% = 523 g CO_2e/kWh Long-run: average of consumption long-run marginal factors, use most recently updated value rather than forecasts (at time of writing). Irregularly updated (Note, long-run refers to interventions that results in sustained change in demand/generation over years. Short-run refers to interventions that result in temporary change in demand/generation.)
Payback%	For demand-turn-down: low carbon London report For demand-turn-up: DSO consensus	Low carbon London Demand-turn-down: From a one-off innovation trial. Assume 21% for reduction services, based on the average of trial events. Assume 100% for load shifting solutions Demand-turn-up: Assume 100%

changes in supply to long-term changes in demand which may be appropriate for long-term scheduled services but not for the more common dispatchable services. A short-run factor should be derived from the emissions factor and efficiency from a gas generator.

Network losses are included in the calculation which, in the case of NESO and Green Book data tables, is included in the grid intensity factor. The NESO time-series grid intensity factor is derived from CO_2 emissions and does not include all Greenhouse Gasses (GHG) unlike the other two sources. At time of publication of this Chapter, there are no plans from the NESO to report on marginal grid intensity or to change the components of GHG.

The benefit of a time-series factor is that it can reflect different grid intensity periods across a day and so better reflects impacts from resources that shift load/exports. A static factor was adopted by the DSOs because there was no marginal grid intensity time-series data available and that it was better to overestimate carbon impact than underestimate to encourage more information to be provided by stakeholders. It was noted that there was limited benefit to adopting a time-series factor at present, but this could be reviewed again when conditions change in future. There are three key reasons:

- A marginal grid-intensity factor is most appropriate for evaluating a change than the average. The marginal grid-intensity does not vary much over time as it is currently driven by gas.
- The implementation cost of moving from a static factor to a time-series factor is relatively high.
- There are subscription-based datasets of time-series marginal grid-intensity available, but these are not openly accessible and therefore not ideal for transparency and replicability.

The carbon impact of energy efficiency is expected to be similar to demand-side-response but with additional complexities. These complexities may relate more to the calculation of the energy saved than to the conversion of energy into a carbon impact, which would also use grid intensity factors. DSOs have limited experience of deploying energy efficiency programmes and therefore it was decided that this should be reviewed in future work as DSOs start developing such programmes.

Payback otherwise known as rebound or bounce back, refers to when demand side response load increases after a turn-down event as the site recovers to pre-event conditions, for instance an increase in cooling load in a building having warmed up during the turn-down event. This could also refer to a similar rebound effect following realisation of economic benefits as a result of an energy efficiency saving measure.

Technology Categorisation

The DSOs have proposed the use of the following technology categorisations, which maps Standard Licence condition (SLC) 31E technology categories by the relevant technology categories required for selecting the appropriate conversion factor (See Table 12.10). This is advisory rather than prescribed to allow for some flexibility

12.3 Approach Adopted by DSOs for Carbon Reporting 271

where the mapping is incomplete. (The SLC31 technology categories may be subject to change pending engagement with Ofgem to add additional granularity.)

12.3.2 Report Submission Format

The reporting format that all DSO presently use, only covers outturn carbon impacts. DSOs can include relative impacts optionally with appropriate narrative explanations to allow the user to understand what assumptions have been used. The calculation of counterfactual and relative impacts have not been standardised.

A table in the format shown in Table 12.11 is included in the Distribution Flexibility Services Procurement Report (as a requirement for SLC 31E) showing energy (requested and delivered) and carbon impact (broken into direct and consequential). An accompanying narrative explains what the data shows.

Energy delivered can differ to energy requested due to under or over delivery. Over-delivery is capped at 150% (captures over-delivery but excludes scenarios where over delivery may not be a consequence of the dispatch). Delivered energy is used to calculate carbon impacts.

In addition, for bioenergy, an additional column to "Direct carbon impact" should be added. In the first of these columns, report on the carbon impact exclusive of biogenic CO_2 using the default emission factors, and in the second column report on the biogenic CO_2 using the "outside of scope" emission factors, from the BEIS/Defra conversion factors (full set). The former calculation nets off the CO_2 released at combustion with the CO_2 absorbed during the growth phase of the bioenergy, whilst the latter calculation reports the CO_2 released at combustion. This is in line with the GHG Protocol guidance for corporate accounting.

Comparable with NESO Methodology

In accordance with the policy objective it was ensured that the methodology was comparable with the NESO approach. This was achieved by adopting a narrow reporting boundary and presenting direct and consequential impacts separately in the report. However, differences in approach remain as it was necessary and desirable for this DSO methodology.

The NESO approach calculates direct emissions from large transmission connected assets in the Balancing Mechanism, presents impacts in terms of the change in grid intensity, publishes information on a half-hourly basis, and uses CO_2 emissions factors (excludes other greenhouse gases).

For DSO reporting, which is on an annual submission cycle, it was necessary to include methodologies for smaller distributed energy resources. It was therefore decided that for storage and demand, the inclusion of consequential impacts of charging and load payback respectively would be more reflective of impacts than just considering impacts at dispatch. It was also decided that CO_2e emissions factors

Table 12.10 SLC 31E technology categories by the relevant technology categories

LC31 technology categorisation	Emissions factor (and grid intensity factor)	Efficiency	Demand payback
Advanced fuel (produced via gasification or pyrolysis of biofuel or waste)		Advanced conversion technologies [a]	
Biofuel—biogas from anaerobic digestion (excluding landfill and sewage)		Anaerobic digestion (AD) [a]	
Biofuel—landfill gas	Biogas—landfill gas		
Biofuel—other	Biogas–biogas	Energy from waste (EfW) [a]	
Biofuel–sewage gas	Biogas–biogas		
Biomass	Biomass—wood logs, wood chips, wood pellets, grass/straw	Biomass—dedicated [a]	
Demand	Grid intensity		Payback (21%), shifting (100%)
Fossil—brown coal/lignite	Figures not available		
Fossil—coal gas			
Fossil—gas	Gaseous fuels—natural gas	OCGT, CCGT [a]	
Fossil—hard coal	Solid fuels—coal (electricity generation)	Coal [b]	
Fossil—oil	Liquid fuel—diesel (average biofuel blend)	Diesel [a]	
Fossil—oil shale			
Fossil—other	Other factors available		
Fossil—peat			
Geothermal			
Hydrogen			
Nuclear			
Solar			
Stored energy (all stored energy irrespective of the original energy source)	Grid intensity	Pumped hydro, CAES, thermal energy storage, lithium ion, zinc, flow, sodium sulphur [c]	
Waste water (flowing water or head of water)			

(continued)

12.3 Approach Adopted by DSOs for Carbon Reporting

Table 12.10 (continued)

LC31 technology categorisation	Emissions factor (and grid intensity factor)	Efficiency	Demand payback
Wind			
Other			

Table 12.11 Template for total carbon impact divided by the total energy delivered

LC31 technology category	Requested energy (MWh)	Delivered energy (MWh)	Direct carbon impact (kgCO$_2$e)	Consequential carbon impact (kgCO$_2$e)
Fossil—gas				
Demand				
Stored energy				
…				
…Total				

Note At the time of the publication of the book, it was proposed to Ofgem to update the data template, to add the carbon impact per dispatch as another column. Technology sub categorisations based on Embedded Capacity Register/G99 was also proposed to be added to reflect different solutions within each SLC31 category, for example if demand is shifted or reduced or the different types of Stored Energy

should be used to capture all greenhouse gases. Currently most DSOs do not calculate and report on grid intensity of their networks which makes reporting the impact in terms of a change in grid intensity more difficult as a starting position, but this could be explored in future work.

Methodology Scope Boundary

Reporting Boundary defines the scope of the calculation. This can include direct impacts (such as combustion of fuel to generate electricity), consequential impacts (such as the carbon emissions from grid generation to produce and transport electricity imported by a battery), or indirect impacts (such as embedded and end of life emissions attributed to the assets). We also describe narrow or wide boundaries to refer to the extent to which the calculation focuses on direct or whole-life emissions respectively.

Following the detailed reviews of other carbon accounting methodologies as discussed in Sects. 12.1 and 12.2, two main types of reporting methodologies being inventory or life-cycle assessments were identified. Carbon inventories tend to have a narrow reporting boundary which records emissions directly attributable to the reporting entity, whilst life-cycle assessments include all emissions with respect to a project or activity.

Figure 12.1 shows different sources of carbon impact from Flexibility Services and how they have been categorised into direct, indirect, and consequential impacts. Direct impact is dispatch (and standby where applicable) whilst consequential impacts are considered as second-order effects such as displacing other grid generation or demand payback. The indirect effects relate to embodied and end-of-life emissions.

It was decided to include direct and some consequential impacts (storage import, demand payback, and grid generation displacement) in the calculation but not indirect impacts for the following reasons.

- This keeps the report comparable with NESO direct-only reporting;
- Direct emissions data sources are standard and readily available which ensures consistency;
- Conversely, indirect emissions data sources and assumptions are less standard and less available, reducing consistency and subject to higher uncertainty;
- For storage and demand, it was considered more reflective of impacts to include the carbon impacts of input energy and payback respectively;
- The Pro Low Carbon project calculated that indirect impacts were a relatively small contributor to overall carbon impacts;

It was recognised however that it would be desirable to include indirect impacts into the methodology as part of future work to capture the full carbon impact.

The inclusion of the displacement of grid generation is the approach taken for quantification of Scope 2 emissions for company reporting as defined under the GHG Protocol where electricity used is at the grid intensity of grid generation to supply the load and hence not used. Increases in grid exports is treated in the same way as reduction in grid imports as it has the same impact on the system. This will also enable calculation of carbon differentials for storage solutions using time-series grid intensities.

It was also discussed whether grid generation displacement would apply to constraint management, which is the main application for DSO Flexibility Services. Load exceeding network limits, allowed to materialise and managed using Flexibility Services, would not be allowed to be supplied from beyond the constrained network as this would breach secure network limits. However, since the constraint is generally an N-1 constraint any excess load would still be physically supplied from outside the constrained network and it was therefore considered appropriate to include this component. This may not apply when the network is islanded. Networks that are not intact were not considered in this methodology.

12.4 Suggested Further Reading

Listed below are the links to Open Networks publications relevant to this chapter in reverse chronological order. Please note, some of the documents/spreadsheets may now by outdated or superseded as the topics have evolved.

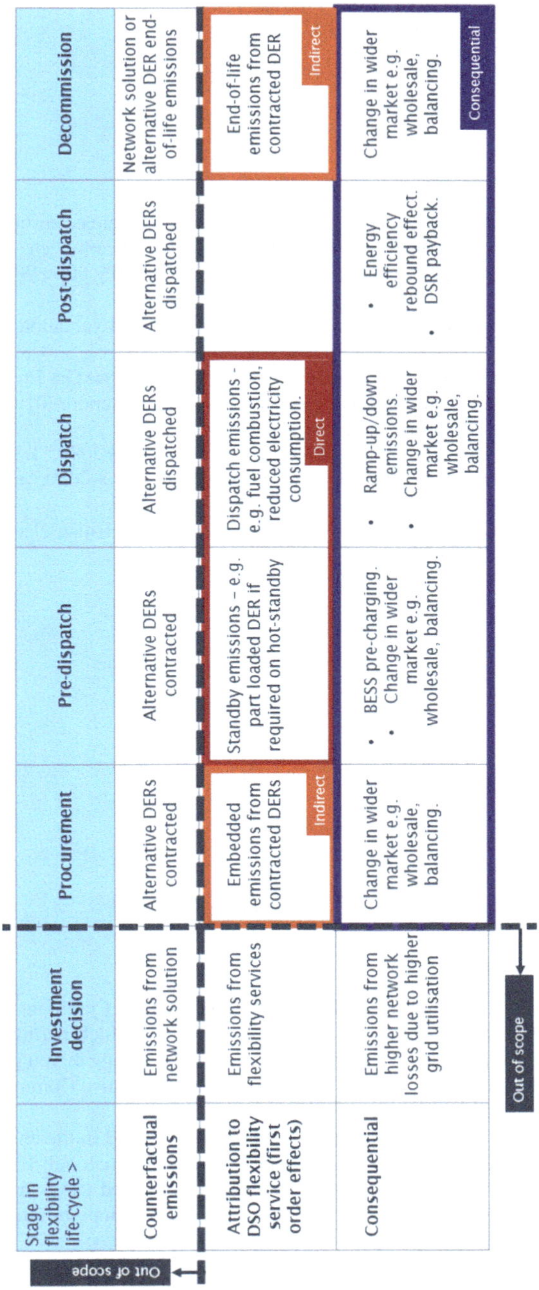

12.4 Suggested Further Reading

Stage in flexibility life-cycle >	Investment decision	Procurement	Pre-dispatch	Dispatch	Post-dispatch	Decommission
Counterfactual emissions	Emissions from network solution	Alternative DERs contracted	Alternative DERs contracted	Alternative DERs dispatched	Alternative DERs dispatched	Network solution or alternative DER end-of-life emissions
Attribution to DSO flexibility service (first order effects)	Emissions from flexibility services	Embedded emissions from contracted DERS	Standby emissions – e.g. part loaded DER if required on hot-standby	Dispatch emissions - e.g. fuel combustion, reduced electricity consumption.		End-of-life emissions from contracted DER
Consequential	Emissions from higher network losses due to higher grid utilisation	Change in wider market e.g. wholesale, balancing.	• BESS pre-charging. • Change in wider market e.g. wholesale, balancing.	• Ramp-up/down emissions. • Change in wider market e.g. wholesale, balancing.	• Energy efficiency rebound effect. • DSR payback.	Change in wider market e.g. wholesale, balancing.

Out of scope ←

Fig. 12.1 Calculation methodology—reporting boundary

1. Carbon reporting methodology (updated) Ver 2.0 (2023).
2. Report on existing carbon reporting methodologies (2022).
3. Carbon reporting methodology Ver 1.0 (2022).

References

1. "EU ETS: Monitoring, reporting and verification," 2024. https://climate.ec.europa.eu/eu-action/eu-emissions-trading-system-eu-ets/monitoring-reporting-and-verification_en
2. "GHG Protocol-Corporate Standard," Green House Gas Protocol, 2015. https://ghgprotocol.org/corporate-standard
3. "The Green Book—Central Government Guidenceon Appraisal and Evaluation," HM Treasury, 2022
4. "2019 Refinement to the 2006 IPCC Guidelines for National Greenhouse Gas Inventories," Intergovernmental Panel on Climate Change, 2019. https://www.ipcc.ch/report/2019-refinement-to-the-2006-ipcc-guidelines-for-national-greenhouse-gas-inventories/
5. "Carbon Management in Infrastructure and Built Environment—PAS 2080," BSI, 2023. https://www.bsigroup.com/en-GB/insights-and-media/insights/brochures/pas-2080-carbon-management-in-infrastructure-and-built-environment/
6. I. Staffell, "Measuring the progress and impacts of decarbonising British electricity," *Energy Policy*, vol. 102, pp. 463-475, 2017.

Open Access This chapter is licensed under the terms of the Creative Commons Attribution 4.0 International License (http://creativecommons.org/licenses/by/4.0/), which permits use, sharing, adaptation, distribution and reproduction in any medium or format, as long as you give appropriate credit to the original author(s) and the source, provide a link to the Creative Commons license and indicate if changes were made.

The images or other third party material in this chapter are included in the chapter's Creative Commons license, unless indicated otherwise in a credit line to the material. If material is not included in the chapter's Creative Commons license and your intended use is not permitted by statutory regulation or exceeds the permitted use, you will need to obtain permission directly from the copyright holder.

Chapter 13
Industry Jargon Buster

This chapter aims to provide more accessible definitions for terms that are used in the networks and energy transition related activities. It includes description for many of the terms that are being used in areas such as the connection of resources to distribution networks and the provision of services from Distributed Energy Resources (DER) to support transmission and distribution network operation.

This chapter is intended to be a reference for stakeholders to provide clarity when discussing the use of the distribution and transmission networks in Great Britain and not as definitive industry definitions. Over 200 terms have been identified by stakeholders, and includes terms identified and used in work that is part of key policy updates.

Note: Updates may be required/issued as further terms are identified for inclusion or as definitions are updated.

13.1 Industry Codes and Sources of Detailed Definitions

Many of the Terms and Definitions used at present are already captured in wider industry code and contractual documents. These industry documents are summarised in Table 13.1. Often the definitions provided in these documents are difficult to interpret quickly, as they use legalistic language, are wordy and they refer to other defined terms. This is necessary to provide a precise understanding of code and contractual obligations.

This chapter provides briefer plain English definitions for terms that are defined elsewhere in code documents, cross-references are also provided.

Table 13.1 Wider Industry Code and Contractual Documents

Codes and contractual documents[1]	Abbrev	Description
Balancing and settlement code [1]	BSC	The BSC covers governance of electricity balancing and settlement in Great Britain
Common connection charging methodology [2]	CCCM	The CCCM is the methodology used to set charges for connection to distribution networks in Great Britain
Competition in connections code of practice [3]	CiC COP	The CIC COP governs how distribution network operators provide services to facilitate competition in the provision of connections to electricity distribution networks in Great Britain
Connection and use of system code [4]	CUSC	The CUSC is the contractual framework for connection to, and use of, the transmission system in Great Britain
Distribution code [5]	D-code	The D-code covers technical aspects relating to the connection to, and use of, distribution networks in Great Britain
Distribution connection and use of system agreement [6]	DCUSA	The DCUSA is a multi-party contract covering use of electricity distribution networks in Great Britain
Electricity safety, quality and continuity regulations [7]	ESQCR	The ESQCR must be met by industry participants to protect the general public and consumers from the dangers of electricity installations
Grid code [8]	G-code	The G-code covers technical aspects relating to the connection to, and use of, transmission networks in Great Britain
National terms of connection [9]	NTC	The NTC are the terms and conditions covering connection to electricity distribution networks in Great Britain
National electricity transmission system security and quality of supply standard [10]	NETS SQSS	The NETS SQSS is s set of criteria covering the planning and operation of the transmission network in Great Britain

[1] *Note* The documents referred in this list/section are often complex, lengthy documents; for example the CUSC is a highly contractual document of over 1,170 pages. Terms and definitions often interact with other multiple terms and conditions link within a topic area. It is therefore not possible to capture all potentially relevant terms within a document of this nature.

13.2 GB Network Voltage Levels

One area where there are different definitions in place across industry documentation, and internationally, is the terminology used to describe transmission and distribution voltage levels. There are several definitions of particular voltage levels in GB industry codes and related documents (e.g. DCUSA, Grid Code, Electricity Safety, Quality and Continuity Regulations). The main voltage levels used for GB distribution and transmission is shown in Table 13.2.

More widely, whilst not used in GB, the international standard IEC60038 uses the following terminology to describe Alternating Current (AC) voltage ranges but these terms are not adhered to in GB documentation.

IEC60038 voltage ranges:

- Low voltage covers nominal voltages from 50 to 1000 V
- Medium voltage covers nominal voltages above 1 kV up to 35 kV
- High voltage covers nominal voltages above 35 kV up to 230 kV
- Extra high voltage covers higher voltages.

13.3 Industry Terms with Common English Description

The following table of terms and definitions includes a column headed "Area" to indicate in what context the particular term is more likely to appear. These areas include charging, connection capacity, connection process, energy trading and settlement, flexible connections (curtailed connections), general contractual terms, industry terms, legal, operations and user commitment. These are further explained below (Table 13.3).

Table 13.2 Main voltage levels used for GB distribution and transmission

Voltage	Terminology	Description
400 kV	Transmission, Supergrid Voltage	The highest AC voltage used on the GB transmission network. Large-scale generators and supply points are connected at this voltage
275 kV	Transmission, supergrid voltage	Much of the early transmission network was constructed at this voltage and there are 275 kV substations and lines in urban areas including London. Generators and supply points are connected at this voltage
132 kV	Transmission/ distribution	Designated as transmission in Scotland. Designated as distribution in England and Wales
66 kV	Distribution, EHV	
33 kV	Distribution, EHV	
11 kV	Distribution, HV	
6.6 kV	Distribution, HV	
3.3 kV	Distribution, HV	
400 V	Distribution, LV	Local street level distribution to domestic/retail properties

13.3 Industry Terms with Common English Description 281

Table 13.3 Industry terms with common English description

Term	Area	D/T	Plain English definition	Code Ref
Abnormal operating conditions	Operations		These are operating conditions where a network is altered from its normal operating conditions, to a different 'abnormal' state. Abnormal operating conditions could include switching in or out of circuits or network assets as a result of faults or other unplanned activities	
Act	Legal		Means the electricity act 1989 (as amended)	Legal requirements
Activation (dispatch)	Industry term—DSO models		The act of instructing a service provider to deliver the service contracted	
Active network management (ANM)	Connection capacity		Active network management is the use of distributed control systems to continually monitor network limits, along with systems that provide signals to DER to modify outputs in line with these limits. Active network management systems are used by DSOs to protect networks that have constraints	Also in SCR charging futures glossary
Additional load	Connection capacity	D/T	In the context of generation connected to the transmission network: Additional load represents any electrical energy used within a site which is not directly related to the production of electrical energy. E.g. a water treatment facility has a diesel generator onsite. The generator requires power to start it up and for its control systems. All other electrical energy demands on the site are classed as Additional Load i.e. heating, lighting, pumps.... In the context of a connection to the distribution network; Additional load is the term relating to requiring more electrical energy than the current supply agreement stipulates. E.g. a factory wishes to install a new pump. The new demand of the pump will cause the factory to exceed its current supply agreement. The difference between the new required supply capacity and the existing agreement is the additional load	CUSC

(continued)

Table 13.3 (continued)

Term	Area	D/T	Plain English definition	Code Ref
Adoption agreement	Connection process	D	If you use an independent connections provider (ICP) to construct the contestable work for your connection, you will have to enter into an adoption agreement. This covers the arrangements for the distribution network operator (DNO) to take over responsibility for the infrastructure installed by the ICP	CiC COP
Agreed export capacity	Industry term		The maximum amount of power (expressed in kW) that is permitted to flow into the distribution system through the connection point, based on the capability of the network in credible demand and generation scenarios	EREC G100
Allowed interruption	Connection capacity	T	An allowed interruption is an interruption relating to some specific circumstances listed within the connection and use of system code (CUSC). Such circumstances include unplanned events (events not on the transmission system), system shutdowns, disconnection or de-energisation (as allowed under certain parts of the CUSC), operation of system to generator operational intertripping scheme as well as some other unlikely circumstances	CUSC
Alternative switched (connection)	Connection capacity		A connection via a substation which will typically have two cables supplying it. In the event of a fault on one cable then full supply can be restored by switching to the alternative cable	
Apparatus	Industry term	D/T	All equipment in which electrical conductors are used, supported or of which they may form a part	CUSC
Application date	Connection process	D/T	The date upon which all required information is received in respect of an application	CCCM

(continued)

13.3 Industry Terms with Common English Description

Table 13.3 (continued)

Term	Area	D/T	Plain English definition	Code Ref
Approved credit rating	General contractual terms		A credit rating is a system that some organisations use to judge how likely it is individuals or businesses will be given credit by a lender. When carrying out business transactions between two companies; businesses will often check the other party's credit rating. The approved credit rating is a pre-defined credit rating that the company is willing to proceed carrying out business with the other company. E.g. a long term debt rating of not less than BB- by standard and poor's corporation	CUSC
Associated DNO construction agreement	Connection process		Where works are required on the national electricity transmission system (NETS) as a consequence of the connection of distributed generation to a distribution system, a construction agreement is required between NGET and the DNO to cover these works. This agreement will be associated with a bilateral connection agreement under CUSC	CUSC Section 10.1
Attributable works	Connection process	T	User Commitment is the means through which National Grid indemnifies itself against the risk of unnecessary transmission investment in the event that a generator terminates its agreements or makes a change which has a material effect on the required works. Attributable Works define the elements of work for which securities are required to be posted by the generator to cover liabilities for these elements. These are defined further within the Connection and use of System Code	CUSC
Authorised persons	General contractual terms	D	Persons authorised by a network company to undertake certain work on the connection equipment, metering equipment and/or the monitoring equipment	NTC
Automatic firm (connection)	Connection capacity	D	An automatic firm connection is an arrangement which, with the exception of a momentary de-energisation resulting from the operation of automatic switching following a fault on any of the circuits forming part of the connection arrangement, will maintain the agreed maximum import capacity or maximum export capacity	

(continued)

Table 13.3 (continued)

Term	Area	D/T	Plain English definition	Code Ref
Balancing and settlement code (BSC)	Energy trading and settlement	D/T	The balancing and settlement code (BSC) covers governance of electricity balancing and settlement in Great Britain. The BSC is a legal document which defines the rules and governance for the balancing mechanism and imbalance settlement processes	Also in SCR charging futures glossary
Balancing mechanism	Energy trading and settlement	D/T	The system of bids and offers relating to the trading of electricity to ensure that supply meets demand in real time, pursuant to the arrangements contained in the balancing and settlement code	CUSC Section 6 (6.8)
Base load	Industry term	D/T	The base load is the minimum level of demand on an electrical network over a defined span of time. E.g. Over a period of 5 years a substations demand fluctuates up and down however never falls below 1 MW. As such the substation base load would be declared as 1 MW	
BEGA	Connection process		A bilateral embedded generation agreement (BEGA) is an agreement type for embedded generators that require access to the transmission network. A BEGA will provide a generator with transmission entry capacity (TEC) and allow it to operate in the energy balancing market. As a result of these rights, a generator with a BEGA must follow the balancing and settlement code (BSC) and pay TNUoS charges if generation is larger than 100 MW	From NG document how to connect to the national electricity transmission system (NETS)

(continued)

13.3 Industry Terms with Common English Description

Table 13.3 (continued)

Term	Area	D/T	Plain English definition	Code Ref
BELLA	Connection process		A bilateral embedded licence exemptible large power station agreement (BELLA) is an agreement type for generators that are classed as 'large' and are smaller than 100 MW. For this reason, it generally applies only in Scotland because generators smaller than 100 MW and larger than 50 MW in England and Wales are classed as 'medium'. A BELLA doesn't give a generator explicit access to the transmission network and it won't be able to get a generation licence, meaning that it won't have to pay TNUoS charges. Generators with a BELLA can take part in the balancing mechanism market if they wish to. They also need a contract with National Grid and a corresponding connection agreement with the DNO	From NG document how to connect to the national electricity transmission system (NETS)
Bilateral agreement	Connection process	D/T	A bilateral agreement is an agreement made between 2 parties. In the context of network use, there are various types of bilateral agreement that cover the arrangements between network parties and connected parties. Examples in CUSC include bilateral connection agreements and bilateral embedded generation agreements	CUSC Section 1, also in SCR charging futures glossary
Budget estimate	Industry term	D	A budget estimate provides an indication of costs and is therefore subject to change. It is not open for acceptance. It can be requested in the early stages of a project, and generally only for larger capital projects. The DNO doesn't require as much information as would normally be available for a formal quote. It is based on a desktop study—the DNO/DSO is unlikely to carry out detailed designs or studies	CCCM

(continued)

Table 13.3 (continued)

Term	Area	D/T	Plain English definition	Code Ref
Bulk supply point (BSP)	Connection process	D	A supply point on the DNO party's distribution system representing an EHV/EHV transformation level e.g. 132/33 kV	NTC
Cancellation charge(s)	User commitment		User commitment is the means through which National Grid indemnifies itself against the risk of unnecessary transmission investment in the event that a generator terminates its agreements or makes a change which has a material effect on the required works. Cancellation charges are applicable on certain elements of work following the cancellation of, or change to, an agreement which results in a material change to the required works. These are defined further within the connection and use of system code (CUSC). There are also a number of related terms defined in the CUSC including: actual attributable works, annual wider cancellation amount statement, attributable works cancellation charge, fixed attributable works cancellation charge, notification of fixed attributable works cancellation charge, pre trigger amount and wider cancellation charge	CUSC
Central volume allocation	Energy trading and settlement		This is a means to determine quantities of active energy to be taken into account for the purposes of settlement in respect of Volume Allocation Units. Any balancing mechanism unit that is connected to the transmission network (e.g. a large coal plant or a steel works) is registered for central volume allocation (CVA). BM units connected to distribution networks (e.g. a supplier's customers in that area or an embedded generator) are normally registered for supplier volume allocation (SVA)	CUSC, BSC
Charging futures	Connection capacity	D/T	Charging futures is a programme to coordinate significant charging reform (on electricity access and charging arrangements), in a way where every stakeholder can equally contribute to change	Also in SCR charging futures glossary
Commercial boundary	Connection process		Normally, this is the boundary between a network company and a user at the higher voltage terminal of the generator step-up transformer	CUSC

(continued)

13.3 Industry Terms with Common English Description 287

Table 13.3 (continued)

Term	Area	D/T	Plain English definition	Code Ref
Commercial services agreement	Energy trading and settlement		An agreement with NGET to govern the provision of and payment for one or more system services	CUSC
Communications outages	Operations		There are instances when due to planned or unplanned circumstances there is a loss of supply or significant reduction in the reliability relating to communications between equipment or sites. E.g. during a storm a radio transmitter becomes damaged resulting in the loss of communications to a generator site	
Company's equipment	General contractual terms	D	Company's equipment means any electrical assets, which are owned/operated and maintained by the network company, and which have been installed to provide an electrical connection between the distribution system and a customer to import and/or export electricity	NTC
Company's premises	General contractual terms	D	A specific reference in the national terms of connection. Land or buildings owned by the network company	NTC
Competition in connections (CIC)	Connection process	D	Most of the work to connect new customers to the existing distribution system (such as housing developments, retail parks or generation connections) can be carried out by either the distributor or an accredited Independent connections provider (ICP). Once completed the new assets will be adopted by a licensed distributor	CCCM

(continued)

Table 13.3 (continued)

Term	Area	D/T	Plain English definition	Code Ref
Complex site	Energy trading and settlement		A term used in metering where energy volumes for a particular site are not simply derived from normal meter readings. A Complex Site is used in BSC documentation to describe a site where the total import volumes or export volumes for a site are derived from the use of a mathematic rule consisting of either: The aggregation of raw metered volumes recorded by multiple settlement meters on the site; or The netting of metered volumes within a site recorded by non-boundary point metering from the metering equipment recording metered volumes at the point of connection to the distributor The aggregation rule is captured on a complex site supplementary information form to enable the data collector to correctly interpret the metered data for the site and ensure that this is converted and submitted into correct settlement values	
Connect	Connection process	D/T	The installation of the required distribution (or transmission) plant and apparatus such that following energisation, electricity may be imported to, and/or exported from, the customer's installation at the connection point	NTC
Connect and manage	Connection process		The connect and manage transmission access regime was introduced by the government in August 2010 and implemented on 11 February 2011. Its aim was to improve access to the electricity transmission network for generators by offering generation customers connection dates ahead of the completion of any wider transmission system reinforcements which may be needed. Any resultant constraint management costs are socialised via BSUoS charges This is part of the arrangements used to manage transmission constraints whereby a customer is permitted to connect to an area of the network ahead of the completion of reinforcement. The connection will then be managed to control the identified constraint(s) until such a time as the reinforcement is completed or the constraint is no longer applicable. Prior to the connection being made there may still be a requirement for enabling works	CUSC Section 1, Also in SCR charging futures glossary
Connected installation	Industry term		This could be a customer installation, a generator Installation or a user installation (as the case may be)	DCUSA

(continued)

13.3 Industry Terms with Common English Description

Table 13.3 (continued)

Term	Area	D/T	Plain English definition	Code Ref
Connected planning data	Connection process	D/T	This is data required pursuant to the grid code planning code which network companies will use in planning network use. The connected planning data replaces data containing estimated values assumed for planning purposes by validated actual values and updated estimates for the future and by updated forecasts for forecast data items	CUSC
Connection	Connection process	D/T	Connection is a term covering a direct connection to an electricity transmission or distribution system. It refers to a network extension and the assets that will connect the network extension to the distribution or transmission system	CUSC
Connection agreement	Connection process	D/T	At distribution, this is an agreement between a customer and a distribution or transmission network company detailing terms and conditions for connecting to and remaining connected to the relevant network	ENA
Connection assets	Connection process	D	Refers to new distribution assets for a new or modified connection to a customer. When a customer applies to be connected to a distribution system, then the assets that get installed are the minimum necessary to cater for the applicant's requirement. These assets which are provided for the sole use of connecting the user are referred to as connection assets and will typically comprise a combination of electrical plant, switchgear, cables and overhead lines. For DNO connections the costs of these assets are charged in full to the customer. These are distinct from reinforcement assets that are only funded in part by the connection customer via apportionment rules	DCUSA

(continued)

Table 13.3 (continued)

Term	Area	D/T	Plain English definition	Code Ref
Connection charges	Connection process		At transmission, connection charges cover the provision of electrical plant, lines and ancillary meters to construct entry and exit points on the national electricity transmission system. They also cover charges in respect of maintenance and repair where these costs are not recoverable as use of system charges, including all charges provided for in the statement of connection charging methodology (such as termination amounts and one-off charges) At distribution level, the full cost of new sole use Connection Assets are charged to the connectee. In addition, the connectee pays for a share of the reinforcement costs under pre-determined apportionment rules Connection charges are paid for by network users and charged by the network operator (transmission or distribution, depending on where the new user connects). (Shallow, shallowish and deep are conceptual terms to describe different principles or approaches for charging for new connections.)	Also in SCR charging futures glossary
Common connection charging Methodology (and statement)	Connection process	D/T	These are the principles and methods by which connection charges are determined	CUSC Section 14 DCUSA CCCM
Connection conditions or "CC"	Connection process		Connection conditions are that portion of the grid code which identify the minimum technical, design and operational criteria which must be complied with by NGET at connection sites and by the generation and demand customers connected to, or seeking connection to, the transmission network. Larger generators or DC converters that are connected to, or seeking connection to, distribution networks must also meet connection conditions	CUSC
Connection entry capacity	Connection capacity		An agreed level of electrical energy generated into the network by a customer. This agreed capacity will be recorded within the connection agreement	CUSC Section 2

(continued)

13.3 Industry Terms with Common English Description

Table 13.3 (continued)

Term	Area	D/T	Plain English definition	Code Ref
Connection equipment	Connection process		A specific, rather than general term, used in the national terms of connection. In this context connection equipment refers to the network company's equipment which has been installed to provide a connection at the connection point	NTC
Connection offer	Connection process	D	This is a formal offer from a network company containing the terms, conditions and charges to make a connection. A connection offer is issued either to a customer or to the independent connection provider (ICP) where applicable	ENA
Connection point	Connection process	D	As distinct from a point of connection. This is located at the entry point or an exit point of the distribution or transmission system. In distribution this is normally at the physical boundary of the Distributor's assets and the customer's electrical equipment. The metering point is normally located here too. Related terms include meter, metering point, MPAN, point of supply, entry point and exit point	D-code
Connection site	Connection process	D/T	Connection sites are locations described in the relevant bilateral agreements at which a user's equipment and connection assets are situated. If two or more transmission system users own or operate plant and apparatus connected at a location, that location is treated as two or more connection sites	CUSC Section 2
Connection works	Connection process	D	Connection works are the works that are required to be undertaken to provide a connection and includes determination of the point of connection	CiC COP
Connections activities	Connection process		These are all the elements of work that are required to be carried out by either the distributor or an independent connections provider (ICP) to provide the connection	CiC COP
Constrained connection	Connection capacity		A constrained connection (or curtailed connection) is a connection where curtailment may be applied to the import and/or export capability of a site. Typically, this is offered to customers to avoid more extensive network reinforcement and it relies on restricting the capacity of the connection under certain network operating conditions to avoid creating conditions outside of the network assets operating limits. This restriction may be imposed on a simple time of day/year basis, or through the use of a more dynamic active network management scheme	

(continued)

Table 13.3 (continued)

Term	Area	D/T	Plain English definition	Code Ref
Constraint managed zones	Connection capacity		These are areas where peaks in demand or distributed generation are managed without needing to reinforce the network	
Constraints (on a network)	Connection capacity		Constraints are a term used for restrictions on the ability of a network to transport energy. For example, due to thermal or voltage limitations. An electricity network is constrained when the required capacity to transport desired electricity flows is higher than the actual capacity on the network. Can also be referred to as network congestion	Also in SCR charging futures glossary
Construction agreement	Connection process	D/T	This is the agreement in place to cover construction of the works required for a connection Each customer who wishes to construct or modify a direct connection to the transmission system, or to commence or modify use of certain types of power station, requires a construction agreement for any construction works that are required for that connection or modification. Distribution companies connecting certain types of power station to their networks also require a construction agreement	CUSC Section 1
Consumer's installation	General contractual terms	D/T	This refers to the electric lines situated upon the consumer's side of the supply terminals together with any equipment permanently connected or intended to be permanently connected thereto on that side	ESQCR
Contestable work	Connection process	D	This comprises connection works that are identified by a DNO/DSO in its connection charging methodology as able to be carried out by an independent connections provider (ICP). In effect, this work is open to competition	CiC COP
Convertible quotation	Connection process		A connection offer that separately identifies the charges for non-contestable works and contestable works and can be accepted by: the recipient in its entirety; or the recipient, or the recipient's duly appointed agent acting on his behalf, in relation only to that part of the quotation relating to the charges for non-contestable works A customer may wish to request this type of quotation if they are looking to compare the price of the network operator carrying out the work to an independent connections provider (ICP). Some DNO/DSOs refer to this type of offer as a dual offer and most offer a convertible quotation as standard	CiC COP

(continued)

13.3 Industry Terms with Common English Description 293

Table 13.3 (continued)

Term	Area	D/T	Plain English definition	Code Ref
Coordinate (between network parties)	Industry term—DSO models		Coordinate in this context is the negotiation and agreement between parties in order that actions and activities on one network does not cause issues on others	
Curtailment	Connection capacity		Curtailment refers to a user's ability to import or export from the network being restricted i.e. the user's access to the network is said to be curtailed The term is usually applicable to generator export but it can also be applied to demand from large industrial sites Typically, curtailment is a temporary reduction in the allowed exports from a generator, below a customer's agreed export capacity. Curtailment is activated in response to a notification or signal that the generator is required to curtail its output	Also in SCR charging futures glossary
Curtailment assessment	Connection capacity		A curtailment assessment is an estimate of the expected curtailment over time, expressed in terms of MWh or the fraction of expected un-curtailed output. Often, this is based on simulation of active network management (ANM) operation across representative time-frames	
Customer	General contractual terms		Customer is a general term used by network companies in connection agreements. It often refers to any person supplied or entitled to be supplied with electricity at any premises within Great Britain	D-code
Customer installation	General contractual terms		In the context of a specific reference in the national terms of connection this refers to any structures, equipment, lines, appliances or devices (not being the DSO's equipment) used, or to be used, at the customer's premises (whether or not owned or used by the customer)	NTC
De-energisation	Operations	D	This is the deliberate movement of any switch or the removal of any fuse or the taking of any other step whereby no electrical current can flow between the transmission or distribution system and the customer's installation	NTC

(continued)

Table 13.3 (continued)

Term	Area	D/T	Plain English definition	Code Ref
Demand control	Operations	D	Demand control refers to the adjustment of customer demand (up or down) to meet wider system needs The term is often used to encompass different methods of achieving demand reduction including customer voltage reduction, customer demand reduction by disconnection, automatic low frequency demand disconnection and emergency manual demand disconnection	D-code
Design variation	Connection process	T	This is a term used to describe a transmission connection design that doesn't meet the full deterministic criteria of the NETS SQSS. Where a customer has opted to be connected using this type of connection design, they may not be entitled to compensation payments in the event of the connection capacity being constrained	CUSC
Developer capacity	Connection capacity	T	This is the MW capacity figure specified on an agreement between NGET and a directly connected distribution system as a consequence of a request for a statement of work	CUSC
DG party	Industry term	D	A DG party is a person, company, or entity that holds a generation licence (or is exempt from requiring one) and has one or more generating stations connected to the distribution or transmission network The term covers the majority of all generator installations connected to the distribution or transmission network with the exception of DNOs, IDNOs, and TOs who have installed generation for network restoration	DCUSA
DGNU payment	Connection capacity	D	This is the mechanism (the distributed generation network unavailability payment) created to make compensation payments for network outages experienced by customers with distributed generation	NTC
Disconnect (or disconnection)	Connection process	D	This means to permanently de-energise an exit/entry point by the permanent physical disconnection or removal of equipment	CCCM/CUSC

(continued)

13.3 Industry Terms with Common English Description 295

Table 13.3 (continued)

Term	Area	D/T	Plain English definition	Code Ref
Disconnection notice	Connection process	D	This is a notice sent by a customer to the network company requesting that the it disconnects one or more connection points	NTC
Distributed generation (DG)	Industry term	D	Also called DG, embedded generation, and distribution-connected generation. These are generators connected to the distribution system, rather than the transmission system. Small and medium sized DG (sub-100 MW) do not pay transmission charges and can receive embedded benefits. Large-sized DG (over 100 MW) do pay transmission charges and do not receive embedded benefits	CUSC, Also in SCR charging futures glossary
Distributed generation connections guide	Connection process	D	This is the guide produced by distribution licensees to provide guidance on the connection process for distributed generation	CCCM
Distribution	Operations	D	Local transportation of electrical energy, typically to customers (demand, generators or storage) across the networks off DNOs, IDNOs or private networks	
Distribution agreement	Connection process	D	This is a connection agreement between a customer and a DNO. This term is used in CUSC to distinguish it from the connection agreement between national grid and a DNO	CUSC
Distribution connection agreement	Connection process	D	This is an agreement between the owner or operator of a distribution system and an owner of a power station to be connected that distribution system	CUSC
Distribution licence	Industry term	D	License conditions place obligations on distributors on how they must operate their businesses. The distribution licence is governed by the regulator, Ofgem	Ofgem
Distribution network	Industry term	D/T	In England and Wales this is the wires, cables and other network infrastructure that typically operate at 132 kV and below, while in Scotland it is the infrastructure that operate below 132 kV. Distribution networks carry electricity from the transmission system and distributed generation to industrial, commercial and domestic users	Also in SCR charging futures glossary

(continued)

Table 13.3 (continued)

Term	Area	D/T	Plain English definition	Code Ref
Distribution network operator (DNO)	Industry term	D/T	DNOs own, operate and maintain the distribution networks. They do not sell electricity to consumers, this is done by the electricity suppliers. There are 14 licensed DNOs in Britain, and each is responsible for a regional distribution services area	Also in SCR charging futures glossary
Distribution services area	Industry term	D	In relation to an electricity distributor, the area (if any) specified as such under its electricity distribution licence	CiC COP
Distribution services provider	Industry term	D	Distribution service providers refers to the distribution network companies that are the former regional electricity companies and who cover 14 separate geographical regions of Great Britain. A distribution service provider has specific distribution licence obligations compared to other distribution network companies	CiC COP
Distribution system	Industry term	D	A distribution system is a system consisting (wholly or mainly) of electric lines owned or operated by an electricity distributor that is used for the distribution of electricity This can be used to describe something slightly different to the distribution network in that it describes more than the physical assets, incorporating also the operation of the system In England and Wales, the distribution system uses lines with a voltage less than or equal to 132 kV. In Scotland and Northern Ireland this is 33 kV and below	CiC COP
Distribution system operator (DSO)	Industry term	D	A distribution system operator (DSO) has a role to monitor, control and actively manage the power flows on the distribution system to maintain a safe, secure and reliable electricity supply As a neutral facilitator of an open and accessible market for network services, a DSO will enable competitive access to markets and the optimal use of DER on distribution networks to deliver security, sustainability and affordability in the support of whole system optimisation. A DSO enables customers to be producers, consumers and storers of energy, enabling customer access to networks and markets, customer choice and great customer service	Open networks

(continued)

13.3 Industry Terms with Common English Description

Table 13.3 (continued)

Term	Area	D/T	Plain English definition	Code Ref
Distributor	Industry term		A distributor (normally a DNO or an IDNO) owns or operates a network for the distribution of electricity. There is a legal definition in the ESQCR. The term distributor is normally used to refer to those distributors that have distribution licences i.e. DNOs and IDNOs. It should be noted that the owners of private networks with electricity's supplier's metered customers on them may also be considered to be distributors. Licenced distributors are required to be party to industry codes	ESQCR
ECCR	Charging	D	ECCR refers to the electricity (connection charges) regulations legislation Also known as 'Second Comer' regulations, the ECCR applies to situations where a customer pays a DNO for a new or modified connection and a subsequent customer's new or modified connection utilises the assets installed for the first customer. When this happens the second customer may be additionally charged a proportion of the costs paid by either the first customer or the DNO to reflect the second customer's requested use. These additional charges may be refunded to the first customer or retained by the DNO If these regulations apply to an applicant, then the quotation letter will specify which version of the legislation is applicable (2002/2017)	CCCM
Embedded generator	Industry term	D/T	In practice, an embedded generator has the same meaning as distributed generation (DG)	Also in SCR charging futures glossary

(continued)

Table 13.3 (continued)

Term	Area	D/T	Plain English definition	Code Ref
Emergency conditions	Industry term		Emergency conditions denote conditions when normal operating arrangements are suspended. Emergency conditions might arise where the condition of an energy system poses an immediate threat of injury or damage, or during a natural disaster or other emergency, or there is an actual or threatened emergency affecting energy supplies	
Enabling works	Connection process		Under the connect and manage methodology, as defined within CUSC, generation projects are allowed to connect to the transmission system in advance of the completion of the wider transmission reinforcement works. Under connect and manage, the works that are required to be completed prior to a generator connecting are classed as enabling works	CUSC Section 13
Energisation	Operations		This is the process which allows electricity to flow between the distribution or transmission network to, or from, the customers' premises. For a domestic dwelling this will be achieved by the insertion of a fuse at the metering position whereas larger connections will involve the closing of switch or circuit breaker	NTC
Engineering recommendations (EREC)	Industry term	D/T	These are the technical standards developed by the energy network association. Engineering recommendations (EREC) are often referred to in industry codes and compliance with certain of these recommendations are often requirements for connection to electricity networks	ENA
Enhanced scheme	Connection process	D	When applying for a new connection, the network operator will provide an offer for the minimum cost scheme, however there are circumstances where the network operator may elect to build a different or enhanced scheme, usually to provide other benefits to the general network. Where this is the case, the connectee is charged the lower of the connection charges applicable to the minimum scheme or the enhanced scheme	CCCM
Entry point	Industry term	D/T	This is the point at which a generator or other users connect to the transmission or distribution system and where power flows into the relevant system under normal circumstances. Refer also meter, metering point, MPAN, point of supply, connection point (exit point)	D-Code

(continued)

13.3 Industry Terms with Common English Description

Table 13.3 (continued)

Term	Area	D/T	Plain English definition	Code Ref
Exempt generator	Connection process		An exempt generator is any generator who, under the terms of the electricity (class exemptions from the requirement for a licence) order 2001, is not obliged to hold a generation licence One class of exempt generator covers generators with a capacity of less than 100 MW who do not provide power of more than 50 MW. Other classes include certain off-shore generators and certain generators who were connected before 30th September 2000	CUSC
Existing capacity	Connection capacity		For existing customers their existing capacity will be either: The maximum capacity used in the calculation of their use of system charges; or For customers who are not charged for use of system on the basis of their maximum capacity the lower of no. of phases × nominal phase-neutral voltage (kV) × fuse rating (A); and the rating of the service equipment	CCCM
Existing network	Industry term		The electricity network in its current form	CCCM
Exit point	Industry term	D/T	An exit point is the point of supply from transmission or distribution system to a user where power flows out from the transmission or distribution system under normal circumstances. Refer also meter, metering point, MPAN, point of supply, connection point (or entry point)	D-code
Export limited connection	Connection capacity		A connection where the power station capacity is greater than the agreed export capacity at that connection point and the export from that power station is limited to the agreed export capacity. This is often the case where there is demand and generation connected on the customer's side of the connection point	
Extension assets	Connection process		These are assets which connect the existing distribution network to a customer's premises and will be charged to the applicant in full	CCCM

(continued)

Table 13.3 (continued)

Term	Area	D/T	Plain English definition	Code Ref
Fault level	Industry term		The fault level is the maximum prospective current or power that will flow into a short circuit at a point on the network, usually expressed in MVA or kA	CCCM
Feasibility study	Industry term	D	A feasibility study may be requested by a customer or project developer to better understand the costs and benefit of differing connection arrangements or capacities, before subsequently making a formal application. It is generally utilised for more complex connections, to consider a number of options for connection and provides estimated costs for each option as appropriate. Costs are purely indicative and are not a binding offer and as such cannot be accepted. The price in any formal connection offer may therefore differ from that given in the feasibility study	CCCM
Firm (connection)	Connection capacity	D/T	The term firm is used to describe a connection that remains available in a first fault scenario. A clear example of a firm connection is a connection of 2 or more circuits to maintain availability in the event of one circuit not being available (Single circuit connections are a clear example of an un-firm/non-firm connection whereby the connection becomes unavailable after a fault and remains unavailable for the duration of the fault repair) A firm arrangement is one which, in the event of a fault on, or the taking out of commission for maintenance or other purposes, any one circuit forming part of the connection arrangement, ensures continued availability of the agreed maximum import capacity or maximum export capacity (assuming that the wider network assets that the connection is connected to are intact and operating normally)	
Firmness	Connection Capacity	D/T	This is the extent to which a user's access to the network can be restricted and, in the case of transmission connections, their eligibility for compensation if it is restricted	Also in SCR Charging Futures Glossary
First circuit outage	Operations		This refers to an operational state of the network where a single item of equipment is out-of-service due to a fault, or due to a planned outage/reconfiguration to enable repairs or maintenance activities	

(continued)

13.3 Industry Terms with Common English Description

Table 13.3 (continued)

Term	Area	D/T	Plain English definition	Code Ref
Flexibility	System services		Flexibility refers to the ability of users on a network to quickly change their operations (e.g. modifying generation and/or consumption patterns) in reaction to an external signal (e.g. change in price) in order to provide system services, such as supporting system balancing and network constraint management. Sources of flexibility are typically demand side response, storage, and dispatchable generation	Also in SCR charging futures glossary
Flexibility market	Industry term—DSO models	D/T	The arena of commercial dealings between buyers and sellers of flexibility services	Also in SCR charging futures glossary
Flexibility service	Industry term—DSO Models	D/T	The offer of modifying generation and/or consumption patterns in reaction to an external signal (such as a change in price) to provide a service within the energy system	Also in SCR charging futures glossary
Flexible connection(s)	Connection capacity		Flexible connections are non-firm, connection arrangements whereby a customer's export or import is managed (often through real-time control) based upon contracted and agreed principles of available capacity. Flexible connections typically allow quicker and cheaper connection to the network but have no defined cap on the extent to which a user's access can be interrupted. Timed connections and connections utilising active network management arrangements are examples of flexible connections. occasionally, flexible connections are also referred to as managed connections. Flexible connections refer to connection capacity and should not be confused with flexibility markets and flexibility services	Also in SCR charging futures glossary
Flexible resources		D/T	Flexible resources, typically distributed generation, storage or demand response, are connected to the electricity network, and are flexible in how they operate and impact the network	Also in SCR charging futures glossary

(continued)

Table 13.3 (continued)

Term	Area	D/T	Plain English definition	Code Ref
GB transmission system	Industry term	D/T	The system consisting of high voltage electric wires owned or operated by transmission licensees within Great Britain. This term is referred to in the CCCM and is similar to the term national electricity transmission system or "NETS" which is defined in CUSC	CCCM, also in SCR charging futures glossary
Generator	Industry term	D/T	This is a person/entity that generates electricity under licence or exemption under the electricity act	D-code
Generator installation	Industry term	D	This refers to any structure, equipment, lines, appliances or devices used or to be used by a generator and connected or to be connected directly or indirectly to a distribution system	DCUSA
Grid supply point ("GSP")	Industry term	D/T	This is a point of delivery from or to the national electricity transmission system to a distribution system or to a non-embedded demand customer	CUSC /D-code
Guaranteed standards of performance	General contractual terms		As part of a network operator's licence they are obligated to provide a minimum standard of service. This is backed by a guarantee and set out in the electricity (standards of performance) regulations 2005. Failure to deliver these standards of service can result in penalties to network operators and in the most severe form result in the loss of licence	CCCM
Independent connection provider (ICP)	Industry term		An independent connection provider (ICP) is an organisation, other than the DNO in whose distribution service region the connection is situated, accredited to undertake contestable works in relation to the provision of a connection to the DNO's distribution system	CiC COP, also in SCR charging futures glossary

(continued)

13.3 Industry Terms with Common English Description 303

Table 13.3 (continued)

Term	Area	D/T	Plain English definition	Code Ref
Independent distribution network operator (IDNO)	Industry term		This means an electricity distributor that is not a distribution services provider (or, if it is, is operating in relation to that part of its distribution system that is outside its distribution services area)	CiC COP, also in SCR charging futures glossary
Indicative maximum generation capability	Connection capacity	D/T	This is a customer's best estimate of its maximum generation	CUSC Section 4
Interactive connection applications	Connection process		These arise where there are two or more applications for connection which would make use of the same part of the existing or committed network or otherwise would have a material operational effect on that network such there is a material impact on the terms and conditions of any connection offer made in respect of such connections. Interactivity at transmission level occurs where a customer connection offer would affect the terms of an outstanding unsigned customer offer, which is reliant upon the completion of the same transmission reinforcement works	CCCM
Interactive connection offers	Connection process	D/T	There are occasions where network companies receive two or more connection applications that will make use of the same part of the existing network and where not all the applicants can be connected. The resulting connection offers are referred to as interactive connection offers. A queue of customers is then formed and the network company will connect as many customers as are technically feasible. Unsuccessful applicants will have their connection offers withdrawn and additional design work will be required	CCCM
Interactivity queue	Connection process	D/T	It is possible for a number of connection offers to have an impact on a common part of the network. Where this part of the network is not capable of allowing all the connections to occur (without reinforcement), the offers become interactive and join an interactivity queue. The queue order provides the contractual priority of the offers which will enter an interactivity process to allow time for each party to consider their offer before acceptance, which will be treated in queue order	CCCM

(continued)

Table 13.3 (continued)

Term	Area	D/T	Plain English definition	Code Ref
Intertrip	Connection capacity	D/T	A system or process to disconnect a generator or demand from the network in very short timescales when a specific event occurs	
Intertripping scheme (categories 1 to 4)	Connection capacity		Intertripping is used as a method of reducing generation capacity to the network, preventing inadvertent back feeding to the network or preventing system overloading. This prevents abnormal system conditions occurring following power system fault(s) such as: over voltage, thermal overload, system instability, etc..... There are four categories of intertripping scheme defined in the CUSC. These differ in the intended application Category 1 schemes are often used as part of a variation to a standard connection design under the NETS SQSS Category 2 schemes are often used to control loading on local circuits when other local circuits are out of service Category 3 schemes are used as an alternative to network reinforcement to alleviate loading on a third party system Category 4 schemes are used to enable the disconnection of the Connection Site from the transmission or distribution system in a controlled and efficient manner	CUSC
Land rights	General contractual terms		All such rights in, under or over land as are necessary for the construction, installation, operation, repair, maintenance, renewal or use of the contestable work or non-contestable work	CCCM
LIFO	Connection capacity		LIFO ("Last In First Out") is a means of allocating network capacity where a network constraint is resolved by curtailing all participating users in the order in which they applied for connection to the network The term LIFO stack refers to the ordered list of participating users. In the context of a multi-customer ANM scheme, a customer recently joining a scheme will be subject to more curtailment that other customers in the scheme who were connected in the scheme earlier	Also in SCR charging futures glossary

(continued)

13.3 Industry Terms with Common English Description 305

Table 13.3 (continued)

Term	Area	D/T	Plain English definition	Code Ref
Local market	Industry term—DSO models		The market(s) for services at local/distribution level e.g. management of constraints on a distribution network	
LV sub	Industry term		A connection to a site with distributor owned voltage transformation where the customer is metered at or very close to the lower voltage of the transformer at LV	
Managed connections	Connection capacity		Please see flexible connections term	
Maximum export capacity	Connection capacity	D/T	In respect of a connection point (or the connection points collectively), the maximum export capacity is the maximum amount of electricity (expressed in kW or kVA) which is permitted by the network company to flow into the distribution system	NTC
Maximum import capacity	Connection capacity	D/T	In respect of a connection point (or the connection points collectively), the maximum import capacity is the maximum amount of electricity (expressed in kW or kVA) which is permitted by the DNO to flow from the distribution system	NTC
Meter	Industry term	D/T	This is a device that measures the amount of energy passing through a given point Refer also metering point, MPAN, point of supply, connection point (or entry/exit point)	NTC, also in SCR charging futures glossary
Meter operator	Industry term	D/T	This is a person who installs, maintains or removes metering equipment used for measuring the flow of energy to or from a network at or near the supply terminals	ESQCR
Meter point administration number (MPAN)	Industry term	D	This is a 21 digit reference to uniquely identify network exit and entry points, such as individual domestic residences. Refer also meter, metering point, point of supply, connection point (or entry/exit point)	CCCM

(continued)

Table 13.3 (continued)

Term	Area	D/T	Plain English definition	Code Ref
Metering point	Industry term		This is the point, determined according to the principles and guidance given at Schedule 9 of the master registration agreement, at which a supply to (export) or from (import) a distribution system is measured. The measurements are used to ascertain a supplier/DG party's liabilities under the balancing and settlement code. The term can also refer to the point where metering equipment has been removed, or was intended to be measured. For an unmetered supply, a metering point can be the point where a supply is deemed to be measured. Refer also Meter, MPAN, point of supply, connection point (or entry/exit point)	DCUSA, also in SCR charging futures glossary
Minimum scheme	Connection process		In the context of a new distribution connection, the minimum scheme is the network design with the lowest overall cost which meets all technical, regulatory and safety requirements in order to provide the capacity required by the applicant	CCCM, also in SCR charging futures glossary
MITS (main interconnected transmission system)	Connection process		The MITS (main interconnected transmission system) refers to the bulk of the GB transmission system. The MITS includes: All the 400 and 275 kV elements of the onshore transmission system; 132 kV elements in Scotland operated in parallel with the remainder of the transmission system; any elements of offshore transmission systems operated in parallel with the remainder of the transmission system. It doesn't include generation circuits, transformer connections to lower voltage systems, interconnections to external systems, and any offshore transmission systems radially connected to the onshore transmission system	
MITS connection works	Connection process		MITS connection works are the transmission works (inclusive of substation works) that are required from the connection site to connect to a MITS substation	CUSC

(continued)

13.3 Industry Terms with Common English Description 307

Table 13.3 (continued)

Term	Area	D/T	Plain English definition	Code Ref
MITS substation	Connection process	D/T	In the context of the definition of MITS connection works, a MITS substation is a transmission substation with more than 4 main system circuits connecting at that substation These are identified in NGET's seven year statement	CUSC
Modification	Connection process	D	The refers to any actual or proposed replacement, renovation, modification, alteration or construction to a customer's plant or apparatus, or the manner of its operation, which materially effects another party Example 1: a customer has an existing connection to the distribution network. due to a requirement for additional capacity it is required to increase the size of the connection and as such a modification to the existing equipment supplying the customer is required Example 2: a customer has accepted a connection offer for a new connection. due to a change in the site layout they request a change to the cable route this results in a Modification to the quotation	NTC, also in SCR charging futures glossary
Mutually exclusive offers	Connection process	D/T	This occurs in situations where multiple offers have been issued for a similar connection to the same customer and are mutually exclusive. Only one of these mutually exclusive offers can be accepted and upon acceptance of one, the other(s) will immediately be withdrawn	
N-1	Connection capacity		N-1 means that is network is planned and operated such that the loss of any one element (e.g. a circuit on an overhead line route, a transformer, an underground cable) still allows the network to operate securely and to continue serving demand	
N-2	Connection capacity		N-2 means that is network is planned and operated such that the credible loss of any two elements (e.g. two circuits on an overhead line route, an underground cable and a separate overhead line circuit) still allows the network to operate securely and to continue serving demand	

(continued)

Table 13.3 (continued)

Term	Area	D/T	Plain English definition	Code Ref
National terms of connection (NTC)	Connection process		The national terms of connection set out the terms and conditions that distribution network operators require users to accept in return for maintaining the connection of the premises to its network	Also in SCR charging futures glossary
National electricity registration scheme (NERS)	Industry term	D	This is the scheme operated on behalf of the DNOs under which independent connection providers (ICPs) may be assessed, audited, surveyed, etc. leading to the issue and maintenance of accreditation for the carrying out of contestable works	
National electricity transmission system or "NETS"	Industry term		This is the system consisting of high voltage electric wires owned or operated by transmission licensees within Great Britain and offshore and used for the transmission of electricity from power stations to sub-stations, or between sub-stations, or to or from any external interconnection This system includes any plant, apparatus or meters that are owned or operated by any transmission licensee, within Great Britain or offshore, in connection with the transmission of electricity, but does not include remote transmission assets This term is referred to in the CUSC and is similar to the term GB transmission system which is defined in the CCCM	CUSC, also in SCR charging futures glossary
National market	Industry term—DSO Models		The market(s) for national services which are not location specific e.g. frequency response, reserve, etc.	

(continued)

13.3 Industry Terms with Common English Description 309

Table 13.3 (continued)

Term	Area	D/T	Plain English definition	Code Ref
NETS SQSS	Industry term	D/T	This is the national electricity transmission system security and quality of supply standard issued under standard condition C17 of the transmission licence. It includes the criteria used to plan and operate the national electricity transmission system	CUSC
New fault level capacity	Connection capacity	D	This is the assessed fault level capacity at the appropriate point on the distribution system following reinforcement It is used in the calculation of the apportioned cost chargeable to the customer in the charging methodology statements The fault level contribution from connection is defined as the incremental increase in fault level caused by the customer Where a customer applies to connect equipment and the Fault Level will cause the network to be reinforced, the new fault level capacity and fault level contribution from connection will be used to calculate the proportion of the costs to be paid by the applicant Where an existing customer requests a change to a connection then the fault level contribution from connection is defined as the incremental increase in fault level caused by the customer	CCCM, also in SCR charging futures glossary
New network capacity	Connection capacity		New network capacity is the assessed network capacity following reinforcement It is used in the calculation of the apportioned cost chargeable to the customer in the charging methodology statements. The new capacity is based on the operator's assessment of the thermal ratings, voltage drop and upstream restrictions and compliance with relevant design, planning and security of supply policies. The equipment ratings to be used are the appropriate operational ratings at the time of the most onerous operational conditions taking account of seasonal ratings and demand	CCCM, also in SCR charging futures glossary
Non-contestable work	Connection process		Non-contestable work is work that can only be undertaken by the host distribution network operator (DNO). It includes those connection Works that are identified by a DNO in its connection charging methodology and statement that may only be carried out by the DNO	CiC COP, also in SCR charging futures glossary

(continued)

Table 13.3 (continued)

Term	Area	D/T	Plain English definition	Code Ref
Normal operating conditions	Operations		Normal operating conditions include a range of conditions under which the system has been designed to operate. Typically, normal operating conditions cover generation variations, load variations and reactive compensation or filter states (e.g. shunt capacitor states), planned outages and arrangements during maintenance and construction work, non-ideal operating conditions and normal contingencies. Abnormal operating conditions could include switching in or out of circuits or network assets as a result of faults or other unplanned activities	
Notification of restrictions on availability	Connection capacity		This is a notification of outage conditions and/or circuit restrictions as applicable. It is usually associated with a design variation. Where a customer is subject to a notification of restrictions on availability, then the customer is not compensated for being constrained off	CUSC, also in SCR charging futures glossary
Operational intertripping	Connection capacity	D/T	This is the automatic tripping of circuit breakers to prevent abnormal system conditions occurring, such as over voltage, overload, system instability etc. after the tripping of other circuit breakers following power system fault(s). Operational intertripping might include generation and demand intertripping schemes	CUSC
Planning limits	Industry term	D/T	This is the result of an assessment of the capacity available to a distribution network operator at a grid supply point (import or export capacity) taking into account the capability of the network, the contracted background and the forecast operation of the network	
Platform/platform market	Industry term—DSO models	D/T	A platform market is a market where user interactions are mediated by an intermediary, the platform provider, and are subject to network effects. As opposed to a marketplace or trading exchange, a platform intermediary must offer inherent value beyond the simple mediation process for the two sides of the market. This added-value usually comes from ICT and the associated complementary innovation that increases utility and attractiveness of the Platform to all user groups [11]	

(continued)

13.3 Industry Terms with Common English Description 311

Table 13.3 (continued)

Term	Area	D/T	Plain English definition	Code Ref
Point of common coupling	Industry term	D/T	This is the point on a distribution network, electrically nearest the Customer Installation, at which other Customers are, or may be, connected. It is often referenced in the assessment of power quality	Engineering Rec G99
Point of connection (POC)	Connection process		As distinct from connection point. For each proposed new connection, this is the point (or points) of physical connection between the DNO's existing distribution system and the new assets for the extended network. I.e. the point on the existing network where the new connection will be connected, whether connected by the DNO or an independent connection provider. This is not a metering point	CiC COP
Point of supply	Industry term		This is the electrical position where the equipment in the customers premises connects to the distribution or transmission network. Usually, the equipment on one side will be owned by the customer and the equipment on the other side will be owned by the network operator. The metering is normally located here Refer also meter, metering point, MPAN, connection point (or entry/exit point)	ENA DG Connection guide
Power ramp rates	Industry term		These are the rates at which a generation or demand site increases or decreases power. Power ramp rates are important where rapid changes in power can affect the stability or operation of the electricity network, and therefore will be used in the design of a network Facilities such as energy storage sites often have capability to increase power export at a rapid rate. This can be a benefit to the system, although needs careful design consideration	
Power station capacity	Industry term		Or power generating facility capacity. The aggregated capacity of all the generating units associated with a single power station	EREC G100

(continued)

Table 13.3 (continued)

Term	Area	D/T	Plain English definition	Code Ref
Pre-qualification	Industry term—DSO models	D/T	Pre-qualification is a process to demonstrate flexibility service providers meet the technical requirements of a flexibility service and that activation does not cause the system to experience additional constraints (at both transmission and distribution levels)	
Principles of access	Connection capacity	D	Principles of access are a methodology or rules by which network access is granted and govern when a curtailment instruction is issued or network capacity released to a user under a flexible connection. They are relevant where non-firm connections are used including active network management (ANM) arrangements	
Pro rata	Industry term	D	In the context of a multi-customer active network management (ANM) scheme, this is an alternative to LIFO for the allocation of curtailment In pro rata allocation, the curtailment levels for customers in a scheme are pro-rated based on their agreed capacities rather than on the date of connection. This means that customers receive equal curtailment as a fraction of their un-curtailed generation	
Pro rata curtailment	Industry term	D	This is associated with curtailment in active network management schemes. Where a constraint occurs and it is necessary to constrain generator output; curtailment is shared equally among all generators in proportion to their capacity and contribution to the constraint Example 1: due to abnormal operating conditions or the pre-determined operating limit for the constraint having been reached it is required to reduce generation output flowing through the network to 15 MW. There are three generators connected to the same point of the network sized 10, 5, 15 MW. A pro-rata disconnection would apply a 50% curtailment to all generators reducing their outputs to 5, 2.5, 7.5 MW respectively Example 2: Same criteria is example 1, however the 5 MW generator is connected in a different area of the network and as such only contributes 1 MW of energy to the curtailed section of network. In this scenario to reduce the generation through the curtailed section of network only the 1 MW of contribution from the 5 MW unit would be pro-rated	

(continued)

13.3 Industry Terms with Common English Description 313

Table 13.3 (continued)

Term	Area	D/T	Plain English definition	Code Ref
Protected import/export capacity	Industry term		In the context of flexible connections, this is a level of capacity that is not subject to being curtailed where curtailment becomes necessary. For example, additional generation added to an existing site may be subject to curtailment, however the original generation on the site remains un-curtailed	
Re-energisation	Operations	D/T	This is the deliberate movement of any switch or the installation of any fuse or the taking of any other step whereby electrical current can flow between the transmission or distribution system and the customer's installation	NTC /CUSC
Regional market	Industry term—DSO models		The market(s) for services at a regional level that may encompass one or more distribution networks and/or an area of the transmission network e.g. management of transmission constraints through the balancing mechanism	
Registered capacity	Connection capacity		This is the maximum amount of active power deliverable by a power station at the entry point as declared by the generator	CUSC
Reinforcement	Industry term		Reinforcement is defined as the work carried out and the assets installed that add capacity (network or fault level) to the existing shared use transmission or distribution system	CCCM
Relevant embedded generator	Industry term	D/T	This is a generator embedded within a distribution network that is reasonably believed to have a significant system effect on the national electricity transmission system A significant system effect could be a change in power flow or fault level on the transmission system such that the operation of the relevant embedded generator needs to be explicitly considered in the planning and operation	CUSC
Relevant section of network	Operations	D	Relevant sections of network (RSNs) are that part or parts of the distribution system that can be used to supply a customer in both normal and abnormal running arrangements. There may be more than one RSN, e.g. at different voltage levels	

(continued)

Table 13.3 (continued)

Term	Area	D/T	Plain English definition	Code Ref
Request for a statement of works	Connection process	D/T	This is a formal request from a DNO to NGET for an assessment of the impact of relevant distribution connected generation upon the transmission network	CUSC Section 6
Required capacity	Connection capacity	D	This is the maximum capacity agreed with a customer. In the case of multiple connections (e.g. a housing development) it may be adjusted after consideration of the effects of diversity. Where an existing customer requests an increase in capacity then it is the increased capacity above their existing capacity	CCCM
Restrictions on availability	Connection capacity	D/T	At times there may be restrictions enforced on a connection due to pre-defined criteria. For transmission connections these will be set out in the relevant notification of restrictions on availability. These restrictions will be in the form of an outage or reduction in capability	CUSC
Second circuit outage	Operations		This refers to operational states of the network where two items of equipment are out-of-service simultaneously. This is usually due to the occurrence of a fault at the same time as a planned outage	
Section 16 (of the Act)	Legal		Section 16 of the Act creates an obligation on DNOs to connect customer's premises and IDNO networks i.e. to provide connection offers where requested to do so	The act
Section 16A (of the act)	Legal	D	Section 16A of the act explains the process for a person who requires a connection and states the minimum information required. It also places an obligation on the distributor to respond as soon as practicable	The act
Section 17 (of the act)	Legal		Section 17 of the act explains the circumstances where the distributor is not required to make a connection. It also explains what the distributor must do before being permitted to disconnect any premises or distribution system	The act
Section 19 (of the act)	Legal	D	Section 19 of the act allows the distributor to recover any reasonably incurred expenses in providing the connection from the applicant	The act

(continued)

13.3 Industry Terms with Common English Description

Table 13.3 (continued)

Term	Area	D/T	Plain English definition	Code Ref
Section 20 (of the act)	Legal	D	Section 20 of the act allows the distributor to require reasonable security from the applicant for the cost of the connection. It also explains the circumstances where the distributor will pay interest to the applicant	The act
Section 21 (of the act)	Legal		Section 21 of the act allows the distributor to require the applicant to accept any reasonable additional terms in respect of making the connection	The act
Section 22 (of the Act)	Legal		Section 22 of the Act the distributor and the applicant to enter into a special agreement with respect to the connection. The rights and liabilities would be subject to the special agreement and not S16 to 21. This does not prevent a person from making an application under S16A	The Act
Section 23 (of the act)	Legal		Section 23 of the act explains the circumstances where a person can refer a dispute to the regulator. No dispute can be referred to the Authority more than 12 months after the connection is made	The act
Secured amount	User commitment		This is the monetary amount that a customer is liable to provide security for against the event of termination of a transmission Bilateral agreement	CUSC, also in SCR charging futures glossary
Service provider	Industry term—DSO models	D/T	Those parties able to offer flexibility services	
Settlement	Industry term—DSO models	D/T	This is the process of measuring and verifying whether a service has been provided and whether there was an imbalance between the contracted position and the outturn	
Single circuit (connection)	Connection capacity	D/T	A single circuit connection arrangement means that in the event of a fault on that circuit or the distribution system feeding that circuit, or the need to take the circuit out of service for maintenance, the customer's connection will remain unavailable for the duration of the necessary works	Also in SCR charging futures glossary

(continued)

Table 13.3 (continued)

Term	Area	D/T	Plain English definition	Code Ref
Site	Industry term	D/T	A site is customer or company premises for which a connection point is made; or for a new connection as defined in the formal connection application site plan	Also in SCR charging futures glossary
Site specific requirements	General contractual terms	D/T	These are works deemed necessary by NGET in accordance with the grid code at an embedded generation site to enable the connection of that generator as identified through the statement of works process	CUSC, also in SCR charging futures glossary
Small-scale embedded generation (SSEG)	Industry term	D	This is defined in EREC G83 as "a generating unit together with any associated interface equipment that can be used independently, rated up to and including 16A per phase, single or multiphase 230/400 V AC and designed to operate in parallel with a public low voltage distribution system". I.e. up to 3.68 kW on a single-phase supply and 11.04 kW on a three-phase supply	ENA
Smart grid	Industry term—DSO models		This is an electrical grid which includes a variety of operational and energy measures including smart meters, smart appliances, renewable energy resources, and energy efficient resources. Electronic power conditioning and control of the production and distribution of electricity are important aspects of the smart grid	
Station demand	Industry term	T	In the context of a generation site; the station demand is the total site combined electrical demand where the site is being supplied by an electricity transmission system or a distribution system. I.e. All the site demand presented when the generation is not operating	CUSC
System losses	Industry term		These are the difference between the energy entering the national electricity system and the net energy exiting the system, split into two categories: technical losses, which occur as an innate by product of the operation of the electricity system i.e. heat generated by electrical plant/equipment Non-technical losses, which occur as a result of illegal abstraction, inaccuracies in unmetered supply inventories and metering conveyance errors i.e. unregistered meters and meter tolerances	

(continued)

13.3 Industry Terms with Common English Description 317

Table 13.3 (continued)

Term	Area	D/T	Plain English definition	Code Ref
Termination amount	User commitment		The monetary amount a customer is liable for in the event of termination of a transmission bilateral agreement. Note this may exceed the secured amount	CUSC, also in SCR charging futures glossary
Thermal rating	Operations	D/T	The current carrying capacity of the cable (or circuit) determined by the heating effect caused by electrical losses	ENA, also in SCR charging futures glossary
Timed connection	Connection capacity	D	A connection arrangement where connection capacity is subject to restrictions within specific time periods	
Transmission	Industry term	D/T	Part of the electricity transmission network transmitting high-voltage electricity from where it is generated to where it is distributed throughout the country. There are 3 transmission operators (TOs) permitted to develop, operate and maintain a high voltage system within their own distinct onshore transmission areas	CUSC, also in SCR charging futures glossary
Transmission circuits	Industry term	D/T	These are onshore or offshore transmission circuits and include the transmission system between two or more circuit- breakers which include, for example, transformers, reactors, cables and overhead lines and DC converters	CUSC
Transmission connection assets	Industry term	D/T	These are the transmission plant and apparatus necessary to connect to the national electricity transmission system at a connection site and which incur connection charges	CUSC
Transmission entry capacity	Connection capacity	T	This is the maximum capacity that shall be accepted into the NETS from a directly connected power station as specified in the relevant agreement	CUSC
Transmission works	Connection process		Transmission works are the works required on the transmission network to either enable a connection, maintain service performance and standards, or to recover equipment where no longer required. In relation to a particular customer, transmission works are specified in Appendix H or identified in the relevant CONSTRUCTION AGREEMent	CUSC, also in SCR charging futures glossary

(continued)

Table 13.3 (continued)

Term	Area	D/T	Plain English definition	Code Ref
Un-firm connections (non- firm)	Connection capacity		As distinct from firm. Un-firm connections are typically single circuit whereby the connection becomes unavailable in the event of a fault or necessary maintenance. The connection remains unavailable for the duration of the necessary works. Un-firm connections have become commonplace for generator connections on distribution networks to reduce connection charges (due to less assets required than for 2 circuit connections)	
Unmetered connection	Connection process	D	This is a connection to the electricity network that is provided without a metering point. A maintained inventory of connected equipment and usage profile will be provided to allow for accurate consumption and maximum capacity charging	Also in SCR charging futures glossary
User commitment methodology	User commitment		The user commitment methodology are the rules by which parties must underwrite works which they trigger on the transmission system. In the event that the party terminates its connection agreement prior to connection (or even if it reduces the capacity at which it eventually connects), it must pay a cancellation charge (the liability) to the network operator. They may also be required to provide security to cover a proportion of the liability prior to the start of any works on the connection	CUSC Section 15, Also in SCR charging futures glossary
Validity period	Connection process		The validity period is the period for which a connection offer or POC offer is open for acceptance	CCCM
Voltage of connection	Connection process	D	This is the voltage at the point of connection (POC) between the existing distribution network and the assets used to provide the new connection. This is not necessarily the voltage of supply to the customer For example, the point of connection may be at high voltage, the new connection asset may include a high voltage to low voltage transformer and consequently the voltage of supply to the customer will be at low voltage i.e. metered at low voltage	CCCM, also in SCR charging futures glossary

(continued)

13.3 Industry Terms with Common English Description 319

Table 13.3 (continued)

Term	Area	D/T	Plain English definition	Code Ref
Whole network	Industry term	D/T	Whole network means taking consideration of both transmission and distribution network costs and impacts	Also in SCR charging futures glossary
Whole system	Industry term	D/T	In the context of open networks, whole system means making optimal network investment and operational decisions for the whole electricity network, not just transmission or distribution networks in isolation from all the equipment connected to the network Whole system is being used in different ways across the electricity and wider energy industries at present Often, in the context of electrical networks, whole system tends to mean the whole electrical system encompassing both transmission and distribution networks, plus all the equipment connected to the networks including generators, demand devices, reactive compensation, energy storage Often, in wider debates about the development of the energy system (rather than just the electricity system) whole system may apply to all aspects of the energy including the whole electricity system, the gas system, fuel transporting and infrastructure, heat networks, and more	Open networks, also in SCR charging futures glossary
Wider transmission reinforcement works	Connection process	D/T	These are transmission reinforcement works (often remote from the connection) other than the enabling works and which are specified in the construction agreement. These works are not required to be completed prior to the user's equipment being energised under a connect and manage arrangement	CUSC, also in SCR charging futures glossary

References

1. I. Staffell, "Measuring the progress and impacts of decarbonising British electricity," *Energy Policy*, vol. 102, pp. 463–475, 2017.
2. Elexon-BSC, "BSC and Codes". https://www.elexon.co.uk/bsc-and-codes/
3. "Inclusion of the Common Connection Charging Methodology and Common Connection Charging Template into the DCUSA," 2011. https://www.dcusa.co.uk/change/inclusion-of-the-common-connection-charging-methodology-and-common-connection-charging-template-into-the-dcusa/
4. "Competition in Connections (CiC) Code of Practice," 2015. https://www.connectionscode.org.uk/
5. "Connection and Use of System Code (CUSC)," National Energy System Operator (NESO). https://www.neso.energy/industry-information/codes/connection-and-use-system-code-cusc
6. The Distribution Code of Licensed Distribution Network Operation of Great Britain. https://dcode.org.uk/
7. Distribution Connection and Use of System Agreement (DCUSA). https://www.dcusa.co.uk/
8. "The Electricity Safety, Quality and Continuity Regulations," 2002. https://www.legislation.gov.uk/id/uksi/2002/2665
9. "Grid Code (GC)," National Energy System Operator (NESO). https://www.neso.energy/industry-information/codes/grid-code-gc
10. ENA, "National Terms of Connection Version 13.0," 2021
11. "Security and Quality of Supply Standard (SQSS)," National Energy System Operator (NESO). https://www.neso.energy/industry-information/codes/security-and-quality-supply-standard-sqss

Open Access This chapter is licensed under the terms of the Creative Commons Attribution 4.0 International License (http://creativecommons.org/licenses/by/4.0/), which permits use, sharing, adaptation, distribution and reproduction in any medium or format, as long as you give appropriate credit to the original author(s) and the source, provide a link to the Creative Commons license and indicate if changes were made.

The images or other third party material in this chapter are included in the chapter's Creative Commons license, unless indicated otherwise in a credit line to the material. If material is not included in the chapter's Creative Commons license and your intended use is not permitted by statutory regulation or exceeds the permitted use, you will need to obtain permission directly from the copyright holder.

Appendix

Appendix I Standard Contract Template (Ver 3.0)

THIS AGREEMENT is made on [·] 20[·].

Between:

[·] (registered number [·]) whose registered office is at [·] (the "**Company**"); and

[·] **LIMITED/PLC**, a company incorporated in [England and Wales] [Scotland] (registered number [·]) whose registered office is at [·] (the "**Provider**"), (together the "**Parties**" and each a "**Party**").

Recitals:

(1) The Company, as owner and operator of the local Network, requires the provision of Flexibility Services (as hereinafter defined) to aid the management and operation of its Network. The Company wishes to contract with providers and/or operators of suitable assets for the provision of such Flexibility Services.

(2) The Provider is the owner and/or operator of assets, or has entered into arrangements for rights in respect of third party owned assets that have the capability to provide Flexibility Services and wishes to make available each Accessible Site for the provision of such Flexibility Services, for example through aggregated or individual assets. The Company will pay the Provider for these Flexibility Services in accordance with this Agreement.

(3) The Company wishes to appoint the Provider to provide the Flexibility Services and the Provider has agreed to provide the Flexibility Services to the Company, on and subject to the terms and conditions contained herein.

It is Agreed:

Glossary and Interpretation

[Note: cross-references to specific Service Terms and/or Annexes will be confirmed by the Company following inclusion of the Specific Service Terms and/or Annexes.]

1 Introduction

1.1 The glossary and rules of interpretation shall apply to any document published or to be published by the company which states (howsoever expressed) that it is governed by or subject to this glossary and rules of interpretation (see definition of associated document).

1.2 Any capitalised term used in the glossary and rules of interpretation shall have the meaning given to it (if any) in the glossary and service glossary as applicable.

1.3 The company may update any of the glossary and rules of interpretation, general terms and conditions, service glossary, service terms, annexes, forms and templates, and other associated documents from time to time by publication of an updated version of the relevant document on its website, and each such updated version shall be effective from the date shown on its front cover provided always that, except with the consent of the provider in writing (which shall include by approved electronic means to the extent permitted by the service terms), any updated version shall not apply to (i) any agreement already in force or (ii) to any service terms already applying to flexibility services currently being provided at the time of publication.

2 Rules of interpretation

2.1 Unless the context otherwise requires:

    2.1.1. The singular includes the plural and vice versa;

    2.1.2. Reference to a gender includes the other gender and the neuter;

    2.1.3. References to an act of Parliament, statutory provision or statutory instrument include a reference to that act of Parliament, statutory provision or statutory instrument as amended, extended or re-enacted from time to time and to any regulations made under it;

    2.1.4. Words denoting persons shall include any individual, partnership, firm, company, corporation, joint venture, trust, association, organisation or other entity, in each case whether or not having separate legal personality; and

    2.1.5. References to a company shall include a corporation or other body corporate and body corporate shall have the meaning given in Section 1173 of the Companies Act 2006.

2.2 A table of contents and headings are for convenience only and shall be ignored in construing the terms of the Agreement.

2.3 Any reference to the words "**including**", "**include**", "**in particular**" or any similar expression shall be construed as illustrative and shall not limit the sense of the words preceding those terms.

2.4 If a term or expression is defined within the Service Terms or Annexes relating to a particular service, the defined term or expression within the Service Terms or Annexes shall apply to the relevant service.

2.5 All references in an Associated Document, General Terms and Conditions, and Glossary to a particular paragraph or Annex shall be a reference to that paragraph or Annex in or to that Associated Document.

Priority of documents

2.6 If there is any conflict between the provisions of any of the documents comprising the Agreement, then the following order of priority between the documents shall apply:

    2.6.1   Associated Documents; and
    2.6.2   General Terms and Conditions and Glossary.

3 Glossary

In the Agreement, unless superseded by additional terms placed within the Service Glossary or Annexes or the context otherwise requires, the following expressions shall have the meaning set out below:

"Accessible site"	A Site that is not a domestic site
"Affiliate"	Any holding company or subsidiary company of a Party, or any company which is a subsidiary of such holding company and "holding company" and "subsidiary" have the meanings given in Section 1159 of the Companies Act 2006
"Agreement"	The general terms and conditions, the glossary, the service terms and service glossary, the annexes, the forms and templates
"Annexes"	The annexes appended to the general terms and conditions
"Apparatus"	All equipment in which electrical conductors are used, supported or of which they may form a part
"Applicable law"	Any applicable law, statute, by-law, regulation, order, regulatory policy, guidance or industry code, rule of court or directives or requirements of any regulatory body (including any health, safety and environmental legislation and approved codes of practice)
"Associated document"	Any document published or to be published by the company which states (howsoever expressed) that it is governed by or subject to this glossary and rules of interpretation in Part 2 above, which includes but is not limited to the service terms, service glossary, annexes and forms and templates
"Authority"	The gas and electricity markets authority
"Availability" or "available"	Means that the flexibility services, in accordance with the service requirements and the utilisation instruction, and where applicable, are available to be delivered to the company for the duration of the service window
"Availability payment!"	Has the meaning given to it in the service terms
"Balancing services activity"	Has the meaning attributed to it in the NESO's transmission licence
"BSC"	Means the balancing and settlement code as administered by Elexon

(continued)

(continued)

"Business day"	Any day other than a Saturday or Sunday or a bank holiday, in England and Wales where the company is located in England and Wales and in the City of Edinburgh where the company is located in Scotland
"Business hours"	Between 9:00 am and 5:00 pm on a Business Day
"Change in ownership"	Means: Any sale, transfer or disposal of any legal, beneficial or equitable interest in fifty per cent (50%) or more of the shares in the Provider (including the control over the exercise of voting rights conferred on those shares, control over the right to appoint or remove directors or the rights to dividends); and/or any other arrangements that have or may have or which result in the same effect as sub-clause a) above
"Charge(s)"	As applicable, the availability payments and the utilisation payments
"CMZ"	Constraint managed zone
"Confidential information"	Any information, however it is conveyed, that relates to the business, affairs, developments, trade secrets, know-how, personnel, customers and/or suppliers of a Party (and/or any its Affiliates) together with all information derived from the above, and any other information clearly designated as being confidential (whether or not it is marked as "confidential") or which ought reasonably to be considered to be confidential
"Connection agreement"	An agreement governing the terms of connection of any plant or apparatus to, and/or any agreement for the supply of electricity to the plant or apparatus or for the acceptance of electricity into, and its delivery from, the company's distribution system or transmission system (as the case may be)
"Connection and use of system code" or "CUSC"	The connection and use of system code designated by the secretary of state for energy security and net zero (DESNZ) as from time to time modified
"Contract award"	The execution and award by the company of a contract for the provision of flexibility services by the provider
"Contract data"	All data other than performance data associated with the agreement

(continued)

Appendix 325

(continued)

"Data protection law"	Any applicable law relating to the processing, privacy, and use of personal data, as applicable to the company, the provider and/or the flexibility services, including in the UK: (i) the privacy and electronic communications (EC directive) regulations 2003 and any current laws or regulations implementing Council Directive 2002/58/EC; and/or (ii) the general data protection regulation (EU) 2016/679 ("GDPR") as retained in the laws of the United Kingdom by the European Union (Withdrawal) Act 2018, and/or any corresponding or equivalent national laws or regulations, once in force and applicable, including the Data Protection Act 2018, and includes any judicial or administrative interpretation of them, any guidance, guidelines, codes of practice, approved codes of conduct or approved certification mechanisms issued by any relevant supervisory authority
"Day"	A calendar day
"DCUSA"	Means the distribution connection and use of system agreement entered into by the DCUSA parties (which includes the Company) and DCUSA limited
"Defaulting party"	Has the meaning given in paragraph 7.1 of the General Terms and Conditions
"Defect"	An issue that may arise with the DER equipment, metering or the communication interface between the Company and Provider which results in non-delivery of Flexibility Services or a misinformed delivery of Flexibility Services
"Development plan"	The defined schedule of design, build and commissioning in respect of a DER project in development
"Distributed energy resources" or "DER"	The electricity generators, electricity storage or electrical loads (both in respect of domestic and non-domestic assets and including, but not limited to, electric vehicle charge points), and other Site equipment, machinery, Apparatus, materials and other items used for the provision of the Flexibility Services as described in the Service Terms
"Distribution code"	The distribution code of licensed distribution network operators of Great Britain
"Distribution licence"	A licence issued under Sect. 6(1)(c) of the Electricity Act 1989
"Distribution limit"	£200,000 (two hundred thousand pounds sterling) or such other amount as may be stated in the Service Terms
"Distribution system"	A distribution network owned and/or operated by the holder of a distribution licence
"NESO"	Means national energy system operator (and any successor to its role)
"Expert"	An independent expert appointed for the purposes of expert determination

(continued)

(continued)

"Flexibility services"	Means, as more particularly described in the service terms, the services to be provided by the provider to the company under and in accordance with this agreement which give the company the ability to manage the load at a specific point of the network at certain points in time
"Force majeure event"	Any event or circumstance which is beyond either the company's or the provider's (as the case may be) reasonable control or its employees and which results in or causes its failure to perform any of its obligations under the agreement, provided that: (a) lack of funds; or (b) any failure or fault in the DER, including insufficient fuel, shall not constitute a force majeure event
"Forms and templates"	Where applicable, the relevant forms and templates associated with the on boarding, procurement, contract award or operation of Flexibility Services
"Fuel security code"	Means the document of that title designated as such by the secretary of state for energy security and net zero as may be amended from time to time
"General terms and conditions"	The general terms and conditions applicable to the provision of flexibility services to be provided under the agreement
"Glossary"	This glossary of terms and interpretation, as applicable to the Agreement
"Good industry practice"	The exercise of that degree of care, skill, diligence, prudence and foresight which would reasonably and ordinarily be expected from a skilled and experienced operator engaged in the same type of undertaking and carrying out services of similar nature, scope and complexity as the Flexibility Services, under the same or similar circumstances or the standard which would reasonably and ordinarily be expected from systems used by a skilled and experienced operator engaged in the same type of undertaking and carrying out services of similar nature, scope and complexity as the Flexibility Services, under the same or similar circumstances
"Grid code"	The technical code for connection and development of the national electricity transmission system as amended from time to time (available at www.nationalgrid.com/uk/electricity/codes/grid-code/code-documents)
"GSP"	Grid supply point
"Industry code"	The BSC, the CUSC, the grid code, transmission code, the distribution code, the DCUSA, the smart energy code, the retail energy code and the fuel security code

(continued)

(continued)

"Insolvency event"	Means any pre-insolvency, creditor protection, or insolvency related actions, events, processes or proceedings, whether in or out of court, including the following (and any proceedings or steps leading to any of the following): any form of bankruptcy, liquidation, administration, receivership, voluntary arrangement, scheme of arrangement, restructuring plan or other compromise or arrangement or scheme with creditors, moratorium, stay or limitation of creditors' rights, interim or provisional supervision by a court or court appointee, winding up or striking off, or any distress, execution, commercial rent arrears recovery or other process levied or exercised; or any similar actions, events, processes or proceedings in any jurisdiction outside England and Wales where the Company is located in England and Wales or alternatively Scotland where the Company is located in Scotland
"Intellectual property rights"	All intellectual property, including patents, trade marks, service marks, domain names, business and trading names, styles, logos and get-ups, rights in goodwill, database rights and rights in data, rights in designs, copyrights and topography rights (whether or not any of these rights are registered, and including applications and the right to apply for registration of any such rights) and all inventions, rights in know-how, trade secrets and Confidential Information lists and other proprietary knowledge and information and all rights under licences and consents in relation to any such rights and all rights and forms of protection of a similar nature or having equivalent or similar effect to any of these that may subsist anywhere in the world for their full term, including any renewals and extensions
"Material adverse effect"	Any event or circumstance which, in the opinion of the Company: Is likely to materially and adversely affect the Provider's ability to perform or otherwise comply with all or any of its obligations under this Agreement; or Is likely to materially and adversely affect the business, operations, property, condition (financial or otherwise) or prospects of the Company
"MPAN"	Meter point administration number
"MSID"	Metering system identifier
"Network"	The electricity network operated by the Company to which the DER is connected
"Non-terminating party"	Has the meaning given in paragraph 7.4 of the General Terms and Conditions
"Party"	Each of the Company and the Provider, together the "Parties"
"Performance data"	Such data relating to the performance of the plant, apparatus and related infrastructure as may be notified by the company to the provider or by the provider to the company from time to time

(continued)

(continued)

"Personal data"	Has the meaning given to it in data protection law[1]
"Plant"	Fixed and movable items used in the generation and/or supply and/or transmission and/or distribution of electricity other than apparatus
"Primacy rules"	Means the primacy rules defined by the energy networks association (as may be updated from time to time)
"Retail energy code"	The retail energy code administered by the Retail Energy Code Company Ltd
"Rules of interpretation"	The rules of interpretation detailed at paragraph 2 above
"Service failure"	As defined in the service terms
"Service glossary"	Any glossary of terms within the service terms as applicable to a particular flexibility service
"Service requirements"	The specification that the flexibility services must be capable of meeting, as defined in the service terms
"Service period"	As defined in the service terms
"Service terms"	The service terms applicable to the provision of flexibility services which form part of the agreement
"Service window"	The time periods during the service period during which the provider agrees to make available, and provide in accordance with the agreement, the flexibility services to the company, as defined in the service terms (if applicable)
"Site"	Means the site on which the DER is located
"Smart energy code"	The smart energy code administered by the smart energy administrator and secretariat
"Statutory requirements"	The requirements placed on the company and/or the provider or affecting or governing the provision and/or use of the flexibility services by applicable law and/or the applicable distribution licence or transmission licence and/or a regulator and/or any relevant codes of practice issued by any government agency or body including in relation to health, safety and environmental matters
"TCM"	Transmission constraint management
"Term"	The duration of the agreement as specified by the company in the service terms
"Terminating party"	Has the meaning given in paragraph 7.1 of the general terms and conditions
"Termination notice"	Has the meaning given in paragraph 7.4 of the general terms and conditions
"Transmission code"	The system operator transmission owner code as required by transmission licences granted under the electricity act 1989
"Transmission licence"	A licence issued under Section 6(1)(b) of the electricity act 1989

(continued)

---

[1] *Drafting note: applicable legislation reference to be confirmed depending on type of data being processed.*

Appendix

(continued)

"Transmission limit"	£500,000 (five hundred thousand pounds sterling) save as provided in the Service Terms
"Transmission system"	The electricity transmission system, as defined in the connection and use of system code
"Unavailability" (or "unavailable")	The flexibility services, in accordance with the service requirements, are not available to be delivered to the company
"Utilisation instruction"	An instruction by the company to the provider to deliver flexibility services
"Utilisation payments"	Has the meaning given to it in the service terms

General Terms and Conditions

[·] 2024

Note: cross-references to specific Service Terms and Annexes will be confirmed.

1 Introduction

1.1 These General Terms and Conditions shall apply to the provision of Flexibility Services by the Provider to the Company.

1.2 References to the "Agreement" in these General Terms and Conditions mean these General Terms and Conditions, the Glossary, the Service Terms and Service Glossary, the Annexes and where applicable, the Forms and Templates.

2 Scope of Flexibility Services

2.1 The Flexibility Services shall be performed in accordance with the Service Terms, these General Terms and Conditions and any other applicable Associated Documents.

3. **Provider's Obligations**

3.1 The Provider will:

  3.1.1 Ensure or procure the Availability of the DER and perform the Flexibility Services in compliance with the terms of the Agreement and all Applicable Laws, Statutory Requirements and Good Industry Practice;

  3.1.2 Ensure that all technical, communication and data provision requirements set out in the Service Terms and Annexes are complied with at all times;

  3.1.3 Act diligently and in good faith in all of its dealings with the Company;

  3.1.4 Ensure that it is available on reasonable notice to provide such assistance or information as the Company may reasonably require in connection with the Flexibility Services;

  3.1.5 At the request of the Company, make available to the Company information in relation to the metering equipment at the DER;

3.1.6 Where reasonably required by the Company in order to inspect and test the DER, or to install, maintain, replace or remove communication equipment belonging to the Company in relation to the provision of Flexibility Services in accordance with the Agreement; grant access to a Site in accordance with paragraph 6.6 of the Service Terms;

3.1.7 Remedy any Defect of the Flexibility Services in accordance with Good Industry Practice and to the satisfaction of the Company;

3.1.8 Disclose the existence of any agreement or arrangement the Provider may have in respect of the DER that provides Flexibility Services under the Agreement that could reasonably impact Availability of the DER or the ability of the Provider to perform its obligations under the Agreement;

3.1.9 Use reasonable endeavours to ensure that a DER that is pre-qualified is not registered with another Provider to provide Flexibility Services to the Company. If the Company identifies that the DER is registered with more than one Provider, the Company will notify both Providers. The DER will remain registered with the existing Provider until sufficient evidence of the Provider to which the Asset is registered has been provided to the Company's satisfaction (acting reasonably).

3.2 The Provider hereby acknowledges that Contract Award does not guarantee that any Flexibility Services will be required by the Company or commit the Company to requiring any, or any particular level of, such Flexibility Services.

4. **Record and Audits**

4.1 The Provider shall keep proper and accurate records of all matters relating to the performance of its obligations under the Agreement.

4.2 The records shall be maintained in a form suitable for audit purposes and shall be retained for any period required by any Applicable Law, and in any event, for the Term of the Agreement and for a period of no less than:

4.1.1 Seven (7) years after expiry or termination of the Agreement where such records contain or relate to financial data and/or Contract Data; or

4.1.2 Unless specified otherwise in the Annexes, four (4) years after expiry or termination of the Agreement where such records relate to Performance Data.

4.3 The Company, or a reputable independent third-party auditor nominated by it, may, on reasonable notice, and in any event on not less than fifteen (15) Business Days' (or such other period as may be specified in the Service Terms or required by Applicable Law) notice, to the Provider and during normal working hours, inspect and review the records, as described in paragraph 4.2, for the purposes of verifying the Provider's compliance with its obligations under the Agreement and/or to meet any other audit or information requirement that may be required by Applicable Law and/or any regulatory body, including the Authority.

4.4 The Provider shall co-operate fully and promptly with any such audit and/or inspection conducted by the Company and provide such reasonable assistance as may be required by the Company in relation to any audit.

General Terms and Conditions 333

4.5 The Provider shall ensure that all paperwork issued by or on behalf of the Provider to the Company (including, without limitation, invoices, correspondence and delivery notes), is complete, accurate and clearly references any other appropriate and necessary information.

5. **Representations and Warranties**

5.1 Without prejudice to its other obligations under and/or pursuant to the Agreement, each Party warrants and undertakes to the other Party at all times that:

5.1.1 It is a duly incorporated and company validly existing under the law of its jurisdiction of incorporation;

5.1.2 It has the right, power, capacity and authority to enter into and perform its obligations under the Agreement;

5.1.3 The entry into and performance by it of the Agreement does not and will not contravene or conflict with any Applicable Law or judicial or official order applicable to it;

5.1.4 It will not be in material breach of any other agreement or arrangement of whatever nature with any person which could or may affect the performance of its obligations under the Agreement;

5.1.5 All information it provides to the other Party will be complete and accurate save to the extent disclosed;

5.1.6 No Insolvency Event is continuing or might reasonably be anticipated; and

5.1.7 No litigation, arbitration or administrative proceedings are taking place, pending, or to the Party's knowledge threatened against it, any of its directors or any of its assets, which, if adversely determined might reasonably be expected to have a Material Adverse Effect.

5.2 Without prejudice to its other obligations under and/or pursuant to the Agreement and in addition to the foregoing, the Provider warrants and undertakes to the Company at all times that:

5.2.1 The DER contracted to provide the Flexibility Services has, as applicable, either:
  (a) Live connection(s) to the Company's Network, associated MPAN or MSID and Connection Agreement(s); or
  (b) A connection offer(s) for a live connection and that the connection(s) can be completed and a Connection Agreement entered into in time to meet the Service Requirements as specified in the Service Terms;

5.2.2 It has, or it will procure that the owner of the DER has, obtained and maintains in force for the Term, either directly or through agreement via its aggregated DER, all licences, permissions, authorisations, consents and permits needed to supply the Flexibility Services in accordance with the terms of the Agreement, including but not limited to any authorisation required pursuant to the regulations, codes, agreements and arrangements referenced in paragraph 5.2.9;

5.2.3 It has neither fixed nor adjusted any Charge under or in accordance with any agreement or arrangement with any other person, and that it has neither communicated to a person (other than its professional advisers) the amount or approximate amount of any Charge in connection with the Agreement (other than in confidence in order to obtain quotations necessary for insurance purposes) nor entered into any agreement or arrangement with any other person to restrain that other person from entering into an agreement for provision of Flexibility Services with the Company;

5.2.4 It shall disclose as soon as reasonably possible any change of circumstances which could affect the delivery of the Flexibility Services;

5.2.5 Where applicable, for each DER project in development, the provider has (or has procured), and, if requested, will promptly provide to the company a copy of the development plan in respect of each DER;

5.2.6 Where applicable, it shall take all reasonable steps to achieve, or procure, the commissioning of each DER project on time and in accordance with the relevant Development Plan;

5.2.7 If, at any time during the Term, the provision of Flexibility Services would cause the Provider to be in breach or non-compliance as described in paragraphs 5.1.3 and 5.2.9, the Provider will not accept or comply with any Utilisation Instruction and will provide notification to the Company as required by the Annexes;

5.2.8 Where any accessible site is occupied by an Affiliate of the Provider or any other third party, the Provider shall be responsible for ensuring that where any provision in the Agreement imposes an obligation on the Provider to do or refrain from doing a particular thing in relation to a Site or any DER at such Site, the relevant Affiliate or third party complies with that obligation as if it were the named "Provider" party to the Agreement; and

5.2.9 The provision of Flexibility Services will not cause it or the DER to be in breach of the Electricity Safety, Quality and Continuity Regulations 2002 (as amended from time to time) (available from the Company on request) or any other enactment relating to health and safety or standards, the Grid Code, Distribution Code, any Connection Agreement, any agreement for the supply of electricity, any restrictions and conditions attaching to relevant authorisations of the Environment Agency

5.3 Without prejudice to any right or remedy, each Party will be entitled to claim damages from the other Party for any breach of representation or warranty set out in the Agreement which causes that Party to incur costs or losses.

General Terms and Conditions

6 **Charges and Payments**

6.1 All Charges and other sums payable under the Agreement shall be paid in accordance with the Service Terms.

7. **Termination**

7.1 Each of the Parties shall have the right, if it is not the Party in breach or in relation to which any of the events concerned occurs ("**Terminating Party**"), to immediately terminate the Agreement on giving written notice of termination to the other Party ("**Defaulting Party**") if at any time during the Term of the Agreement:

   7.1.1 Subject to paragraph 7.3, the Defaulting Party is in material and/or persistent breach of the Agreement;
   7.1.2 An Insolvency Event occurs in relation to the Defaulting Party;
   7.1.3 Paragraph 11.6 of these General Terms and Conditions applies.

7.2 Either Party shall have the right to immediately terminate the Agreement on giving written notice of termination to the other Party under paragraph 9.4 of these General Terms and Conditions.

7.3 For the purposes of paragraph 7.1.1 and without limitation, the following shall be deemed to be a material breach by a Party of the Agreement:

   7.3.1 The Defaulting Party fails to pay (other than by inadvertent error in funds transmission which is discovered by Terminating Party, notified to the Defaulting Party and corrected within thirty (30) Business Days following such notification) any amount properly due or owing from it pursuant to paragraph 6, and such non-payment continues unremedied and not disputed in good faith and upon reasonable grounds at the expiry of thirty (30) Business Days immediately following receipt by the Defaulting Party of written notice from the Terminating Party of such non-payment;
   7.3.2 Paragraphs 8.3 or 15.10 of these General Terms and Conditions apply; or
   7.3.3 Any other material breach by the Defaulting Party of any of its obligations under the Agreement which, if capable of remedy, the Defaulting Party fails to remedy within ten (10) Business Days after service of a written notice from the Terminating Party specifying the breach and requiring it to be remedied.

7.4 Either Party (the "**Terminating Party**") may at any time on providing no less than ninety (90) Days prior written notice to the other Party (the "**Non-Terminating Party**") terminate the Agreement. Where the Non-Terminating Party fails to respond to a Termination Notice in accordance with this paragraph 7.4, the Non-Terminating Party shall be deemed to have accepted the Termination Notice.

Accrued liabilities

7.5 On termination, the rights and liabilities of the Parties that have accrued before termination shall subsist.

Surviving provisions

7.6 This paragraph and the following provisions of the Agreement shall survive termination or expiry:

   7.6.1 Paragraph 4 (*Records and Audit*);
   7.6.2 Paragraph 6 (*Charges and Payment*);
   7.6.3 Paragraph 7 (*Termination*);
   7.6.4 Paragraph 8 (*Service Failure*);
   7.6.5 Paragraph 10 (*Indemnity, Liability & Insurance*);
   7.6.6 Paragraph 12 (*Confidentiality*);
   7.6.7 Paragraph 13 (*Intellectual Property Rights*);
   7.6.8 Paragraph 14 (*Data Protection*);
   7.6.9 Paragraph 17 (*Dispute Resolution*);
   7.6.10 Paragraph 21 (*Waiver*);
   7.6.11 Paragraph 24 (*Governing Law and Jurisdiction*);
   7.6.12 Glossary; and
   7.6.13 Any other provision of the Agreement that expressly or by implication is intended to come into, or continue in force, on or after termination or expiry of the Agreement.

Consequences of termination or expiry

7.7 Where requested by the other Party, on termination or expiry of the Agreement each Party shall delete or return Confidential Information provided by the other Party for the purpose of the Agreement.

7.8 Following termination or expiry of the Agreement, the Provider shall promptly at the Provider's cost:

   7.8.1 Deliver to the Company for approval a final invoice detailing all monies due to it under the Agreement;
   7.8.2 Submit to the Company within thirty (30) Business Days all invoices with supporting documents for payment of all outstanding sums in connection with the provision of the Flexibility Services.

7.9 Where the Company terminates the Agreement as a result of a material and/or persistent breach by the Provider pursuant to paragraph 7.1.1, the Company may recover from the Provider any and all costs, losses and expenses reasonably incurred by the Company as a result of such termination, including where relevant such costs, losses and expenses associated with appointing a replacement Provider. Such costs, losses and expenses shall be payable by the Provider to the Company provided that the liability of the Provider in respect of this paragraph 7.9 shall not exceed (as applicable):

7.9.1 The Transmission Limit where such costs, losses and expenses are in connection with, or relate to, DER connected to the Transmission System; or

7.9.2 The Distribution Limit where such costs, losses and expenses are in connection with, or relate to, DER connected to the Distribution System.

7.10 The Parties agree that any costs, losses and expenses incurred by the Company pursuant to paragraph 7.9 shall be deemed direct losses and costs of the Company and accordingly not be subject to paragraph 10.3.

8. **Service Failure**

8.1 Notwithstanding its obligations under paragraph 8.2, the Provider shall notify the Company as soon as reasonably practicable upon becoming aware of the inability of the Provider to provide the Flexibility Services in all or any part of any contracted Service Window (if applicable) as set out in the Service Terms.

8.2 In the event of a Service Failure by the Provider, the Company may require the Provider to:

8.2.1 Provide the Company with a written explanation as to the cause of the failure of service delivery;

8.2.2 Implement a rectification plan for improving performance and/or reducing the number of occurrences of Unavailability, which may include at the Company's discretion, a repeat of any commissioning tests undertaken on initial installation and commissioning of the DER;

8.2.3 Propose a variation to the Service Requirements as specified in the Service Terms; or

8.2.4 Take any other action that may be agreed with the Company in order to alleviate a Service Failure (as reasonably required in the circumstances).

8.3 In the event that:

8.3.1 The Provider fails to comply with the terms of paragraph 8.2;

8.3.2 The Provider's proposals are not accepted by the Company (acting reasonably);

8.3.3 The Parties (acting reasonably) fail to reach agreement on any rectification actions; or

8.3.4 The Provider's performance in respect of the Service Failure notified by the Company does not significantly improve within thirty (30) Days of the date of the notice,

such failure will be deemed a material breach of the Agreement for the purposes of paragraph 7.1.1 of these General Terms and Conditions and paragraph 7.9 shall apply.

9. **Force Majeure**

9.1 A Party shall not be in breach or default of the Agreement to the extent that it is prevented from performing any of its obligations under the Agreement as a result of a Force Majeure Event, for so long as the Force Majeure Event continues to prevent such performance.

9.2 If a Force Majeure Event occurs, the following process will apply:

9.2.1 The affected Party will notify the other Party as soon as reasonably practicable of:
(a) The occurrence and description of the Force Majeure Event;
(b) The date on which the Force Majeure Event commenced and its likely duration (if known); and
(c) The effect of the Force Majeure Event on the Party's ability to perform its obligations under the Agreement;

9.2.2 As soon as is reasonably practicable following notification pursuant to paragraph 9.2.1, the Parties shall meet to discuss how best to continue their respective obligations under the Agreement; and

9.2.3 The affected Party will use reasonable endeavours to mitigate the impact of the Force Majeure Event on its ability to perform its obligations under the Agreement.

9.3 For the avoidance of doubt the non-performance of either Party's obligations under the Agreement arising prior to the Force Majeure Event, shall not be excused as a result of the Force Majeure Event.

9.4 If a Force Majeure Event prevents, hinders or delays a Party in performing its obligations under the Agreement for a continuous period of at least two (2) calendar months, either Party may terminate the Agreement with immediate effect.

10. Liability, Indemnity and Insurance

10.1 Subject to paragraph 10.2, and save where any provision of the Agreement provides for an indemnity, the Parties acknowledge and agree that neither Party nor any of its officers, employees or agents shall be liable to the other Party for loss arising from any breach of the Agreement other than for loss directly resulting from such breach and which at the date of formation of the Agreement was reasonably foreseeable as not unlikely to occur in the ordinary course of events from such breach in respect of:

10.1.1 Physical damage to the property of the other Party, its officers, employees or agents; and/or

10.1.2 Any liability arising under paragraph 5.3 and/or

10.1.3 The liability of such other Party to any other person for loss in respect of physical damage to the property of any person subject, for the avoidance of doubt, to the requirement that the amount of such liability claimed by such other Party should be mitigated in accordance with general law,

and provided further that the liability of any Party in respect of all claims for the losses referred to in this paragraph 10.1 shall not exceed (i) the Transmission

Limit where such claims are in connection with, or relate to, DER connected to the Transmission System or (ii) the Distribution Limit where such claims are in connection with, or relate to DER connected to the Distribution System, in each case per incident or series of related incidents.

10.2 Nothing in this Agreement shall exclude or limit the liability of either Party for death or personal injury resulting from the negligence of that Party or any of its officers, employees or agents, and each Party shall indemnify and keep indemnified the other Party, its officers, employees and agents from and against all such and any loss or liability which such other Party may suffer or incur by reason of any claim on account of death or personal injury resulting from the negligence of that Party or its officers, employees or agents.

10.3 Subject to paragraph 10.2, and save where any provision of the Agreement provides for an indemnity or otherwise, neither Party nor any of its officers, employees or agents shall in any circumstances whatsoever be liable to the other Party for:

10.3.1 Any loss of profit, loss of revenue, loss of use, loss of data, loss of contract or loss of goodwill; or

10.3.2 Any indirect or consequential loss; or

10.3.3 Loss resulting from the liability of the other Party to any other person howsoever and whenever arising save as provided in paragraphs 10.1.3 and 10.2.

10.4 Subject to paragraph 10.2, and save where any provision of the Agreement provides for an indemnity, the liability of any Party in respect of all claims for the losses referred to in paragraph 10.1 shall be subject to an aggregate cap of two million pounds sterling (£2,000,000).

10.5 The Provider shall procure (and on request provide evidence to the Company of) appropriate insurances as required by law and necessary for the safe and efficient performance of the Agreement to cover the liabilities set out in paragraph 10, with a reputable insurance company.

10.6 If the Provider appoints a sub-contractor in connection with the provision of the Flexibility Services, the Provider shall ensure that the sub-contractor maintains appropriate insurance to the extent set out in paragraph 10. If the Provider acts as an aggregator in connection with the provision of the Flexibility Services to Accessible Sites, it shall, where it is reasonably practicable to do so, ensure that the DER owners and operators for which it acts maintain appropriate insurance to the extent set out in paragraph 10.

10.7 The Provider's liabilities under the Agreement shall not be deemed to be released or limited by the Provider taking out the insurance policies referred to in paragraph 10.

11. **Transfers, sub-contracting and Change in Ownership**

11.1 Where pursuant to paragraph 24:

11.1.1 The governing law of this Agreement is English law, any reference to "assign" shall be construed as relating to an "assignment"; or

11.1.2 The governing law of this Agreement is Scots law, any reference to "assign" shall be construed as relating to an "assignation".

11.2 Save as provided for in paragraph 11.3, the Agreement is personal to the Parties and neither Party shall assign, transfer, mortgage, charge, sub-contract or deal in any other manner with any or all of its rights and obligations under the Agreement without the prior written consent of the other Party (such consent not to be unreasonably withheld, conditioned or delayed).

11.3 The Company may without the consent of the other Party assign, novate or transfer the benefit or burden of the Agreement or any other rights and/or obligations pursuant to these General Terms and Conditions to: (i) the holder of a Distribution Licence; (ii) the holder of a Transmission Licence with responsibility for carrying out the Balancing Services Activity; or (iii) to an Affiliate of the Company but only where such Affiliate of the Company holds a Distribution Licence or a Transmission Licence.

11.4 If either Party sub-contracts any part of the provision or obligations of Flexibility Services, then the responsible Party shall be fully responsible for the acts, omissions or defaults of any sub-contractor (and its employees) as if they were the acts, omissions or defaults of the responsible Party.

11.5 If ownership, occupancy or use (for the purpose of providing the Flexibility Services) of any Accessible Site changes, or may change, during the Term, the Provider shall promptly notify the Company of the same. Where (i) the ownership, occupancy or use (for the purpose of providing the Flexibility Services) of any Accessible Site changes during the Term; or (ii) the use (for the purpose of providing the Flexibility Services) of any domestic Site changes during the Term, the Provider shall update its records and ensure that such records are reflective of such changes. The Company and the Provider shall if required, and at the reasonable request of the Company discuss the implications of the change and the options available to minimise any disruption that may be caused by the change.

11.6 The Company reserves the right to terminate the Agreement in accordance with paragraph 7.1.3 if a Change in Ownership of the Provider occurs and the new owner of the Provider fails to meet any of the Company's reasonable due diligence checks as notified to the Provider.

12 **Confidentiality**

12.1 The Company is required to disclose certain information in accordance with this Agreement under obligations within its Distribution Licence or Transmission Licence (as applicable), or an Industry Code. Information shared will include but may not be limited to provider names, awarded prices, volumes,

General Terms and Conditions 341

GSP and asset locations, and contract durations. Pursuant to the Primacy Rules, the Company, as applicable, shall be entitled to share information relating to the Agreement for the purpose of industry initiatives in relation to network or system constraint management and electricity network optimisation and the Company shall be entitled to make publicity releases and/or announcements regarding either this Agreement and/or the Company's activities under the Agreement. It shall not be a breach of this paragraph 12 where the Company discloses any such information. Such information shall include but is not limited to:

12.1.1 CMZ locations;
12.1.2 CMZ requirements;
12.1.3 A list of TCM generators;
12.1.4 An agreed form of 'risk of conflict forecast';
12.1.5 NESO planning outputs;
12.1.6 Company outages;
12.17 Transmission outages; and
12.18 Any additional Company related information as may be required, as may be updated from time to time on agreement from the Company or the NESO.

12.2 Subject to paragraphs 12.1, 12.3.4 and 12.3.5, no public announcement or statement regarding the completion, performance or termination of the Agreement shall be issued or made by the Provider without the Company's prior written approval (such approval not to be unreasonably withheld or delayed). Neither Party shall be prohibited from issuing or making any such public announcement or statement to the extent expressly permitted or if it is necessary to do so in order to comply with any Applicable Law or the regulations of any recognised stock exchange upon which the share capital of such Party is from time to time listed or dealt in.

12.3 Save as permitted by paragraph 12.1, each Party shall treat as strictly confidential and shall not disclose any Confidential Information relating to the other Party received or obtained as a result of entering into or performing this Agreement. The restrictions imposed by this paragraph 12.3 shall not apply to the disclosure of any Confidential Information:

12.3.1 Which is in or becomes part of the public domain otherwise than as a result of a breach of paragraph 12.3, or which either Party can show was in its written records prior to the date of disclosure of the same by the other Party, or which it received from a third party independently entitled to disclose it;

12.3.2 Which is required to be disclosed by law, an Industry Code or pursuant to any licence of the Party concerned;

12.3.3 To a court, arbitrator or administrative tribunal in the course of proceedings before it to which the disclosing Party is a party;

12.3.4 To any parent, subsidiary or fellow subsidiary undertaking on a "need to know" basis only. In this paragraph 12.3.4, the words "parent", "subsidiary" and "undertaking" shall have the meanings as provided in Sections 1159, 1161 and 1162 of the Companies Act 2006;

12.3.5 By the Provider to any owner and/or operator of relevant Plant and Apparatus to the extent necessary to enable the Provider to submit an offer or tender to provide Flexibility Services pursuant to the Agreement and fulfil its obligations under the Agreement.

12.4 Save as permitted by paragraph 12.1, neither Party shall use the name, brands and/or logos of the other Party for any purpose without the other Party's prior written approval (such approval not to be unreasonably withheld or delayed).

13 Intellectual Property Rights

13.1 The Agreement does not transfer any interest in Intellectual Property Rights.

13.2 All Intellectual Property Rights owned by or licensed to either Party shall at all times both during the Term of the Agreement and after its termination or expiry, belong to or be licensed to the Party providing that intellectual property and neither Party shall make any use of the other Party's intellectual property other than to the extent reasonably necessary in performing its obligations pursuant to the Agreement, provided that nothing in this paragraph 13.2 shall operate so as to exclude any non-excludable rights of either Party.

14 Data Protection

14.1 Each Party shall, at its own expense, ensure that it complies with all applicable Data Protection Law.

14.2 The Parties acknowledge that as at the date of the Agreement, neither Party acts as a processor on behalf of the other. If at any point during the Term, either Party considers that one Party is acting as processor on behalf of the other, then the Parties shall promptly meet to negotiate in good faith a separate data processing agreement to cover the matters required by the Data Protection Law.

15 Modern Slavery, Anti-bribery and Living Wage

Modern slavery

15.1 The Parties undertake, warrant and represent that:

15.1.1 Neither Party nor any of its officers, employees, agents or subcontractors:
(a) Has committed an offence under the Modern Slavery Act 2015 ("**MSA Offence**");
(b) Has been notified that it is subject to an investigation relating to an alleged MSA Offence or prosecution under the Modern Slavery Act 2015; or

General Terms and Conditions 343

   (c) Is aware of any circumstances within its supply chain that could give rise to an investigation relating to an alleged MSA Offence or prosecution under the Modern Slavery Act 2015;

 15.1.2 They shall comply with all applicable anti-slavery and human trafficking laws, statutes, regulations and codes from time to time in force including but not limited to the Modern Slavery Act 2015;

 15.1.3 They shall notify the Company immediately in writing if they become aware or has reason to believe that they, or any of its officers, employees, agents or subcontractors have breached or potentially breached any of the Provider's obligations under this paragraph 15.1. Such notice to set out full details of the circumstances concerning the breach or potential breach of Provider's obligations;

 15.1.4 They shall include in their contracts with subcontractors and suppliers' anti-slavery and human trafficking provisions that are at least as onerous as those set out in this paragraph 15.1; and

 15.1.5 They will respond to all reasonable requests for information required by the other Party for the purposes of completing other Party's annual anti-slavery and human trafficking statement.

15.2 The Provider shall indemnify the Company against any losses, incurred by or awarded against the Company as a result of any breach of anti-slavery and human trafficking laws, statutes, regulations and codes or the Modern Slavery Act 2015.

15.3 The Provider will permit the Company and its third party representatives, on reasonable notice during normal Business Hours, but without notice if there are reasonable grounds to suspect an instance of slavery and human trafficking, to access and take copies of records and any other information held at the Provider's premises (which shall be the Provider's office premises and other business premises) and to meet with personnel and more generally to audit compliance with its obligations under this paragraph 15. The Provider shall give all necessary assistance to the conduct of such audits during the term of the Agreement.

Anti-bribery

15.4 The Provider shall have suitable controls and compliance procedures in place and shall not engage in any activity, practice or conduct which would constitute an offence under the Bribery Act 2010 and shall promptly report to the Company any request or demand for any undue financial or other advantage of any kind received or offered by the Provider in connection with the Agreement.

15.5 The Provider shall immediately notify the Company if a foreign public official exerts a direct or indirect influence over the performance of the Agreement.

15.6 The Provider shall not:

 15.6.1 Offer or agree to give any person working for or engaged by the Company or any other Affiliate of the Company any gift or other consideration which could act as an inducement or a reward for any act or failure to act connected to the Agreement, or any other agreement between the Provider and the Company or any Affiliate of the Company, including its award to the Provider and any of the rights and obligations contained within it; nor

15.6.2 Enter into the Agreement if it has knowledge that, in connection with the Agreement, any money has been, or shall be, paid to any person working for or engaged by the Company or any other Affiliate of the Company by or for the Provider, or that an agreement has been reached to that effect, unless details of any such arrangement have been disclosed in writing to the Company and has been approved by the Company before execution of the Agreement.

15.7 The Provider shall indemnify the Company against any losses, incurred by or awarded against the Company as a result of any breach of anti-corruption and anti-bribery laws, statutes, regulations and codes or the Bribery Act 2010.

15.8 The Provider agrees to provide the Company with such reasonable assistance as it may require from time to time to enable it to perform any activity required by any relevant government, agency or competent authority in any relevant jurisdiction for the purpose of compliance with any anti-slavery laws or anti-bribery laws (including but not limited to the Modern Slavery Act 2015 and the Bribery Act 2010).

Living wage

15.9 Where applicable the Provider agrees to:

15.9.1 pay all of its personnel who are directly employed by it in respect of the provision of the Flexibility Services used within the UK not less than the real living wage (as defined at https://www.livingwage.org.uk/ as may be updated from time to time) for the Term of the Agreement; and

15.9.2 ensure all employees of its contractors and subcontractors performing the provision of the Flexibility Services used within the UK are paid not less than the real living wage (as defined at https://www.livingwage.org.uk/ as may be updated from time to time) for the Term of the Agreement.

15.10 Any breach of this paragraph 15 by the Provider shall be deemed a material breach of the Agreement for the purposes of paragraphs 7.1.1 and 7.9.

16 Notices

16.1 Unless otherwise specified in the Service Terms, all notices shall be submitted in accordance with the processes, and to the relevant addresses, set out in the Service Terms.

16.2 A notice shall be deemed to have been received:

16.2.1 If delivered by hand or recorded delivery post within Business Hours at the time of delivery or, if delivered by hand outside Business Hours, at the next start of Business Hours;

16.2.2 If sent by first class post, at 9.00 a.m. on the second Business Day after posting.

16.3 E-mail communications may be valid for notices the purposes of the Agreement, where agreed between the Parties. Such email notices shall be deemed to have been received on the Day of sending, or where outside of Business Hours on the first Business Day thereafter.

16.4 In verifying service of a notice, it shall be sufficient to prove that delivery was made or that the envelope containing the notice was properly addressed and posted.

16.5 This paragraph 16 does not apply to the service of any legal proceedings, or other documents in any legal action or other method of dispute resolution.

17. **Dispute Resolution**

17.1 The Parties shall use good faith efforts to resolve any operational issue, dispute, claim or proceeding arising out of or relating to the Agreement.

17.2 In the event that a dispute cannot be resolved within thirty (30) Days of written notice of the dispute, the dispute shall be escalated to the Parties' senior representatives (named in the Service Terms, or as otherwise notified by either Party to the other) who have authority to settle the same and/or may refer the dispute to the forms of dispute resolution in accordance with paragraph 17.3.

17.3 If thirty (30) Days following such an escalation the Parties have still not resolved the dispute, then either Party shall have the right to refer the dispute to either:

17.3.1 Arbitration; or

17.3.2 An Expert for determination; or

17.3.3 Such other process as is agreed between the Parties.

17.4 For the avoidance of doubt, paragraphs 17.2 and 17.3 shall not preclude a Party from raising arbitration proceedings (or where other processes have been agreed under paragraph 17.3.3 court proceedings) in the event a claim is considered to be nearing the end of a prescription and/or limitation period pursuant to the Limitation Act 1980 or the Prescription and Limitation (Scotland) Act 1973 (as applicable) or where determination is required in the event of an emergency where the time periods set out in this paragraph 17 would not be suitable.

17.5 In the event that the Parties cannot agree any other process under paragraph 17.3.3, then either Party may refer any dispute to the courts of: (i) England and Wales if the Company is incorporated in England and Wales; and (ii) Scotland if the Company is incorporated in Scotland (as applicable).

Arbitration

17.6 Where any dispute is referred in accordance with paragraph 17.3.1 to arbitration, the following provisions shall apply:

    17.6.1 If the Company is incorporated in England and Wales, the seat of arbitration shall be London. If the Company is incorporated in Scotland, the seat of arbitration shall be Edinburgh;

    17.6.2 The number of arbitrators shall be one. Where no arbitrator is named or where the named arbitrator is not able or unwilling to act the appointer of the arbitrator (and of any replacement) shall be The Chartered Institute of Arbitrators;

    17.6.3 Whatever the nationality, residence or domicile of either Party and wherever the dispute or difference or any part thereof arose, (i) the laws of England and Wales shall be the proper law of any reference to arbitration if the Company is incorporated in England and Wales or (ii) the laws of Scotland shall be the proper law of any reference to arbitration if the Company is incorporated in Scotland, and in particular (but not so as to derogate from the generality of the foregoing) the rules and provisions of (i) the Arbitration Act 1996 (notwithstanding anything in Section 108 thereof) shall apply if the Company is incorporated in England and Wales or (ii) the Arbitration (Scotland) Act 2010 shall apply if the Company is incorporated in Scotland, to any such arbitration wherever the same or any part of it shall be conducted;

    17.6.4 For the avoidance of doubt, both Parties confirm and agree that nothing in the Agreement to arbitrate prevents a Party:

        (a) Challenging the award of an arbitral tribunal as provided for under the Arbitration Act 1996 and the Arbitration (Scotland) Act 2010;

        (b) Seeking the remedy of specific performance or any other power or remedy that would be available to the English court or Scottish court (as the case may be) from the arbitral tribunal in accordance with the Arbitration Act 1996 and the Arbitration (Scotland) Act 2010;

        (c) Seeking interim relief from the English court or Scottish court (as the case may be) under the Arbitration Act 1996 and the Arbitration (Scotland) Act 2010, or from any other court with competent jurisdiction; or

        (d) Seeking to enforce any arbitral award in the English court or Scottish court (as the case may be) or any court of competent jurisdiction.

    17.6.5 Without prejudice to any other mode of service allowed under any relevant law, where a Provider is not incorporated in any part of Great Britain, the Provider agrees that if it does not have, or shall cease to have, a place of business in Great Britain it will promptly appoint, and shall at all times maintain and identify to the Company, an agent for the service of process in Great Britain to accept service of process on its behalf in any proceedings commenced in support of, or in relation to arbitration, in the courts of England and Wales or Scotland (as the case may be).

Expert determination

17.7 Where any dispute is referred in accordance with paragraph 17.3.2 to an Expert for determination, the following provisions shall apply:

17.7.1 The Expert shall act as an expert and not as an arbitrator and shall decide those matters referred to them using their skill, experience and knowledge, and with regard to all such other matters as they in their sole discretion consider appropriate;

17.7.2 If the Parties cannot agree upon the selection of an Expert, the Expert shall be determined by (i) the President for the time being of the Law Society of England and Wales, if the Company is incorporated in England and Wales or (ii) the President for the time being of the Law Society of Scotland, if the Company is incorporated in Scotland;

17.7.3 All references to the Expert shall be made in writing by either Party with notice to the other being given contemporaneously, and the Parties shall promptly supply the Expert with such documents and information as they may request when considering any referral;

17.7.4 The Expert shall be requested to use their best endeavours to give their decision upon the question before them as soon as possible in writing following its referral to them, their decision shall, in the absence of fraud or manifest error, be final and binding upon the Parties;

17.7.5 If the Expert wishes to obtain independent professional and/or technical advice in connection with the question before them:
   (a) The Expert shall first provide the Parties with details of the name, organisation and estimated fees of the professional or technical adviser; and
   (b) The Expert may engage such advisor with the consent of the Parties (which consent shall not be unreasonably withheld or delayed) for the purposes of obtaining such professional and/or technical advice as they may reasonably require;

17.7.6 The Expert shall not be held liable for any act or omission, and their written decision will be given without any liability on the Expert's part to either Party, unless it shall be shown that they acted fraudulently or in bad faith;

17.7.7 Save to the extent otherwise expressly provided herein pending the determination by the Expert, any subsisting Agreement shall continue to the extent possible for the Parties to perform their obligations; and

17.7.8 The Expert shall at their discretion be entitled to order that the costs of the reference of a dispute to them shall be paid by the Parties in whatever proportions they think fit.

18. **Severance**

18.1 If any provision of the Agreement becomes or is declared invalid, unenforceable or illegal by a judicial or other competent authority, such invalidity, unenforceability or illegality shall not prejudice or affect the remaining provisions of the Agreement, which shall continue in full force and effect notwithstanding such invalidity, unenforceability or illegality.

18.2 The Company and the Provider each acknowledge that it has entered into the Agreement on an arm's length basis and that it has taken independent legal advice in so doing.

19 **Third Party Rights**

19.1 For the purposes of the Contracts (Rights of Third Parties) Act 1999 or where appropriate the Contracts (Third Party Rights) (Scotland) Act 2017, the Agreement is not intended to, and does not, give any person who is not a Party to it any right to enforce any of its provisions.

20. **No Agency or Partnership**

20.1 Nothing in the Agreement shall be deemed to constitute a partnership or joint venture or contract of employment between the Parties nor constitute either Party the agent of the other.

20.2 Neither Party shall act or describe itself as the agent of the other, nor shall it make or represent that it has authority to make any commitments on the other's behalf, including but not limited to the making of any representations or warranty and the exercise of any right or power.

21. **Waiver**

21.1 No failure or delay by any Party to exercise any right, power or remedy under the Agreement will operate as a waiver of it nor will any partial exercise preclude any further exercise of the same, or of some other right, power or remedy.

22. **Entire Agreement**

22.1 The Agreement and the Associated Documents referred to in it together constitute the entire agreement and understanding of the Parties relating to the matters contemplated by the Agreement and those documents, and supersede any previous drafts, agreements, understandings or arrangements between any of the Parties relating to the subject matter of the Agreement and those documents, which shall cease to have any further effect.

23. **Counterparts**

23.1 Where executed in counterparts:

23.1.1 The Agreement shall not take effect until all of the counterparts have been delivered; and

23.1.2 Delivery will take place when the date of delivery is agreed between the Parties after execution of the Agreement as evidenced by the date at the top of the Agreement.

23.2 Where not executed in counterparts, the Agreement shall take effect after its execution upon the date agreed between the Parties as evidenced by the date at the top of the Agreement.

24 **Governing Law and Jurisdiction**

24.1 The validity, construction and performance of the Agreement and any claim, dispute or matter (whether contractual or non-contractual) arising under or in connection with the Agreement or its enforceability shall be governed by and construed: (i) in accordance with English law if the Company is incorporated in England and Wales; and (ii) in accordance with Scots law if the Company is incorporated in Scotland.

[Drafting note: Signature documents can be added by the DSO as per its process]

Signed by the duly authorised representatives of the Parties as an agreement on the date first written above

Signed)
)
)
for and on behalf of
[COMPANY])   ....
Director/Duly Authorised Signatory

Signed)
)
)
for and on behalf of:
[PROVIDER])   ....
Director/Duly Authorised Signatory

Flexibility Services Service Terms—Company Active Services

[·] 2024

1 **Introduction**

   [Each DSO to add description of Services and/or terms of the Flexibility Services to be included.]

   These Service Terms relate to the Company's procurement of Flexibility Services on its electricity distribution network.

   [Parameters table to be included by DSOs with description of Company Active Services.]

2 **Changes to Service Terms**

   [To be reviewed for formal governance requirements following the establishment of Market Facilitator.]

3 Service Terms Glossary

   [Each DSO may add or remove definitions as applicable to their individual population of the Service Terms Template]

These additional terms placed within the Service Terms are applicable to all Associated Documents and shall supersede terms within the General Terms and Conditions and Glossary. The following expressions shall have the meaning set out below:

"Accepted [MW/MVAR]"	The [MW/MVAR] accepted in accordance with [this Annex]
"Accepted availability window"	Where services have been contracted to include variable availability, the accepted availability window is the period required for service provision to be made available following the agreement between the Company and Provider during the Availability Refinement Period. If a service does not have an Availability Refinement Period, then this Accepted Availability Window is defined within the Contract Award
"Accepted end time"	The date and time (to the nearest minute) as notified in accordance with the Service Terms at which the Accepted [MW/MVAR] is no longer required to be delivered

(continued)

(continued)

"Accepted start time"	The date and time (to the nearest minute) as notified in accordance with the service terms at which the accepted [MW/MVAR] shall be delivered
"Active power"	The product of voltage and the in-phase component of alternating current measured in units of Watts and standard multiples thereof i.e. 1000 Watts = 1 kW, 1000 kW = 1 MW, 1000 MW = 1GW, 1000 GW = 1TW
"Agreed availability capacity"	The volume of capacity required to be made available for the provision of services following the agreement between the company and provider during the availability refinement period, where applicable
"Asset point metering"	The metering measured directly from the DER and is downstream of the boundary point metering
"Availability fee"	The fee payable in consideration for the provider making the DER available and calculated in accordance with the provisions of the service terms
"Availability payments"	Means the payments made by the company in respect to the accepted availability windows
"Availability refinement period"	Means the period defined within the product parameters where a refinement of the availability window and agreed availability capacity is agreed
"Availability status"	Available or unavailable
"Boundary metering point"	The metering measured at the point of supply from the Company network
"Demand"	The demand (in MW) of active power consumed by plant and/or apparatus
"Demand response active power code"	As defined in the Grid Code
"Demand response provider"	As defined in the Grid Code
"Generation"	The electrical output (in MW) of a [Unit]
"Monthly utilisation performance factor"	The calculation of the impact on the Availability Payment, for the relevant month, of how the Provider performs where Utilisation Instructions have been issued
"Output"	Active power output (in MW) achieved by plant and/or apparatus
"Performance report"	Means a report in relation to the flexibility services provided by a DER, or groups of DER responding to utilisation instructions in accordance with the service terms
"Power requirement"	Means the level of power injection or demand reduction required by the company within a specified service window (if applicable) and delivered by the provider following a utilisation instruction
"Recovery time"	The minimum time required between the end of a flexibility service delivery window and the commencement of the next flexibility service delivery window, as defined in the service terms

(continued)

Flexibility Services Service Terms—Company Active Services 353

(continued)

"Requested end time"	The date and time (to the nearest minute) as notified in accordance with this annex at which the requested MW is no longer required to be delivered
"Requested MW"	The MW requested by the company in accordance with the service terms
"Requested start time"	The date and time (to the nearest minute) as notified in accordance with the service terms at which the requested MW shall be delivered
"Service meter"	The measuring equipment, as defined by the company in the service terms, that shall be used to determine delivery of the flexibility services
"Service meter data"	The meter data recorded at the service meter at the site(s) listed in the service terms
"Service period"	The period as specified in the service terms
"Service requirement"	The relevant service requirements detailed in the notification of contract award
"Service window"	The relevant service window detailed in the notification of Contract Award
"Stop instruction"	An instruction from the Company to the Provider, instructing the Provider to cease delivery of the Flexibility Services, as more particularly described in the Service Terms
"Utilisation fee"	The amount payable by the Company to the Provider for the utilisation of any Flexibility Service, as defined in the Service Terms
"Utilisation payments"	The payments made by the Company in respect to the capacity or energy delivered over a specified period of time, in response to a Utilisation Instruction
"Zone"	The feeding area of the DERs being managed or where the Flexibility Services will be provided and to which the Flexibility Services will be delivered
[Additional company terms…]	
[…]	

4 Service Details

4.1 Service Parameters

    4.1.1 Details of the Service Parameters shall be provided in the notification of Contract Award and examples shall be available within the relevant tools and templates schedule.

4.2 Service Windows

    4.2.1 Details of the Service Windows shall be provided in the notification of Contract Award and examples shall be available within the relevant tools and templates schedule.

4.3 Service Requirements

    4.3.1 Details of the Service Requirements shall be provided in the notification of Contract Award and examples shall be available within the relevant tools and templates schedule.

5 Invoicing and Charges

5.1 All invoices should reference the statement number and be sent to the Nominated Person as identified in Part 1 of this Agreement.

[Note: Each DSO to amend and to provide any further commentary in respect of specific payment processes/invoices]

5.2 Charges

[Each DSO to provide details of contracted pricing here or use this section to signpost to where pricing details are found]

5.3 Calculation of Charges

    5.3.1 There shall be two types of Flexibility Services payments: Utilisation Payments and Availability Payments. The application of the payment type depends on the Flexibility Service product being delivered.

Utilisation Payments

    5.3.2 Utilisation Payments are made when a Utilisation Instruction is issued by the Company. Utilisation Payments can be in terms of:
        (a) For every metered time period, energy (MWh) delivered supplied by the Provider and multiplied by the Utilisation Fee (£/MWh); or,
        (b) The capacity (MW) delivered multiplied by the Utilisation Fee (£/MW) over a period-of-time.

    5.3.3 Where the Provider has not fully met the Utilisation Instruction, an additional performance calculation is applied to the Utilisation Payments. The Monthly Utilisation Performance Factor determines how much under delivery is eligible for payment.

Availability Payments

    5.3.4 Where Availability is applicable to a Flexibility Service, payments are paid for every Accepted Availability Window in respect of the contract DER groups. Availability Payments are subject to a Monthly Utilisation Performance Factor.

    5.3.5 Availability is determined by:
        (a) For every metered time period, the Agreed Availability Capacity (MW) multiplied by the Availability Fee.

    5.3.6 Where a Service Provider declares Unavailability, or was not Available at the time of delivery, then no Availability Payment will be made for that metered time period.

    5.3.7 Availability Performance is calculated monthly and Availability Payments are recovered should the delivered capacity be lower than the agreed delivery capacity. A performance factor is applied to the

Availability Payment and such performance factor shall consider the Provider's Monthly Utilisation Performance Factor.

[Note: Insert details on where to find documentation: Detailed Payment Calculations Vx, Specific Parameters Table, etc.]

5.4 Payment Terms
[DSOs to confirm agreement]

5.4.1 In consideration of the provision by the Provider of the Flexibility Services in accordance with the terms of this Agreement, the Company shall pay to the Provider the Charges.

5.4.2 All invoices shall be paid within thirty (30) Days of the date of invoice (the "Due Date for Payment").

5.4.3 If the Company intends to pay less than the sum stated as due in the self-billing invoice it shall, not later than five (5) Business Days before the Due Date for Payment, give the Provider notice of that intention by issuing a notice which shall specify both the sum that it considers to be due to the Provider at the date the notice is given, or the sum which it considers is due from the Provider to the Company, and the basis on which that sum is calculated. *[DSOs to confirm details of any platform specific /ITT specifics etc.]*

5.4.4 Unless otherwise agreed in writing between the Parties, payment of invoices shall be made by the Company either (at the Company's option) by BACS payment to a bank account nominated in writing by the Provider or by cheque sent to an address nominated in writing by the Provider (or, where no such address is nominated in writing by the Provider then to the Provider's registered office).

5.4.5 All sums payable under this Contract shall be exclusive of VAT. The payor of any sums shall pay an amount equal to such VAT to the payee in addition to any sum or consideration on receipt of a valid VAT invoice from the payee.

5.4.6 If the payor fails to pay to the payee any undisputed amount payable by it under this Agreement, the payee may charge the payor interest on the overdue amount from the due date up to the date of actual payment at the rate of two per cent (2%) per annum above the base rate of the Bank of England. Such interest shall accrue from day to day from the due date until actual payment of the overdue amount, whether before or after judgment. The relevant Party shall pay the interest together with the overdue amount. The Parties acknowledge that their liability under this paragraph 5.4.6 is a substantial remedy for the purposes of Section 9(1) of the Late Payment of Commercial Debts (Interest) Act 1998.

5.4.7 The payor may, without limiting any other rights or remedies it may have, withhold or set off any amounts owed to it by the payee against any amounts payable by the payor to the payee under this Contract.

[Each DSO to provide performance factor table for reduction of charges if applicable]

[Each DSO to provide information in relation to recovery and withholding of payment if applicable]

6 Sites and DER

[Note: each DSO to signpost relevant information where applicable.]

6.1 Details of the Provider's DER are to be submitted to the Company through the [·], as further described in [·]. *[Note: each DSO to provide relevant details]*

6.2 Where a DER forms part of a successful [Trade], the Company will confirm such trade within the [Trade Award]. *[Note: each DSO to provide a description of the relevant Trade/Trade Award]*

6.3 Should the Provider wish to change its DER post [Trade Award], this can be accommodated through the [·], as further described in [·]. *[Note: each DSO to provide relevant details]*

6.4 The Company will allow the inclusion of additional sites at any time throughout the Term. Such changes to DER shall take effect not earlier than the following operational period.

6.5 Participation above the Accepted [MW/MVAR] stated in the [Trade Award] cannot be exceeded.

[Note: each DSO to include details of access requirements for Sites.]

7 Communications

7.1 Senior Representatives

[Note: Insert details of senior reps for each Party]

Escalations process

Escalation level	Company representative	Service provider representative
1	Relevant company authorised person	*[·]*
2	Relevant company manager/ commercial manager	*[·]*

7.2 Process and systems for communications

[Each DSO to provide details of Process and systems here or use this section to signpost to where Process and systems are found.]

Flexibility Services Service Terms—Company Active Services   357

Utilisation instructions	Stop instructions	Unavailability notices
[Delete as appropriate: • *E-mail* • *Telephone* • *API]* The utilisation instruction must specify for a DER: • The zone to which the utilisation instruction relates; • The requested start time; • The requested end time; and • The requested MW	*[Delete as appropriate:* • *E-mail* • *Telephone* • *API]*	*[Note: Set out process for issuing Unavailability notices—update to refer to the form and who it should be sent to]*

7.3 Acceptance of instructions

[Each DSO to amend as required]

The Provider may accept the instruction by responding (by any method as approved by the Company) to the Utilisation Instruction within [thirty (30) minutes] from the time of the request, setting out:

- The Accepted Start Time, which cannot be earlier than, but must be no later than [thirty (30) minutes] from, the Requested Start Time;
- The Accepted End Time, which can be no later than the Requested End Time but otherwise has to be at least *[thirty (30) minutes]* from the Accepted Start Time; and
- The Accepted MW, *[which shall be at least [0.1MW]] [which shall be at least [·] % of the Requested MW]* and can be no greater than the Requested MW.

8 **Performance Monitoring**

8.1 Metering Standards

For Asset Point Metering, the Provider will ensure compliance with the following metering standards set out within the most recent published relevant Balancing and Settlement Code of Practice Eleven: code of practice for the metering of balancing services assets for settlement purposes:

- The metering 'accuracy requirements';
- The 'asset meter calibration test certification';
- The 'limits of error';
- The 'sealing' requirements.

For Boundary Point Metering, the Provider should be compliant with Balancing and Settlement Codes of Practice 1, 2, 3, 4, 5 and 10 as applicable.

[Note: DSOs to include optional, hyperlink to additional information]

If requested by the Company, the Provider shall provide evidence of compliance with the above standards. This may be in the form of certification, photo, or written confirmation.

8.2 Submission of Performance Report
[Note: each DSO to provide.]

8.3 Testing and monitoring

Processes	Standards	Timetable of testing
[Note: Provide information as appropriate]	*[Note: Provide information as appropriate]*	*[Note: Provide information as appropriate]*

8.4 Service Meter

Measuring equipment	Service meter data		Standards
[Note: Provide details of the measuring equipment]	Minute-by-minute and half hourly data, will be accepted for settlement purposes. Certain products rely on minute-by-minute metering granularity for accurate performance monitoring and settlement. Where an alternative to minute-by-minute granularity is provided the data may be disaggregated. As such, this could result in performance monitoring and calculation inaccuracies *[Note: DSO option to insert table to specify granularity preference for products]*		

8.5 Service Failure
[Note: DSOs to agree wording but it shall cover:]
Each of the following shall constitute a Service Failure:

8.5.1 Reduced Capacity: if a unit providing Flexibility Services fails to deliver Flexibility Services in accordance with Utilisation Instructions at a delivery performance of at least 60% over 2 months or for [·] dispatches, whichever is sooner.

8.5.2 Unavailability level: if a unit providing Flexibility Services is Unavailable and has more than 30% of [agreed availability] of the Service Period in a month;
Unavailability Notification: if the Provider fails to notify the Company that a unit providing Flexibility Services is Unavailable within [·]

8.6 Monitoring of DER development projects

Details of development milestone	Expected date of delivery	Comments
[Note: Provide information as appropriate]	*[Note: Provide information as appropriate]*	*[Note: Provide information as appropriate]*

[Note: to be [N/A] if there are no planned assets.]

8.7 Auditing

Process	Requirements
[Note: Provide information as appropriate]	*[Note: Provide information as appropriate]*

[Note: The Company may repeat the assessment process on an annual basis.]

8.8 Additional performance obligations

[Each DSO to set out any additional performance obligations required of the Provider]

8.9 Non-delivery and under-delivery

[Each DSO to set out any further points relating to non-delivery or under-delivery]

9 Data Protection

[Each DSO to set out any points relating to Data Protection including links to any standalone data protection documents]

10 Details of Flexibility Provider and Special Conditions

Provider	Provider's company number and registered office
Requested MW	[·]
Contract number	[[to be completed by the Company post award]]
Provider's addresses for notices	[·] Address: [·] Contact Number: [·] For the attention of: [·]
Company's addresses for notices	[·] Address: [·] Contact Number: [·] For the attention of: [·]
Provider's Nominated Person	[·]
Company's Nominated Person	[to be split into operation and commercial details]

10.1 Cyber Security

Providers are required to comply with the Company's applicable Cyber Security policy or terms and a copy can be accessed at the following link *[Note: DSO to provide relevant link]*

Annexes to Flexibility Services Service Terms—Company Active Services

[·] 2024

Definitions

The additional terms placed within the Service Terms shall also apply to these associated Annexes.

Annex 1—Flexibility Management Systems/Technical Requirements

A1.1   Company Flexibility Management System Details

[Each DSO to add here details of Flexible Power/alternative Operational Requirements, System specifics i.e. API interfaces or external references (websites, documentation etc.]

A1.2   Dispatch Principles

[Each DSO to provide details of Dispatch Principles here, or signpost to where available.]

Annex 2—Auction/Tender/Trade Guidelines

[Guidance and rules for post contract award auctions/trades. Most relevant to DSOs using a Call-Off or Framework approach]

[Delete if not relevant]

Annex 3—Special Requirements

[Note: Annex to be amended/deleted as appropriate]

[Note: Annex included for any specific variations, specific HSE requirements or obligatory additions of the Company]

Forms and Templates to Flexibility Services Service Terms—Company Active Services

[·] 2024

1 **Introduction**

Where a Company adopts Forms and Templates as part of its process to contract for Active Services these shall be detailed here.
[Each DSO to amended/deleted as appropriate.]

2 **Unavailability/Remedy Template**

[Form of unavailability notification /remedy notification]
[in accordance with paragraph 7.2, this is a notification of Unavailability of Flexibility Services.]

Company Name:	
Zone ID:	
Flexible Unit:	
From Date/Time:	[Unavailable from]
To Date/Time:	[Unavailable to]
Reason:	
Name:	[of individual making notification]
Date:	[of notification]

3 **Performance Report Template**
   [Each DSO to amend as required]

4 *[Other Templates as Required]*
   [Each DSO to amend as required]

The manufacturer's authorised representative in the EU is Springer Nature Customer Service Centre GmbH, Europaplatz 3, 69115 Heidelberg, Germany. If you have any concerns regarding our products, please contact ProductSafety@springernature.com

Printed and bound by CPI Group (UK) Ltd, Croydon, CR0 4YY

02/01/2026

02028205-0006